Sixth Edition

Fire Service Rescue

Edited By
Carl Goodson

Validated By
The International Fire Service Training Association

Published By
Fire Protection Publications • Oklahoma State University

Cover Photo:
Rescuers search for victims in the rubble of the Alfred P. Murrah Building in Oklahoma City, OK, which was bombed April 19,1995.

ISBN 0-87939-129-4
Library of Congress 96-85275

Sixth Edition
First Printing, June 1996

Printed in the United States of America

Dedication

This manual is dedicated to the members of that unselfish organization of men and women who hold devotion to duty above personal risk, who count on sincerity of service above personal comfort and convenience, who strive unceasingly to find better ways of protecting the lives, homes and property of their fellow citizens from the ravages of fire and other disasters . . . ***The Firefighters of All Nations.***

Dear Firefighter:

The International Fire Service Training Association (IFSTA) is an organization that exists for the purpose of serving firefighters' training needs. Fire Protection Publications is the publisher of IFSTA materials. Fire Protection Publications staff members participate in the National Fire Protection Association and the International Association of Fire Chiefs.

If you need additional information concerning our organization or assistance with manual orders, contact:

Customer Services
Fire Protection Publications
Oklahoma State University
930 N. Willis
Stillwater, OK 74078-8045
1 (800) 654-4055

For assistance with training materials, recommended material for inclusion in a manual, or questions on manual content, contact:

Technical Services
Fire Protection Publications
Oklahoma State University
930 N. Willis
Stillwater, OK 74078-8045
(405) 744-5723

THE INTERNATIONAL FIRE SERVICE TRAINING ASSOCIATION

The International Fire Service Training Association (IFSTA) was established as a "nonprofit educational association of fire fighting personnel who are dedicated to upgrading fire fighting techniques and safety through training." This training association was formed in November 1934, when the Western Actuarial Bureau sponsored a conference in Kansas City, Missouri. The meeting was held to determine how all the agencies interested in publishing fire service training material could coordinate their efforts. Four states were represented at this initial conference. Because the representatives from Oklahoma had done some pioneering in fire training manual development, it was decided that other interested states should join forces with them. This merger made it possible to develop training materials broader in scope than those published by individual agencies. This merger further made possible a reduction in publication costs, because it enabled each state or agency to benefit from the economy of relatively large printing orders. These savings would not be possible if each individual state or department developed and published its own training material.

To carry out the mission of IFSTA, Fire Protection Publications was established as an entity of Oklahoma State University. Fire Protection Publications' primary function is to publish and disseminate training texts as proposed and validated by IFSTA. As a secondary function, Fire Protection Publications researches, acquires, produces, and markets high-quality learning and teaching aids as consistent with IFSTA's mission. The IFSTA Executive Director is officed at Fire Protection Publications.

IFSTA's purpose is to validate training materials for publication, develop training materials for publication, check proposed rough drafts for errors, add new techniques and developments, and delete obsolete and outmoded methods. This work is carried out at the annual Validation Conference.

The IFSTA Validation Conference is held the second full week in July, at Oklahoma State University or in the vicinity. Fire Protection Publications, the IFSTA publisher, establishes the revision schedule for manuals and introduces new manuscripts. Manual committee members are selected for technical input by Fire Protection Publications and the IFSTA Executive Secretary. Committees meet and work at the conference addressing the current standards of the National Fire Protection Association and other standard-making groups as applicable.

Most of the committee members are affiliated with other international fire protection organizations. The Validation Conference brings together individuals from several related and allied fields, such as:

- Key fire department executives and training officers
- Educators from colleges and universities
- Representatives from governmental agencies
- Delegates of firefighter associations and industrial organizations
- Engineers from the fire insurance industry

Committee members are not paid nor are they reimbursed for their expenses by IFSTA or Fire Protection Publications. They come because of commitment to the fire service and its future through training. Being on a committee is prestigious in the fire service community, and committee members are acknowledged leaders in their fields. This unique feature provides a close relationship between the International Fire Service Training Association and other fire protection agencies, which helps to correlate the efforts of all concerned.

IFSTA manuals are now the official teaching texts of most of the states and provinces of North America. Additionally, numerous U.S. and Canadian government agencies as well as other English-speaking countries have officially accepted the IFSTA manuals.

Table Of Contents

Tables

Preface

This sixth edition of **Fire Service Rescue** is intended to serve as a primary text for firefighters and rescue squad members to guide them in the safe and efficient performance of basic rescue operations. The information contained in this manual reflects the current state of the art in rescue practices and is consistent with the third edition of IFSTA **Essentials Of Fire Fighting** and NFPA 1001, *Standard for Fire Fighter Professional Qualifications*, 1992 Edition.

Acknowledgement and special thanks are extended to the members of the validating committee who contributed their time, wisdom, and knowledge to this manual.

Chairman
George Dunkel
Tualatin Valley Fire and Rescue
Aloha, OR

Secretary
Bob Anderson
Spokane County Fire District #9
Spokane, WA

Wes Kitchel
Santa Rosa Fire Department
Santa Rosa, CA

Bryan Neilson
Calgary Fire Department
Calgary, Alberta

John Norman
New York Fire Department
New York, NY

Duane McKay
Prince Albert Fire Department
Prince Albert, Saskatchewan

John Ryan
Texas A&M University
College Station, TX

Tim Gallagher
Phoenix Fire Department
Phoenix, AZ

Jim Kellam
Virginia Beach Fire Department
Virginia Beach, VA

Mike McGroarty
La Habra Fire Department
La Habra, CA

Others who generously contributed their time and expertise to the committee were:

Robert Watters
Rochester Fire Department
Rochester, NY

Bruce Walz
University of Maryland
Baltimore, MD

Mike Carpenter
Lamar University
Beaumont, TX

Jim Mendonza
Columbia College
Columbia, CA

The following individuals and organizations contributed information, photographs, or other assistance that made the completion of this manual possible:

AIM Safety USA, Inc., Austin, TX
Angel-Guard Products, Inc., Worcester, MA
Biomarine, Inc., Malvern, PA
Ron Bogardus, Albany, NY
Calgary (Alta) Fire Department
Clemens Industries, Inc., Brookeville, MD
DELSAR, Inc., Chapel Hill, NC
East Greenville (PA) Fire Company #1

Emmaus (PA) Fire Department
Fire Wagons, Inc., Stillwater, OK
Flexlite, Inc., Edison, NJ
Keith Flood, Santa Rosa (CA) Fire Department
Steve George, Midwest City (OK) Fire Department
Bobby Henry, Altus (OK) Fire Department
Hughes Missile Systems Co., Rancho Cucamonga, CA
Innovative Safety Systems, Inc., Peabody, MA
Ron Jeffers, New Jersey Metro Fire Photographers Assn.
Los Angeles (CA) Fire Department
New York (NY) Fire Department
Laura Mauri, Spokane County (WA) Fire District #9
NiteRider Technical Lighting Systems, San Diego, CA
Paratech Incorporated, Frankfort, IL
Phillips 66 Oil Co., Borger, TX
Phoenix (AZ) Fire Department
Paul Robbins, Air Line Pilots Assn.
SKEDCO, Inc., Portland, OR
Stanley Hydraulic Tools, Milwaukie, OR
Stillwater (OK) Fire Department
Vespra (Ont) Fire Department
Warrington Group Ltd., North Hampton, NH
Joel Woods, Maryland Fire & Rescue Institute (MFRI)
Richey Wright, Roco Rescue, Baton Rouge, LA

Last, but not least, gratitude is extended to the following members of the Fire Protection Publications staff whose contributions made the final publication of this manual possible:

Barbara Adams, Senior Publications Specialist
Marsha Sneed, Senior Publications Specialist
Mike Wieder, Senior Publications Editor
Carol Smith, Senior Publications Editor
Susan S. Walker, Instructional Development Coordinator
Don Davis, Publications Production Coordinator
Ann Moffat, Graphic Design Analyst
Rick Arrington, Graphic Designer
Connie Burris, Senior Graphic Designer
Desa Porter, Senior Graphic Designer
Ben Brock, Graphics Assistant
Don Burull, Graphics Assistant
Eric Barnum, Research Technician
Michael Huskey, Research Technician
Christian Jaehrling, Research Technician
Mark Slight, Research Technician
Mike Spini, Research Technician
Ben Warren, Research Technician

Lynne C. Murnane

Lynne Murnane
Managing Editor

Introduction

RESCUE APPLICATIONS

Situations to which rescuers may respond range from the very simple to the extremely complex. Rescue incidents are often extremely dangerous for rescue personnel and require the application of knowledge, skill, and teamwork. This knowledge, skill, and teamwork are developed and maintained through training and experience.

Because the lessons learned during actual emergency incidents may or may not apply to other incidents and because emergencies occur at random intervals, rescuers cannot rely upon experience alone to prepare them to deal with all types of rescue emergencies. Rescuers must have a solid foundation of classroom and hands-on training that is periodically reinforced by realistic and challenging training exercises. Many states, provinces, and regional training systems offer excellent courses in various rescue disciplines, and rescue personnel are encouraged to take advantage of these training opportunities.

Another significant part of many rescue incidents is the emergency medical care needed by the victims. Emergency medical care and rescue often must be done concurrently at the same incident and require equal consideration. However, this manual does not address emergency medical skills. To learn more about emergency medical care, see the IFSTA **Fire Service First Responder** manual. That manual provides in-depth information on all aspects of basic emergency medical care.

PURPOSE AND SCOPE

The purpose of this manual is to present the principles and practices of basic fire service rescue in a manner that will enable the discerning student to safely and effectively participate in rescue operations. The scope of this manual is limited to those situations to which most firefighters and rescue squad members may be called. Some situations and environments, such as wilderness search/rescue and underwater search/rescue, have been omitted because they are beyond the range of responsibility for most fire departments and rescue squads. However, by the time a firefighter or rescue squad member completes a course of study based on this manual, he or she will have acquired much of the knowledge needed to safely and efficiently perform basic rescue operations under appropriate supervision.

1

Introduction To Rescue Services

Chapter 1

Introduction To Rescue Services

INTRODUCTION

There are two general elements to consider when examining the task of rescue: needs assessment and incident management and control. Each department should assess the potential for rescue situations occurring within its district and then evaluate the capabilities of the department for handling these types of rescues.

This chapter identifies the need for specialized rescue equipment and services and describes various potential rescue situations. It identifies the other agencies and organizations with which the rescuers may have to interact and discusses the means by which existing rescue and extrication resources may be evaluated.

ASSESSING THE NEED FOR RESCUE SERVICES

Evaluating The Response District

An important part of assessing the need for rescue services is to determine the frequency of and potential for rescue incidents. The questions that must be answered are as follows:

- Are rescues performed regularly or infrequently?
- Do rescue situations occur more frequently during a particular time of the day or on a particular day of the week?
- How severe are the rescue incidents that usually occur in the district? Do they involve simple, quick actions or long, tedious operations?
- Do rescue problems involve natural hazard areas (caves, lakes, or cliffs) or constructed areas (highways, railroads, or large buildings)?
- Are parts of the district more likely than others to generate rescue calls?
- Are there areas where rescues will be more difficult to perform than others?
- What is the current average response time for a rescue company, and how well-equipped and prepared are they?
- What is the potential for extrication situations, aircraft incidents, railroad accidents, agricultural and/or industrial accidents, structural collapse, cave-ins, rescues from either elevated areas or confined spaces, water-related incidents, subterranean rescues, incidents involving either public utilities or hazardous materials, or other rescue situations?

One of the best ways to find this information is by searching fire department alarm records for the district in question (Figure 1.1). The records should reflect the kinds of problems typical to the area. Law enforcement agency records may also reveal useful data.

Figure 1.1 Alarm records can reveal much about past emergency activity within the district.

Communicating with people who are familiar with the area may also be helpful. Local physicians or hospital administrators can provide information on the types of injuries and problems they have encountered. Another good source of background information may be the local road supervisor or state or federal highway personnel from the Department of Transportation (DOT). These individuals can provide information on the amount of traffic in the area and on the types of vehicles that commonly travel through the area.

Do not overlook members of the fire department or the general public who can assist in compiling a list of the hazards in the area. These people will be familiar with problem areas such as natural hazards, hazardous road curves or intersections, and so on. All hazards should be considered and noted; however, anecdotal evidence should be verified.

Figure 1.2 Rescue units may be valuable resources at fires.

Potential Rescue Situations

FIREGROUND SEARCH AND RESCUE

One of the most common situations in which a rescue unit is needed is for rescue operations on the fireground (Figure 1.2). A well-trained and well-equipped rescue unit can be an extremely valuable resource on the fireground, and if possible, one should be assigned to each working fire. The rescue personnel should be trained and equipped to perform all of the routine truck company operations such as forcible entry, ventilation, search, and control of utilities. If necessary, they can also be used to advance a hoseline from an engine. The specialized equipment carried on many rescue units can give the incident commander (IC) more options for fire attack than would otherwise be available. For example, pavement breakers (jackhammers) and oxyacetylene cutting equipment can be used to breach floors or walls for ventilation or to provide access for hose streams (Figure 1.3). Winches may be used to move obstructions, and longer-duration self-contained breathing apparatus (SCBA) may allow firefighters to reach the seat of a stubborn blaze.

Figure 1.3 The specialized equipment carried on rescue units is often needed on the fireground. *Courtesy of Stanley Hydraulic Tools.*

While searching a fire building for civilian occupants is the task most frequently assigned to a rescue company, the primary reason for having a rescue unit at a working fire is to rescue trapped or overcome firefighters (Figure 1.4). When firefighters are in trouble, it is often the specialized training and equipment that the rescue unit has that are needed to save them. Examples of the special situations for which rescue units are uniquely suited are confined spaces, structural collapses, and high-angle environments. Having personnel on scene who are rescue trained and equipped can sometimes mean the difference between a firefighter surviving or not surviving.

Depending upon the capabilities of a particular rescue unit, rescue personnel may be able to pro-

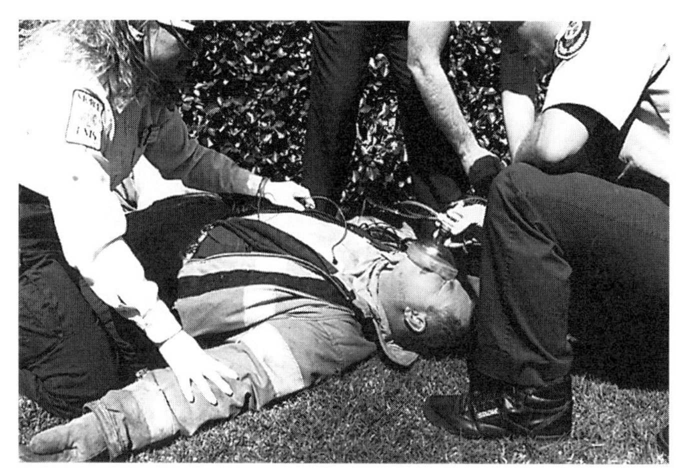

Figure 1.4 Rescue units are sometimes needed to treat injured firefighters.

vide a variety of other services on the fireground (Figure 1.5). These could include providing light and power, refilling SCBA cylinders from a cascade system, providing medical treatment for both firefighters and civilians, and monitoring and/or mitigating hazardous materials situations. But because of the need to keep the rescue unit available for its primary mission — firefighter rescue — some departments do not use rescue personnel for overhaul or other nonlife-threatening situations on the fireground.

Figure 1.5 Rescue personnel can provide support to fire suppression personnel.

VEHICLES

The most common rescue situations involve transportation accidents on public roadways (Figure 1.6). Many factors within each district influence both the frequency and the severity of vehicle accidents. The main factor is the type of transportation routes that traverse the district. A greater frequency of accidents would be expected on a heavily traveled interstate highway than on a rural dirt road (Figure 1.7). However, the potential for rescue situations occurring on back roads with relatively little traffic should not be ruled out. Various aspects of the roadway must be considered: the contour of the road, dangerous or blind intersections, types of vehicles using the roads, and the area's climate. All of these factors combine to increase or decrease the potential for motor vehicle accidents within the district. For more information on traffic-related rescues, refer to the IFSTA **Principles of Extrication** manual.

Figure 1.6 Vehicle extrications are the most common rescue incidents.

Figure 1.7 The type and volume of traffic on roads within the district must be assessed.

AIRCRAFT

Almost no area is without air traffic, so it must be considered in every evaluation of potential rescue needs within a given area (Figure 1.8). Even departments without airports or aircraft facilities in their areas should consider the minimum rescue capabilities needed for aircraft-related rescues.

Figure 1.8 The potential for aircraft-related rescues must also be gauged. *Courtesy of Paul Robbins.*

Obviously, the potential is much greater if the district is responsible for an airport or landing strip. The size of the airport, the amount of air traffic, and the type of aircraft using the facility are important factors that need to be known when determining the rescue and extrication capabilities necessary. The level of protection provided by the airport, if any, should also be taken into account. For more information on aircraft-related rescues, refer to the IFSTA **Aircraft Rescue and Fire Fighting** manual, DOT publication HS805-703, *Air Ambulance Guidelines*, or the appropriate *Aeronautical Information Publications* of Transport Canada Aviation.

RAILROADS

Other potential transportation emergencies are those involving railroads. If the district contains rail lines, they should be examined to see what characteristics they possess that would increase the potential for rescue situations. Important factors include the number of rail lines going through the area, the amount of rail traffic, the types of trains using the rails, the conditions of the rails and roadbeds, the speeds used by the trains going through the area, and the means by which junctions within the area are controlled (Figure 1.9). If passenger trains use the lines, the possibility of a multicasualty incident must be assumed. For more information on railroad-related rescues, refer to the IFSTA **Principles of Extrication** manual.

AGRICULTURE

The potential for rescue problems involving agricultural machinery must be considered because it is a frequent source of injury and entrapment (Figure 1.10). If the district contains agricultural operations, a number of different types of accidents may occur. Such accidents include vehicle rollovers; people caught within large machinery; grain bin entrapments; and exposure to toxic gases, pesticides, and other agricultural chemicals. These incidents may require both special equipment and training to be handled properly. Because of the rugged construction of agricultural equipment, normal tools and techniques used in automobile extrication and rescue may not be effective. To

Figure 1.9 Planners must consider the potential for railroad rescue incidents.

Figure 1.10 Farming operations can produce rescue incidents.

further complicate matters, the accident may occur far from a surfaced road, requiring the use of four-wheel-drive vehicles to reach the scene. For more information on agricuture-related rescues, refer to Chapter 11, Special Rescues, and the IFSTA **Principles of Extrication** manual.

INDUSTRY

Industries within the district must be evaluated because industrial machinery may cause many of the same types of rescue problems as agricultural machinery. Common industrial machinery, such as chain or screw conveyors, punch presses, and woodworking equipment, is notorious for inflicting injuries to and/or entrapping workers. All industries within the district should be visited to identify the types of machines being used (Figure 1.11). Earthmovers, cranes, derricks, mining equipment, and other heavy equipment can pose serious problems in rescue situations if the department is not prepared to deal with them. For more information on industrial rescues, refer to the IFSTA **Principles of Extrication** manual.

Figure 1.11 The potential for industrial rescue incidents must be considered.

STRUCTURAL COLLAPSE

Incidents involving structural collapse can challenge the capabilities of even the most prepared rescue unit. The response district should be evaluated for buildings of inferior construction as well as for those that are old and deteriorating into dangerous condition (Figure 1.12). Buildings still under construction should be visited regularly during construction. Even dilapidated, vacant buildings, especially in urban areas, must be considered because they may be inhabited by vagrants, thus creating an unexpected rescue problem. For more information, refer to Chapter 6, Structural Collapse Rescue.

Figure 1.12 Structural collapse is possible in almost every response district. *Courtesy of Steve George.*

CAVE-INS

The possibility of excavation cave-ins must be considered, even though excavations are usually temporary situations. If there is a large amount of pipeline and/or underground cable installation in the area, rescue capabilities may be needed. The soil conditions in an area will also influence the potential for cave-in incidents. Loose, sandy soil is more susceptible to instability than soil with a high shale or clay content. Rescuers should be familiar with the types of soil within their jurisdiction.

Any large amount of underground construction will dictate the need for a rescue unit to be prepared for trench rescues (Figure 1.13). Cave-ins necessitate the use of heavy shoring equipment, which is different from the standard multipurpose rescue equipment typically carried by engine and truck companies. Trench incidents also require rescuers to have knowledge of proper trenching techniques.

Figure 1.13 The amount of trenching within the district can indicate the level of preparedness needed.

Taking either incorrect or careless actions at a cave-in could aggravate the situation and even turn rescuers into more victims. For more information, refer to Chapter 10, Trench Rescue.

In addition to trench incidents, similar types of problems can occur at a number of different locations, including quarries and lumberyards. Quarries store large piles of stone, sand, coal, or other materials. Some lumberyards also store these materials in large piles (Figure 1.14). Grain-storage facilities and cement-production plants are also potential problem areas. For more information, refer to Chapter 11, Special Rescues.

Figure 1.14 The potential for engulfment in loose material must be considered.

ELEVATED AREAS

Rescues involving elevation differences can also require a specially equipped rescue unit. The elevation differences could be man-made, such as tall buildings, transmitter towers, or elevated water storage tanks; or they can involve natural terrain features such as canyons, cliffs, or tall trees (Figure 1.15). A survey of the district must be made to identify both the type and the approximate height of the structures, identify areas frequented by rock or mountain climbers, and give an estimate of the likelihood of rescue situations developing in these areas. For more information, refer to Chapter 4, Rope Rescue.

Elevators and escalators must also be evaluated for potential problems in multistory buildings of all types and occupancies. Elevator rescues most often involve cars stalled between floors, and even though an elevator mechanic is best-suited to handle this situation, the fire department is likely to be called first to assist in removing trapped occupants (Figure 1.16). The survey should identify all passenger, freight, and service elevators, as well as

Figure 1.15 Elevation differences can indicate a need for rope rescue capability.

Figure 1.16 A potential for elevator rescues exists in almost every district.

large dumbwaiters. The overall appearance and condition of the building is often a good indicator of the condition of its elevators. For more information, refer to Chapter 7, Elevator Rescue.

CONFINED SPACES

Rescue operations involving confined spaces can be especially difficult. Typical examples of confined spaces include empty storage tanks, silos, rail tank cars, utility manholes, wells, septic tanks, compost pits, mines, and caves (Figure 1.17). In addition to the problems of small, cramped spaces with limited access, the atmosphere in these spaces is often oxygen-deficient and may contain toxic gases. Special rescue equipment and SCBA or supplied-air respirators are necessary to perform rescues in these situations. Railroad yards, tank farms, industrial facilities, and farming operations are also possible sources of confined space problems. Chemical plants in particular should be scrutinized not only for the types of tanks they have but also for the properties of their contents. For more information, refer to Chapter 8, Confined Space Rescue.

Figure 1.17 Confined spaces exist in many forms and in every district.

WATER HAZARDS

The response district should also be surveyed for bodies of water that might represent a drowning hazard. Streams, rivers, aqueducts, canals, swimming pools, ponds, lakes, and storage reservoirs, as well as the ocean, should be considered. Even small, normally calm bodies of water can be the scene of a water rescue or recovery incident. Areas used for recreation, such as swimming, boating, and fishing, should be considered high-probability areas (Figure 1.18).

Another consideration is whether any of the local waterways contain low-head dams. These dams are often called "drowning machines" because once someone becomes entrapped in the downwash, it is nearly impossible for them to extricate themselves without assistance. Low-head dams are especially hazardous during periods of high water levels or flooding conditions (Figure 1.19). If a small number of these hazards exist, only minimal equipment, such as ropes, dragging equipment, and life rings, may be necessary. If there is a high potential for a greater frequency of these types of incidents, more equipment, such as boats and diving equipment, may be justified. The rescuers must receive specialized training before placing this type of equipment into service. The potential for water-related incidents also occurs during periods when the body of water is frozen. Such activities as skating or ice fishing can also create difficult rescue situations, and personnel likely to encounter such emergencies should receive special training for ice/water rescue. For more information, refer to Chapter 9, Water And Ice Rescue.

Figure 1.18 Any body of water can be the source of rescue incidents.

Figure 1.19 Rescues from low-head dams require highly trained and specially equipped rescue units.

MINES/CAVES

Rescue situations involving man-made shafts or natural caves must be included in the list of possible rescue problems if there are any subterranean spaces within the response district. Mines are a common source of entrapment caused by cave-ins or explosions. Caves more often involve lost or injured spelunkers (recreational cave explorers). In most cases, firefighters and rescue squad members are not sufficiently trained and equipped to perform these rescues. Regular turnout gear is not effective and may even be hazardous. Specially trained mine/cave rescue personnel should be consulted if these structures exist within the district. For more information, refer to Chapter 11, Special Rescues.

PUBLIC UTILITIES

Virtually no populated area is immune to the possibility of a rescue situation involving public utilities. Depending on the needs of the area being served, these utilities can include electrical distribution, municipal water supply, sewers and storm drains, natural gas, telephone services, and others (Figure 1.20). The care and maintenance of these utilities require people to work in dangerous areas, including confined spaces, trenches, hazardous atmospheres, and elevated areas. Rescue personnel should be familiar with the types of utilities and facilities that serve their area. Of special interest to rescue personnel are underground vaults, high-voltage transmission lines, pumping and treatment plants, electrical substations, and any other utility installations that might present unusual rescue problems. Improper handling of these incidents can put both the victim and the rescuer in greater jeopardy than they were initially.

Figure 1.20 Public utilities can be the source of rescue incidents.

HAZARDOUS MATERIALS

Being prepared to handle hazardous materials emergencies is an essential function of every rescue unit. The response district should be surveyed to see what hazardous materials are used by local industries. It is important to identify all these materials, review their Material Safety Data Sheets (MSDS), note amounts and storage methods, and record locations. If possible, a survey should be conducted to determine the types of materials that are commonly transported through the area. At the present time there is no nationwide or international system in place to assist with this effort. Knowing exactly what materials are likely to be encountered will promote the safe and effective handling of any incident. Extensive training is necessary to handle an incident of this nature. For more information on dealing with hazardous materials, refer to the IFSTA **Hazardous Materials for First Responders** manual.

SPECIAL SITUATIONS

Any fire department, with or without a trained and equipped rescue unit, will be expected to respond to just about any rescue situation imaginable. Special rescues can range from a child with his head caught in a porch railing to an obese adult stuck in a bathtub. Specialized rescue equipment may sometimes be needed in hospital emergency rooms. Countless possibilities exist in any locality; however, most potential rescue situations are peculiar to the location. Therefore, these types of situations can be anticipated and evaluated, resulting in the formulation of pre-incident plans.

INTERACTING WITH OTHER ORGANIZATIONS

A properly trained and equipped rescue unit can provide extremely valuable assistance to other fire department units, to other governmental agencies, and to the private sector. However, one of the main concerns whenever a rescue unit is requested at an incident is that the incident is one that will benefit from the rescue unit's intervention and not one that is just more likely to add rescue personnel to the list of victims.

Fire Response

The services of a well-prepared rescue unit can be invaluable when the unit is assigned to assist on the fireground. Ideally, a rescue company should be assigned to all alarms. On the fireground, rescue units can employ their specialized training and equipment, perform truck company functions, or supplement the available manpower for other nonrescue functions.

Other Government Agencies

A rescue unit can assist law enforcement agencies in a variety of ways. These ways may include forcible entry to assist an investigation, emergency lighting to illuminate a crime scene, or a body recovery operation. Similar services can be provided to agencies concerned with protecting the environment, dealing with flood control, operating the municipal water supply, and others.

Private Sector

In some areas, rescue units work very closely with private ambulance companies to provide the public with a greater variety and higher level of emergency care. Properly trained and equipped rescue units are well-suited to extricate traffic accident victims and to provide basic life support and initial treatment. Their first aid capabilities make rescue units extremely valuable at mass casualty incidents. Rescue units also work directly with hospitals and with utility companies when needed.

Emergency Medical Services

Rescue personnel should establish a close working relationship with emergency medical service (EMS) personnel. Since one of the major functions of a rescue unit is the removal (and sometimes the initial treatment) of people trapped in wrecked vehicles and similar situations, it is important for rescue personnel to have an appropriate level of first aid training (Figure 1.21). The level of training needed depends on the local EMS system and the general operating guidelines. In many jurisdictions, the main purpose of rescue personnel is to perform rescue and extrication functions. Beyond that, they only provide first responder medical treatment. For the purpose of this manual, the level of medical service provided by rescuers will be assumed to be first responder.

Rescue companies frequently arrive at an incident before emergency medical units. In this case, fire and rescue vehicles should be positioned to allow ambulances direct access to the scene (Figure 1.22). Basic patient assessment and primary treatment should be initiated immediately. To the extent possible, any hazards that might further endanger victims or emergency personnel working on the scene, such as downed electrical wires or leaking gasoline, should be eliminated. Any obstacles or obstructions near the victim(s) should be removed. These procedures are discussed further in later chapters dealing with particular types of incidents.

Figure 1.21 Rescue personnel can be an essential part of the EMS delivery system.

If the ambulance is delayed, rescue personnel may start "packaging" the patient for transport. Again, it is important to assess the skills of the rescue personnel versus those of the responding ambulance personnel. For example, if the rescue unit's level of training is basic life support and the medical unit responding is also basic life support, rescue personnel may start splinting, bandaging, and doing anything else that may be necessary. However, if the responding medical unit is an advanced life-support (ALS) unit, it may not be appropriate for rescue personnel to start these activities. They might hinder the efforts of the paramedics who may need to start intravenous solutions (IVs), insert airways, or begin other advanced life-support procedures. If the rescue unit has ALS capabilities, it may initiate full treatment and then coordinate with the medical unit when it arrives. This is why it is imperative that rescue personnel and ambulance crews know each other's capabilities and establish basic protocols for patient handling. Each unit or agency should know

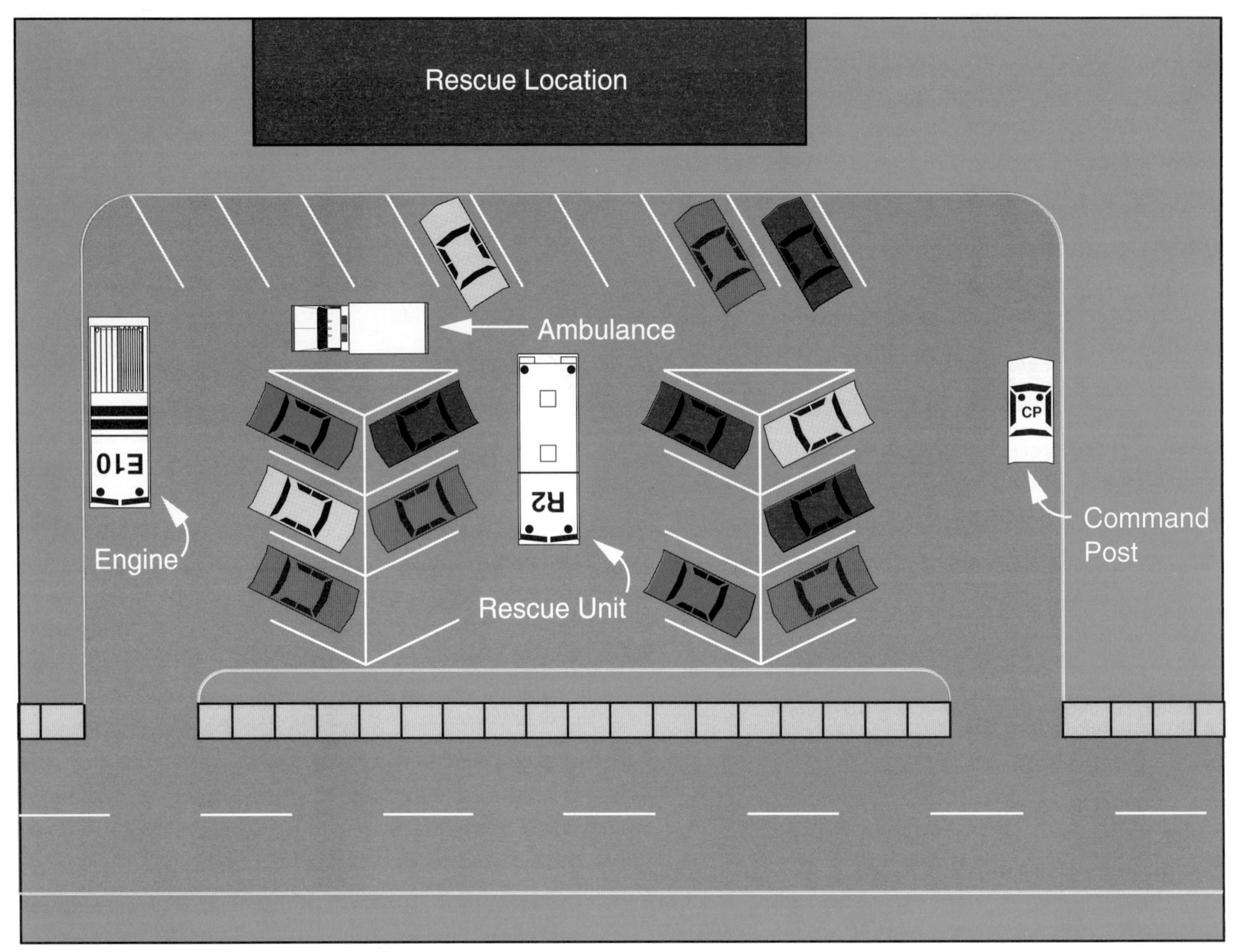

Figure 1.22 Emergency vehicles must provide access for EMS units.

what its duties are if it arrives before the other unit and what each unit's duties will be once both groups are on scene.

In most jurisdictions, once rescue and EMS units are on the scene, EMS personnel are responsible for treating patients, and rescue personnel are responsible for freeing entrapped victims and for scene safety. Close coordination between the two groups is very important to avoid working against each other, wasting valuable time, and perhaps further endangering victims and rescuers. Because rescue and EMS personnel will be concentrating on dealing with victims, their attention will be focused there and not on their surroundings. This can make them more vulnerable to being struck by vehicles driving by the scene, so emergency vehicles should be positioned to form a barrier between the traffic and the emergency scene.

Hospital Personnel

In some unusual incidents, hospital personnel may be called to the scene of an emergency (Figure 1.23). This situation is most likely to occur during a mass casualty incident; in such cases, hospital personnel are needed on-scene to assist in performing triage (sorting victims by the severity of their injuries) or conducting primary treatment of more seriously injured victims. It does not necessarily have to be a large incident to require hospital personnel on the scene. Although quite rare, in some areas where EMS personnel are not trained to provide advanced life support, hospital personnel may be called to the scene to perform such functions as starting IVs while extrication operations are in progress. Another example where hospital personnel are needed is a serious industrial or an agricultural equipment entrapment where major medical

Figure 1.23 Hospital personnel may sometimes respond to the scene. *Courtesy of Mike Wieder.*

procedures (such as amputation of a limb) may be the only way the victim can be freed.

Most medical doctors and nurses are not accustomed to working outside a hospital. Thus, it is important for rescue personnel to assist them in functioning safely at an emergency scene. Someone should be assigned to stay with them while they are on the scene to protect them from injury.

Prior to an incident occurring, rescue personnel should establish a dialogue with hospital personnel with whom they might be working at some point (Figure 1.24). Rescue personnel should become familiar with what the hospital has to offer as well as its limitations, what hospital personnel are and are not prepared to do, how they can and cannot assist on the scene, what they will need once they arrive, and any other information that might be helpful.

Figure 1.24 Rescue personnel should get to know those with whom they will work during emergencies.

Air Operations

Many jurisdictions use helicopters for the transportation of seriously injured patients, for transporting rescue personnel and equipment, and as rescue platforms (Figure 1.25). These units can greatly reduce travel time of both rescuers to the scene and patients to a hospital. However, operating helicopters is very costly and adds an increased element of risk to an emergency operation. Therefore, the decision to use these units must take into account the cost and the potential risks as well as the potential benefits.

If helicopters are available in their area, rescue personnel must prepare to work with them. Rescuers should be aware of the helicopter's availability, capabilities, and limitations. To promote the safest and most efficient operations, rescue personnel and helicopter crews should jointly develop protocols and procedures and train together to refine them. Only those who have been trained to do so should be allowed to operate near a rescue helicopter (Figure 1.26). For more information about working with helicopters, refer to DOT publication HS805-703, *Air Ambulance Guidelines* or to applicable *Aeronautical Information Publications* of Transport Canada Aviation.

Helicopters are best used as a means of transporting injured victims from the scene to a medical facility after they have been rescued, but some situations may make the use of helicopters in the actual rescue feasible. The key to making good decisions regarding the use of helicopters is to take into account their inherent limitations as well as their unique capabilities. The main limitations on

Figure 1.25 Helicopters can be essential to the timely transportation of victims to medical facilities.

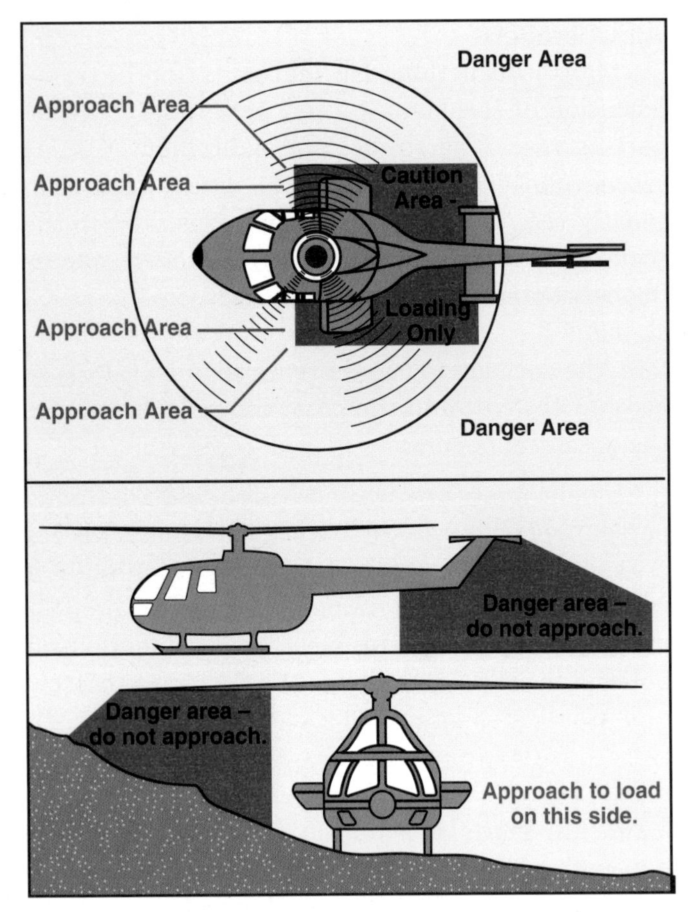

Figure 1.26 Danger zones around a helicopter.

the use of helicopters in rescue are response time, operating space requirements, and communications difficulties.

Conditions at the scene may make it tempting to call for a helicopter instead of using other rescue techniques, especially if a helicopter is within a few miles (kilometers) of the scene. But a 15-minute flight can mean a one-hour response time when the time needed for assembling the flight crew, getting required preflight clearances, and readying the helicopter for flight are considered.

When victims are stranded in narrow gorges or other areas with limited open space, there may not be sufficient room for helicopters to safely operate. Bridges, electrical transmission lines, and cliffs or other natural formations at the rescue scene may preclude the use of helicopters in the rescue effort.

Even when conditions are such that helicopters could operate safely and effectively, it may be difficult to find a common radio frequency that will allow rescuers on the ground to communicate with the helicopter crew. Hand signals may be used if both ground and flight personnel are familiar with them.

Law Enforcement

Law enforcement personnel may be present at the rescue scene and may be part of the rescue effort; however, they have their responsibilities at a rescue scene just as any other unit does. They are responsible for maintaining the flow of traffic during rescue operations on highways, roads, and streets and for investigating traffic accidents on public roadways (Figure 1.27). When victims are either unconscious or otherwise unable to provide information, law enforcement personnel can often secure the needed information by using resources such as computer data bases.

It is important for law enforcement and rescue personnel to understand each other's functions and what each can expect from the other at the scene. Control of the rescue scene will be the responsibility of the IC, but scene security and crowd/traffic control usually will be assigned to law enforcement (Figure 1.28). Rescue personnel should be aware of the need to preserve evidence because every traffic accident is a potential crime scene. Rescue personnel should also work with local law enforcement personnel to develop guidelines for positioning rescue vehicles at emergency scenes so that the vehicles do not create an additional traffic hazard. The guidelines should be based on the following general principles:

- Emergency vehicles should be positioned wherever they will be most useful in the rescue operation.

Figure 1.27 Traffic control is usually left to law enforcement.

Figure 1.28 Law enforcement personnel are trained and equipped to handle crowd control. *Courtesy of Mike Wieder.*

- If necessary, emergency vehicles can be positioned to form a barrier to protect the scene from oncoming traffic.
- Whenever possible, emergency vehicles should be parked off the roadway with their emergency lights turned *OFF*.

Utility Companies

Many incidents involve utilities in some way, so it is important for rescue personnel to have a good working relationship with local utility company personnel. It is also important for rescue units to coordinate with the utilities on mutual responses and to know what to do until the utility crews arrive. In addition, utility companies may have specially trained and equipped emergency response teams that can greatly assist in rescue efforts (Figure 1.29).

There are several guidelines that are important to follow when dealing with any utility:

- The utility should be notified immediately upon determining that its equipment or system is involved and that its services are required.
- Rescuers should not touch anything with which they are unfamiliar prior to the arrival of utility personnel.
- Rescue personnel should work with and become familiar with the various utilities within the district and develop general operating guidelines acceptable to all parties.

Figure 1.29 A public utility emergency response team.

ELECTRIC/TELEPHONE/CABLE COMPANIES

Probably the most common utility-related emergencies are those involving downed power lines, often caused by severe weather or by vehicles striking utility poles. These power lines can be very dangerous to victims and rescuers. The lines involved can be of several types, including electrical, telephone, or cable television. All should be considered dangerous. Even though telephone and cable television lines themselves are not dangerous, they may be in contact with electrical lines at some other point and may be conducting current. The rescue company should have a clear understanding with all of the utility companies on how these incidents are to be handled under local protocols. Regardless of the rescue unit's other actions, rescue personnel should notify the appropriate utility *immediately* when a problem is found. Once utility company personnel arrive, rescue personnel should be ready to work with them to mitigate the problem.

WATER/SEWER DEPARTMENTS

Rescue personnel should also be prepared to work with local water and sewer authorities. Incidents that might involve these agencies include drownings in reservoirs or holding ponds; rescues from sewers, storm drains, or underground aqueducts; damaged fire hydrants; and trench collapses. Water utility personnel may be able to provide maps and other information about the layout of these systems. They can warn of other hazards that may be present, and they may be needed to shut down broken mains or damaged fire hydrants that are flowing water.

GAS COMPANIES

A good working relationship with local gas utilities is also very important for rescue personnel. Gas company personnel may be needed to shut off gas leaks that occur for a variety of reasons, ranging from an automobile striking a meter to a main being excavated during construction or to an explosion with building collapse (Figure 1.30). It is important for rescue personnel to be trained to deal with gas-related emergencies, and the local supplier can often provide the training. In some cases, the supplier may be able to supply combustible gas detectors during emergencies. The local gas supplier should be contacted for information peculiar to the rescue district in question; it may be able to supply maps of the distribution system and valve locations. In most areas, gas distribution systems have far fewer valves than do water systems. For this reason, escaping gas often can be shut down only by the gas company excavating on both sides of a break and crimping the main.

Figure 1.30 Natural gas explosions can produce rescue incidents.

Other Agencies

In addition to the agencies mentioned, there are a number of other entities with which the rescue company may come in contact. These include public health departments, coroner/medical examiner's offices, and the Environmental Protection Agency (EPA), to name a few. Any possible contacts should be identified and relationships established so that these agencies will be able to work more effectively with the rescue company when an incident occurs.

Support Of Rescue Units

Up to this point, the discussion has focused on how the rescue unit can work with and benefit fire companies and others. It is also important to remember that there are many ways in which others can help the rescue unit. Suppression companies can assist in the control of hazards at the emergency scene. This may include controlling traffic, providing hoselines or foam for fire control, and/or providing additional personnel and equipment. Fire suppression personnel should also be trained in basic rescue and extrication techniques so that they may be used more effectively to augment rescue unit personnel.

EVALUATING EXISTING RESCUE RESOURCES

The keys to successful rescue operations are adequate pre-incident planning before the emergency and dispatching a sufficient number of properly trained and equipped rescue personnel to the emergency when it occurs. In most cases, this requires striking a balance between acquiring all of the rescue resources that may be needed and the fiscal resources of the jurisdiction.

Pre-Incident Planning

As in any other type of emergency, the more that rescue personnel know about the locations to which they may be sent, the more they will be able to anticipate problems and devise safe and effective solutions. There is no substitute for rescue personnel conducting pre-incident planning surveys of their assigned district (Figure 1.31). The fewer surprises there are when dealing with a rescue problem, the better.

Manpower

While there is no definite number of personnel required to be assigned to a rescue unit, there are two major factors that influence staffing levels. These two factors are: (1) the type and level of

Figure 1.31 A typical rescue target hazard.

service being provided (which also influence the type and amount of training and equipment needed) and (2) the number of personnel available.

If the types of rescue services to be provided are limited to those that can be handled by an engine company with the tools and equipment carried on an engine, the staffing needs will be quite different than those involving technical rescue operations. If a particular fire department cannot provide the necessary manpower and other resources to perform technical rescues, the department should identify where those resources can be obtained in an emergency.

In a career department, the number of personnel available will be fairly definite because it will be dictated by the funding available in the department's budget. In a volunteer rescue company, it is often difficult to predict how many members will respond when an alarm is sounded, but some minimum number should be established without which the unit will not respond. For efficiency and safety, a limit should be set for both a minimum and a maximum number of personnel allowed to respond on a given rescue vehicle.

When evaluating personnel for possible assignment to a rescue unit, consider their maturity and emotional stability as well as their physical and mental capacities. Rescue personnel must be emotionally stable enough to deal with highly stressful situations in a calm and professional manner. If an individual lacks the motivation and/or the aptitude to acquire the necessary skills through training, the individual should not be assigned to a rescue unit.

To be effective, the rescuer must be able to apply the four components of a successful rescue: (1) *knowledge* of the techniques available, (2) *skill* necessary to perform the techniques, (3) *physical fitness* needed to apply the skills, and (4) *judgment* to decide which techniques to apply and when. Rescuers can only acquire, develop, and maintain these components through training.

Vehicles And Equipment

The types and levels of service that the department is committed to providing dictate the types of vehicles and equipment that are needed. The most motivated and trained personnel cannot perform rescues safely and effectively without the specialized equipment they need. The amount and types of vehicles and equipment needed, as well as staffing and training needs, should be a direct result of the sort of district survey described earlier in this chapter.

Once a survey of the district is complete, existing vehicles and equipment should be evaluated to determine their suitability for the types of service that are to be provided. The vehicles and equipment should be evaluated using both the general information provided in Chapter 3, Rescue Vehicles And Equipment, and the specific information provided in the chapters dealing with the types of rescue problems anticipated within the district.

It may be determined that a rescue vehicle is not even necessary. If only light rescue capabilities are justified, it may be more cost-effective to outfit an engine company as a rescue-engine; or a truck company may be able to carry the rescue equipment and perform the work when needed. On the other hand, the evaluation may determine that there is a need for one or more rescue vehicles. All of these considerations are a part of the evaluation of the district and the department.

Training

Maintaining an effective rescue unit requires an extensive amount of initial and ongoing training for the personnel assigned to the unit. The department must decide whether to provide this training in-house or to seek sources outside of the department. In either case, the training must be delivered by fully qualified trainers who emphasize safety as well as the technical skills. Regardless of how initial rescue training is provided, it is still necessary to have a good training program within the department because the acquired skills must be maintained through practice. All training, whether initial or maintenance, should be fully documented.

It is recommended that departments send their rescue personnel to recognized schools for appropriate training whenever possible. These schools may be at the local, county, state, or national level. They are often better equipped to deliver the necessary training than are individual fire departments, and the opportunity to meet and interact with rescue personnel from other areas is extremely valuable. Sending personnel to recognized schools is also important when training them in specialized areas such as diving or hazardous materials incidents.

If rescue training is to be delivered locally, existing fire training facilities should be assessed for application to rescue situations, and the possibility of adding special rescue props should be considered. Training resources may also be available from local utility companies, construction and demolition companies, transportation companies, and even junkyards. They may be able to provide instruction as well as surplus equipment, facilities, or junked vehicles for training purposes.

The initial stage in the training process for rescue personnel should be a presentation of rescue theory in the classroom (Figure 1.32). A well-qualified trainer who is knowledgeable on the specific subject, as well as with the roles, responsibilities, and limitations of this particular rescue unit, should be selected to present the material. Subjects that should be outlined to new members include:

- An incident command/management system
- Roles of the rescue unit
- Duties of the rescue unit
- General operating guidelines
- Familiarization with the equipment to be used
- Procedures for interacting with other organizations

Figure 1.32 Rescuers should learn rescue theory.

Once the trainees have been sufficiently oriented in the basics, they can begin hands-on training (Figure 1.33). Every member should become thoroughly familiar with the capabilities, limitations, and safe operation of every piece of equipment carried on the unit. This phase of training is where contact with other organizations can be valuable. Outside training facilities or sites, such as junkyards, swimming pools (water evolutions), large buildings (elevator/high-angle rescue), or abandoned structures (structural collapse, smoke house, search and rescue), should be used if available. All training evolutions should be well planned and organized, with safety being the foremost consideration (Figure 1.34). The training should build toward a series of very realistic and intense hands-on exercises that encompass all the various roles identified for the unit.

Once the skills training has been completed, it will be necessary for rescue personnel to continue to practice their skills through regular drills and major exercises at least annually. The exercises should be carefully designed to require only those actions that the rescue company is able to provide. Initially, the problems should be both small and relatively simple — for example, a mock vehicle accident, a person stranded on a ledge, or people clinging to a tipped canoe.

As the personnel's skills and confidence increase, more complex problems can be staged. Eventually, problems beyond the capabilities of the rescue unit alone can be created. These exercises will provide an opportunity for the rescue unit to

Figure 1.33 Rescuers in hands-on training.

Figure 1.34 Safety officer observing rescue training.

call and interface with other agencies. This interaction under controlled conditions increases the capabilities of all involved and helps develop good working relations. As the ability to handle these incidents improves, larger staged incidents or disaster drills may be attempted. These drills are excellent ways of improving how the various organizations work both independently and together in large multicasualty incidents.

A very important part of any exercise of this type is an objective postexercise critique (Figure 1.35). This review enables personnel to analyze the strong and weak points of the operation and to identify areas for further training.

Figure 1.35 A critique of the exercise will help rescuers refine techniques and improve performance.

Administration And Organization

The fire department administration must decide how the rescue unit fits into the overall emergency services delivery system, taking into account the unit's capabilities and limitations and the funding available. The administration must anticipate those situations that will be beyond the unit's current capabilities and be prepared to either upgrade those capabilities or to make sure that the necessary resources will be available from outside sources when they are needed. Beyond that, the most important element in providing for safe and effective rescue operations is the routine use of an incident management system in all drills, exercises, and actual incidents. Proper rescue scene organization will promote safe and effective operations. Such organization is extremely important because the rescue unit's work routinely involves life-threatening situations, and any lapse in organization could have serious consequences for victims and rescuers alike.

Chapter 1 Review

Directions

The following activities are designed to help you comprehend and apply the information in Chapter 1 of **Fire Service Rescue**, Sixth Edition. To receive the maximum learning experience from these activities, it is recommended that you use the following procedure:

1. Read the chapter, underlining or highlighting important terms, topics, and subject matter. Study the photographs and illustrations, and read the captions with each.
2. Review the list of vocabulary words to ensure that you know the chapter-related meaning of each. If you are unsure of the meaning of a vocabulary word, look up the word in the glossary or a dictionary, and then study its context in the chapter.
3. On a separate sheet of paper, complete all assigned or selected application and review activities before checking your answers.
4. After you have finished, check your answers against those on the pages referenced in parentheses.
5. Correct any incorrect answers, and review material that was answered incorrectly.

Vocabulary

Be sure that you know the chapter-related meanings of the following words:

- anecdotal *(6)*
- augment *(18)*
- breach *(6)*
- critique *(21)*
- dilapidated *(9)*
- dumbwaiter *(11)*
- entrapment *(8)*
- evolution *(20)*
- gorge *(16)*
- imperative *(13)*
- interface *(21)*
- intervention *(12)*
- intravenous *(13)*
- jurisdiction *(9)*
- low-head dam *(11)*
- packaging *(13)*
- preclude *(16)*
- protocol *(13)*
- scrutinize *(11)*
- spelunker *(12)*
- subterranean *(5)*
- susceptible *(9)*
- terrain *(10)*
- trenching *(9)*
- triage *(14)*

Application Of Knowledge

1. List specific factors within your district that influence the frequency and severity of vehicle accidents. *(Local protocol)*
2. List the three most hazardous intersections or sections of roadway within your district. *(Local protocol)*
3. Perform a rescue needs assessment for your district. Or, read and revise your district's present assessment. Be sure to include each of the following areas: *(Local protocol)*
   - Fireground
   - Vehicle
   - Aircraft
   - Railroads
   - Agriculture
   - Industry
   - Structural collapse
   - Cave-ins
   - Elevated areas
   - Confined spaces
   - Water hazards
   - Mines/caves
   - Public utilities
   - Hazardous materials
4. Survey your district for low-head dams. Write down the locations of each. *(Local protocol)*

Review Activities

1. Identify each of the following abbreviations and acronyms:
   - DOT *(6)*
   - MSDS *(12)*
   - SCBA *(6)*
   - IC *(6)*
   - EMS *(13)*
   - ALS *(13)*
   - IV *(13)*
   - EPA *(18)*
2. Explain the differences among the following procedures:
   - Rescue *(5)*
   - Recovery *(11 & 13)*
   - Extrication *(7)*
3. Explain the differences between the following pairs of procedures:
   - Basic patient assessment/primary treatment *(13)*
   - Basic life support/advanced life support *(13)*
4. List the questions that must be answered when performing a rescue needs assessment. *(5)*
5. Name six sources of information for a needs assessment. *(6)*
6. List advantages of having a rescue unit on the fireground. *(6)*
7. State the primary reason for having a rescue unit at a working fire. *(6)*
8. Name the most common type of rescue. *(7)*

9. Describe characteristics a rescue unit should look for when assessing a district for potential railroad rescue situations. *(8)*
10. List types of agricultural accidents that may require rescue operations. *(8)*
11. Name types of industrial equipment notorious for inflicting injuries to or entrapping workers. *(9)*
12. List types of buildings that the response team should evaluate for potential structural collapse. *(9)*
13. Name conditions that can result in trench rescue situations. *(9)*
14. Explain why quarries, lumberyards, grain-storage facilities, and cement-production plants are prime sites for potential trench rescue. *(10)*
15. Name typical areas involving elevation differences that may require a rescue unit. *(10)*
16. List the types of elevators that should be identified in a rescue needs assessment. *(10)*
17. List typical examples of confined spaces. *(11)*
18. Explain why low-head dams are called "drowning machines." *(11)*
19. Describe hazards associated with public utilities. *(12)*
20. Explain the rescue unit's responsibilities regarding hazardous materials incidents. *(11)*
21. Explain the assistance and benefits a rescue unit can offer each of the following:
    - Fire response *(12, 13)*
    - Government agencies *(13)*
    - Private sector *(13)*
    - Emergency medical services *(13)*
    - Hospital personnel *(14)*
    - Air operations *(15)*
    - Law enforcement *(16, 17)*
    - Utility companies *(17)*
    - Water/sewer departments *(17, 18)*
    - Gas companies *(18)*
22. Explain why it is imperative that rescue personnel and ambulance crews know each other's capabilities and establish basic protocols for patient handling. *(13, 14)*
23. Sketch the most effective and safest position for EMS and rescue vehicles at an emergency scene. *(14)*
24. Describe limitations to the use of helicopters in rescue operations. *(15, 16)*
25. List rules of thumb for rescue and law enforcement personnel in developing guidelines for positioning vehicles at the emergency scene. *(16, 17)*
26. List guidelines for working with a utility company. *(17)*
27. Explain why *all* downed power lines — including telephone and cable television lines — should be considered dangerous. *(17)*
28. Explain ways in which suppression companies can assist and support rescue units. *(18)*
29. Name two basic keys to successful rescue operations. *(18)*
30. Name the two major factors that influence staffing levels. *(18, 19)*
31. Describe guidelines for volunteer personnel allowed to respond to an incident. *(19)*
32. Name the four attributes that a rescuer must have and use to effect a successful rescue. *(19)*
33. Explain what dictates the vehicles, equipment, and staffing needed for any given district. *(19)*
34. Compare and contrast in-house and outside rescue unit training. *(19, 20)*
35. List the basic theory areas that rescue personnel should learn before they can begin hands-on training. *(20)*
36. Explain the importance of a postexercise critique during the training of rescue personnel. *(21)*

Questions And Notes

Rescue Scene Management

Chapter 2
Rescue Scene Management

INTRODUCTION

The efficient handling of rescue incidents demands a well-organized rescue operation. This includes providing an appropriate response and using a clearly defined incident command/management system to manage the scene and the available resources. The keys to effective response, organization, and management are planning and training. Pre-incident planning and realistic training enable emergency crews to become familiar with general operating guidelines, interfacing with other agencies, and the incident command/management system. Well-trained rescuers know what can and cannot be done with the resources available. Knowledge of the capabilities and limitations of equipment and personnel will improve the "decision making" of those in charge. This chapter examines some of the basic principles involved in emergency response, implementing an incident command/management system, and rescue scene management.

Figure 2.1 A dispatcher receives a rescue call.

EMERGENCY RESCUE RESPONSE

Receiving Information And Dispatching Calls

An important part of effectively operating a rescue unit is making sure that it arrives at the correct location as quickly and safely as possible This type of response can be achieved in part through good communication practices and procedures. The rescue system is usually activited by a dispatcher who receives the call for help (Figure 2.1) The dispatcher should be familiar both with the district and with field operations in order to communicate effectively with both the caller and the responding units. If an excited or confused caller provides incorrect information about the location of an incident, such as mistakenly describing the location to be at an intersection of two parallel streets, a dispatcher who is familiar with the district can often recognize the discrepancy and question the caller further to get accurate information. A dispatcher who is familiar with field operations can sometimes anticipate the need for additional resources and can notify the affected crew(s) so they can be better prepared to respond when called.

Effectively handling the initial response is the first step toward maximizing efficiency in organizing the rescue scene. In order to determine the nature and extent of the emergency and, therefore, what type of response is most appropriate, dispatchers should follow a standardized procedure or checklist for taking information from the reporting party.

The dispatcher receiving the call should first determine the nature and extent of the problem because it may not be an emergency. The problem may involve more than one agency, or it may not be an incident that is appropriate for the rescue unit. The next piece of essential information is the location of the problem — street address, business name, or some landmark. And finally, the dispatcher should find out how many people are involved and what their status is — injured, trapped, or in immediate jeopardy.

Although obtaining this information may seem to be a cumbersome and time-consuming process, it is the best way of determining the services and resources required at an incident. It may require skill on the dispatcher's part because the caller may be excited, hysterical, or incoherent (Figure 2.2). The dispatcher should try to get as much information as possible, giving priority to the nature of the problem and the location of the incident. The dispatcher can then use this information to dispatch emergency crews based on predetermined response levels and on what appears to be needed.

Figure 2.2 Callers may have difficulty describing their location.

Planning The Response

A major part of pre-incident planning is determining what the initial response for each type of rescue incident should be. Many fire departments have predetermined responses for structure fires such as two engines and one ladder or three engines and two ladders. To the extent possible, responses to rescue calls should also be standardized so that each responding unit knows which other resources can be expected to arrive at the scene. First-arriving units can use this knowledge to call for additional help or to cancel units already en route. A predetermined, standardized response has many benefits and is an important step in organizing the emergency scene.

In determining the standard response for a rescue call, the resources most likely to be needed for dealing with such incidents should be included. Standardizing the response helps to ensure that all units that may be needed will arrive soon enough to be used effectively. Primary elements that should be included are emergency medical services, rescue and extrication capabilities, fire protection, law enforcement, and other outside agencies. Each of these entities should be aware of its role and how it will be expected to participate in emergency activities.

Taking Charge

Any rescuer, regardless of rank, may have to assume command of a rescue incident. All rescue personnel should be prepared to accept this responsibility. The major functions of the incident commander (IC) will be making decisions and managing resources, not the hands-on activities of making the rescue.

In addition to knowing rescue procedures, the potential IC must also be mentally and emotionally prepared to assume command. Making decisions involving life-threatening situations can produce significant stress, and the stress is multiplied if the rescuer has not been properly trained and prepared for the pressures of command. Realistic drills and simulations will help rescuers develop the confidence and mental toughness needed to command rescue incidents. An effective leader must remain calm even if there is havoc at the scene. This calmness may appear to be indifference, but it is necessary to reassure victims and subordinates.

Initial Decisions

Upon arriving at the scene, the rescuer in charge will have to make two critical decisions: (1) whether it is safe and/or feasible to attempt a rescue and (2) whether the resources on scene or en route are sufficient or if more need to be called. As part of the first decision, the IC must decide whether the operation about to be undertaken is a rescue or a body recovery. Body recoveries should be handled more cautiously and with less risk than rescues or extrications of live victims. There will always be some risks associated with any rescue operation. However, the IC must decide when those risks are great enough to warrant limiting the actions of rescue personnel. This may be a difficult decision to make, and the rescuers may feel frustrated because they cannot help the victim as much as they would like. It is necessary to weigh these feelings against the potential for rescuers becoming additional victims and the likelihood of the rescue being successful.

It is essential for all rescuers to remember that they did not cause the accident, they are not responsible for the victim being in that situation, and they are not obligated to sacrifice themselves in a heroic attempt to save the victim — especially not in an attempt to recover a body. In fact, it is irresponsible and unprofessional for rescuers to take unnecessary risks that might result in their being incapacitated by an injury and therefore be unable to perform the job for which they have been trained. It is not the function of the fire/rescue service to add victims to the situation. The IC's first priority must be rescuer safety; the second priority is the victim's safety. The IC should never choose a course of action that will require rescuers to take unnecessary risks.

INCIDENT COMMAND/MANAGEMENT AT RESCUE OPERATIONS

Perhaps the most important factor in managing an efficient and organized emergency scene is using a formal incident command/management system. Staffing levels, equipment availability, and other general operating guidelines determine how the command/management system will be implemented within any given organization. Other agencies that may be responding should be familiar with the type of system used so they can function properly within its structure. This can best be accomplished through regular, joint training exercises held at least annually.

It is important that the system be established as part of the operational routine and that it be used at every incident so that all department personnel become familiar with it. Incident command/management systems are useful to large and small departments alike, but they can be especially beneficial to smaller departments that are not accustomed to working large-scale incidents.

Many departments have established incident command/management systems that specify who is in command at all times. These generally specify that the first member on the scene, such as a firefighter or a company officer, assumes command by advising the dispatcher that he or she is "in command." If more than one person arrives simultaneously, the senior person usually has this responsibility. The person assuming command has the *full authority* that goes with the position and remains in command until command is formally transferred or the incident is terminated.

Initially, establishing a formal command post (CP) may not be necessary; however, at least designating the first-arriving emergency vehicle as the command post (until it is determined whether a formal CP is needed) is recommended. If the incident involves multiple units and it appears that it will be a protracted operation, a formal, easily identified command post (CP) should be established (Figure 2.3).

Figure 2.3 A clearly identified command post.

Implementing The System

The incident command/management system should be initiated by the first rescuer arriving on the scene of an emergency. This person should immediately begin to evaluate the situation with the following questions:

- What has occurred?
- What is the current status of the victim(s)?
- Is the situation stable or is it getting worse?
- Can the rescue be handled with the resources on scene or already en route?

The person making the evaluation is at least temporarily in command of the incident. If the evaluation reveals that the actions required to mitigate the emergency are beyond the scope of this person's training, command should be transferred at the earliest opportunity to someone more qualified. In the meantime, the individual should do whatever he or she is qualified to do, such as initiating the IMS by naming the incident and announcing the location of the command post. The actual announcement might be as follows:

"Dispatch, Engine 31."

Dispatch answers, "Go ahead, 31."

"We are at the low-head dam in Harding Park where there is an overturned canoe and the two occupants are missing. I am establishing Harding Command. Dispatch the dive rescue team and an ALS ambulance to the command post at the west end of the dam."

Dispatch replies, "Harding IC, Dispatch copies you need a dive rescue team and an ALS ambulance at the west end of the dam in Harding Park."

If the company on scene is not equipped to begin the rescue, they should begin to do whatever they can. This may be limited to cordoning off the area and isolating any witnesses to the accident. The IC should begin to formulate an incident action plan that reflects the following priorities:

1. Providing for rescue personnel safety and survival
2. Rescuing those who can be saved
3. Recovering the remains of those beyond saving

Whenever the command/management system is implemented, there should be only ***ONE*** incident commander, except in multijurisdictional incidents when a unified command is appropriate. In rescue incidents, a multijurisdictional incident may be one that is located on the boundary between two jurisdictions or is in a body of water that lies in or flows into two or more districts (Figure 2.4). Even when a unified command is used, the chain of command must be clearly defined, and all orders should be issued by one person through the chain of command to avoid the confusion caused by conflicting orders. In industrial rescue incidents, company policy may dictate that an employee, such as the facility manager, must be in charge of anything on company property. But when the expertise and resources of the rescue unit are clearly needed, the manager must defer to the IC for the strategic and tactical decisions needed to mitigate the emergency. The manager can act on behalf of the company by advising the IC in the decision-making process.

The IC should amass enough resources to handle the incident and organize them in a way that will ensure that orders can be carried out promptly, safely, and efficiently. Having sufficient resources on scene will help to ensure the safety of all involved. The organization must be structured so that all available resources can be utilized to achieve the goals of the Incident Action Plan. If necessary, the IC can appoint a command staff to help gather, process, and disseminate information.

All incident personnel must function according to the Incident Action Plan. Personnel should function according to the department's general operating guidelines, but every action should be directed toward achieving the goals and objectives specified in the plan.

Decision Making

The decision-making process for a rescue operation is the same as for any other type of emergency. The facts are gathered, the problem is defined, a strategy is developed, alternatives are considered, tactics are chosen and implemented, and progress is evaluated. If the strategy or tactics do not pro-

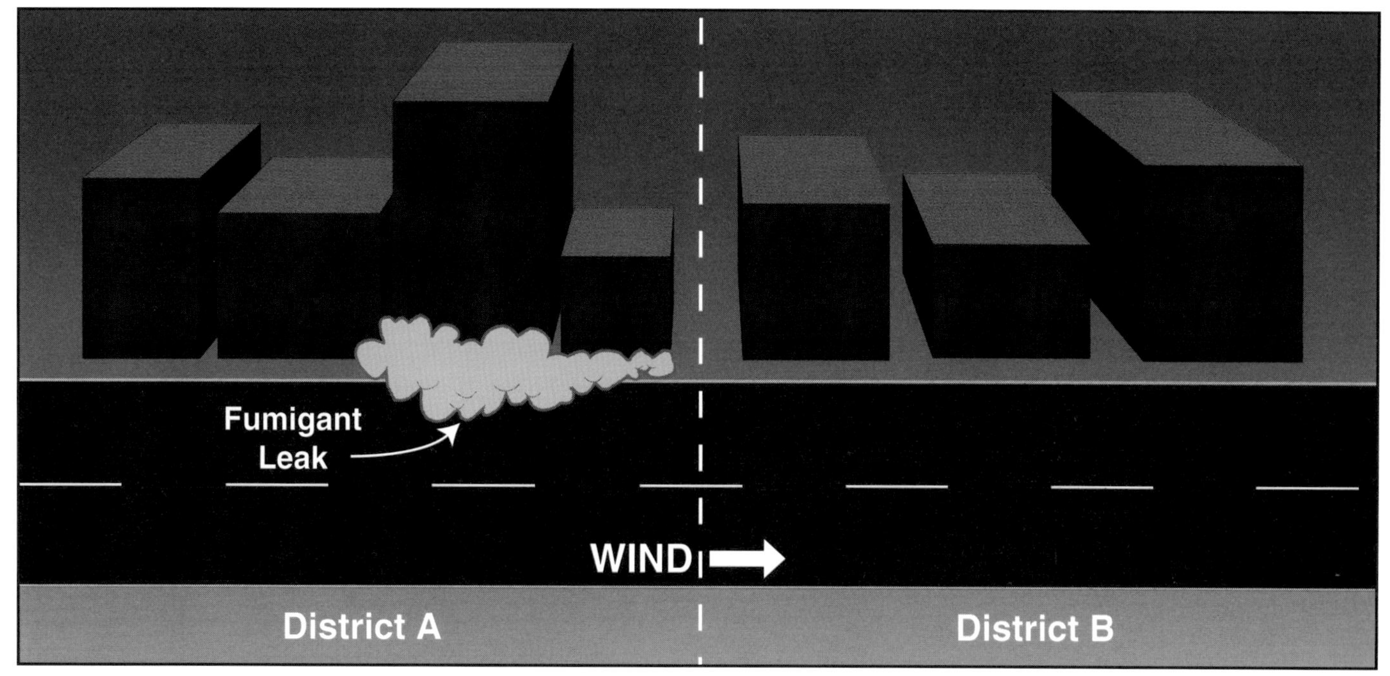

Figure 2.4 A typical multijurisdictional rescue incident.

duce a satisfactory result, an alternative is chosen. Effective leaders know that they can make mistakes and are capable of changing strategies or tactics to achieve the desired result.

PHOENIX MODEL

An excellent example of a decison-making model is the one developed by the Phoenix (AZ) Fire Department. Their emergency response risk management model is used to guide their officers in the decision-making process. The essence of the model is as follows:

- Each emergency response is begun with the assumption that they *"can protect lives and property."*
- They will *"risk their lives a lot, if necessary, to save savable lives."*
- They will *"risk their lives a little, and in a calculated manner, to save savable property."*
- They will *"**NOT** risk their lives at all to save lives and property that have already been lost."*

Obtaining Expert Assistance

The IC cannot be expected to be an expert on every aspect of every situation encountered, so he or she should obtain expert assistance when necessary. Following are some examples of the types of experts whose advice may be needed by the IC:

- Structural or mechanical engineers
- Chemists or hazardous materials specialists
- Railroad officials
- Farmers or agricultural extension agents
- Industrial plant maintenance or engineering personnel
- Elevator mechanics or building engineers
- Mine or cave rescue experts

Experts or professionals in a specialized field can be brought to the CP to advise the IC. They may be better equipped to predict the consequences of a particular action and to suggest alternatives. When dealing with unfamiliar situations, an effective IC will take advantage of every resource available.

Building The Organization

Rescue situations can be as simple as a baby locked inside a car or as complex as the collapse of a heavily occupied structure. Depending on the nature and the scope of the incident, different levels of incident management will be needed. The entire incident command/management organization need not be implemented on every incident. Only those parts of the system that are needed to handle that particular incident safely and efficiently — and that make managing that incident easier — should be used.

When a relatively simple rescue situation occurs, a formal command/management system may not be needed to handle the incident safely and effectively. However, it is good practice for the officer in charge of the first-arriving (and perhaps only) unit to formally assume command as part of the initial report of conditions upon arrival. This keeps incident command/management in the forefront and makes for a smooth transition from handling a single-unit incident to managing a more complex incident with a command/management system. The more that people practice using a command/management system on the simple incidents, the more likely they are to use it on the big ones, when it really counts.

Not all rescue situations will be simple, and the size and complexity of the incident organization should reflect the size and complexity of the situation. When a complex situation develops, command may have to be transferred several times as the organization grows to meet the need. It is important that the transitions be made as smoothly and as efficiently as possible.

COMPONENTS OF A RESCUE OPERATION

As part of establishing command, a systematic size-up must be made that includes certain critical decisions. In general, every rescue incident includes two common elements: (1) scene assessment and (2) establishing command and control. The IC is responsible for seeing that these critical steps are taken; how well they are initially performed will set the tone for the entire operation.

Scene Assessment

Much of the success of a rescue operation can depend upon the proper assessment of the situation (Figure 2.5). The first-arriving unit should assume command and begin scene assessment immediately. While the majority of rescue incidents can be handled by the initial response units, the first important decision to be made is whether the initial response is sufficient to handle the situation. If not, it is essential that the IC immediately call for additional resources. If there is any doubt about the number of resources that will be needed, the IC should call for too many resources rather than too few. Responding units that are not needed can be canceled while still en route and can be returned to quarters.

Figure 2.5 First-in officer interviews a witness.

Command And Control

After an initial size-up, the IC must decide how to utilize the available resources. Under the rescue command and control category, there are several ways in which the IC can subdivide the resources into functional groups/sectors within the chain of command. These groups/sectors may vary from incident to incident or from agency to agency, but in general, those most commonly used are as follows:

- Rescue
- Triage/treatment
- Transportation

The typical rescue incident command chart looks similar to other incident command charts, except that the titles of some of the components are different. The system by which the functional groups/sectors and geographic divisions/sectors are organized should meet the specific needs of the particular rescue operation. The typical rescue scene organization may have to be modified according to the demands of a specific incident and to the needs, capabilities, and limitations of the agencies involved.

Within the limits of span-of-control, the IC is responsible for the actions and coordination of these groups/sectors. On relatively small incidents, the IC can and should stay in constant contact with the supervisors of the groups/sectors and make certain they are performing and interacting properly (Figure 2.6). On larger, more complex incidents, an Operations Section Chief (Ops) should be appointed to coordinate the operations of the various functional groups/sectors (Figure 2.7).

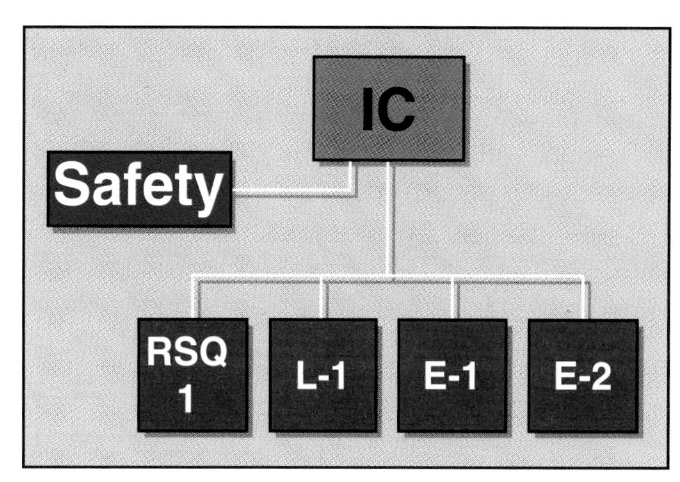

Figure 2.6 Organization of a small rescue incident.

Group/Sector Operations

Good management strategies must be employed in order to handle both small and large incidents, and the use of an incident command/management system to assign functional groups/sectors should facilitate the task. Preparation is vital in order to achieve good results. It is important that rescue personnel train and work with this system regularly so that company members become and stay familiar with it.

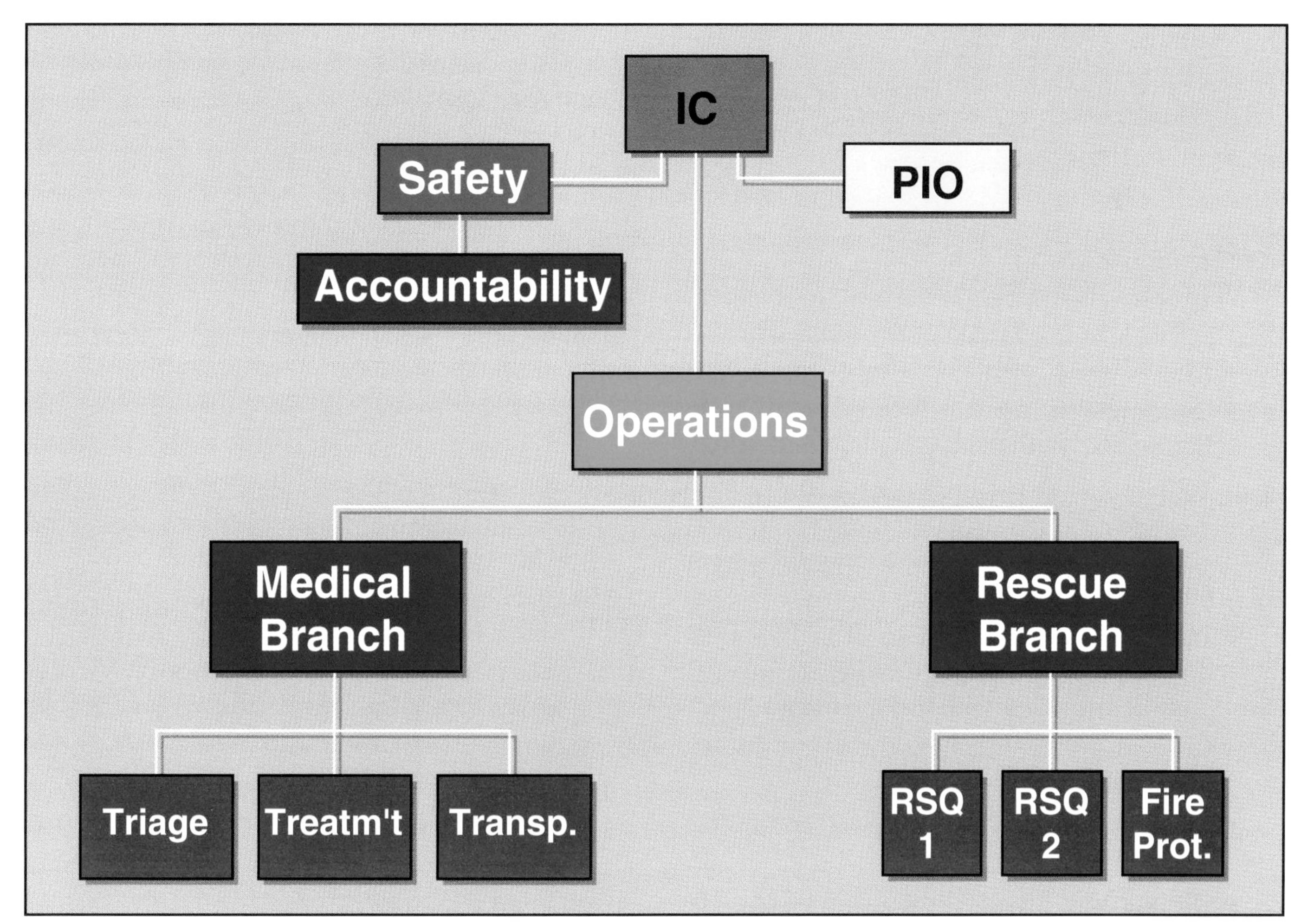

Figure 2.7 Organization of a more complex rescue incident.

Each group/sector should have one person in charge — a group/sector supervisor — to see that all of its assigned responsibilities are completed. The group/sector supervisor is also responsible for coordinating through channels with the IC and with other groups/sectors. Group/sector supervisors, as well as the IC and all other position specialists, should be easily identifiable to personnel on the scene (Figure 2.8). Wearing colored vests (labeled with the position titles) over protective clothing and marking a specific location or vehicle as the command post are the generally accepted ways of promoting easy identification.

Figure 2.8 A clearly identified position specialist.

Duties Of Each Group/Sector

RESCUE GROUP/SECTOR

The responsibilities of the Rescue Group/Sector will vary with the type, magnitude, and complexity of the situation. In general, its duties are as follows:

- Determine the number, location, and condition of the victim(s), both alive and otherwise.
- Evaluate the resources required for the rescue of trapped victims and/or the recovery of bodies.
- Determine whether treatment is necessary and if it can be safely delivered on-site or if victims will have to be moved before treatment. If necessary, move victims to the triage/treatment area(s).
- Advise the IC (through channels) of resource requirements.
- Allocate and supervise resources assigned to the rescue function.
- Report progress to the IC, and give an "all clear" signal when all of the victims have been removed.
- Coordinate with other groups/sectors through channels.

TRIAGE/TREATMENT GROUP/SECTOR

The responsibilities of the Triage/Treatment Group/Sector are to perform triage and begin initial treatment of the victims. Those requiring treatment should be stabilized and continually monitored until they are transported to a medical facility. Separate areas should be set up for those with the most serious injuries (immediate treatment), those with less serious injuries (delayed treatment), and those with minor injuries (minor treatment). The Triage/Treatment Group/Sector Supervisor should advise the Transportation Group/Sector Supervisor (if one has been appointed) of the number of victims in the immediate-, delayed-, and minor-treatment categories so that appropriate transportation can be arranged. The responsibilities of the Triage/Treatment Group/Sector can be summarized as follows:

- Triage victims and continually evaluate their condition.
- Determine the resources needed to treat and transport victims, and advise the IC through channels.
- Identify and establish suitable treatment areas for immediate-, delayed-, and minor-treatment categories. Locate these areas near an easily accessible pickup point for transport. Advise the IC of these areas.
- Assign and coordinate resources to provide suitable treatment for victims.
- Determine transportation priorities, and communicate this information to the Transportation Group Supervisor and/or the IC. Victims in the minor-treatment category may be able to provide their own transportation or use public transportation.
- Maintain an accurate record of victims and where they are transported in the absence of a Transportation Group/Sector.
- Keep the IC informed of progress/problems.

- Coordinate with other groups/sectors through channels.

TRANSPORTATION GROUP/SECTOR

The Transportation Group/Sector is responsible for taking stabilized victims to the appropriate medical facilities (Figure 2.9). In order for it to do its job properly, the Transportation Group/Sector must coordinate with the Treatment Group/Sector. Generally, transportation will not be the job of the rescue unit but most likely will be handled by the emergency medical organization within the jurisdiction of the incident. This does not mean that fire/rescue personnel should not be involved in the operation of the Transportation Group/Sector. On the contrary, appropriate fire department personnel should be assigned to this group/sector to coordinate with other groups/sectors and with those providing the transportation. The responsibilities of the Transportation Group/Sector are as follows:

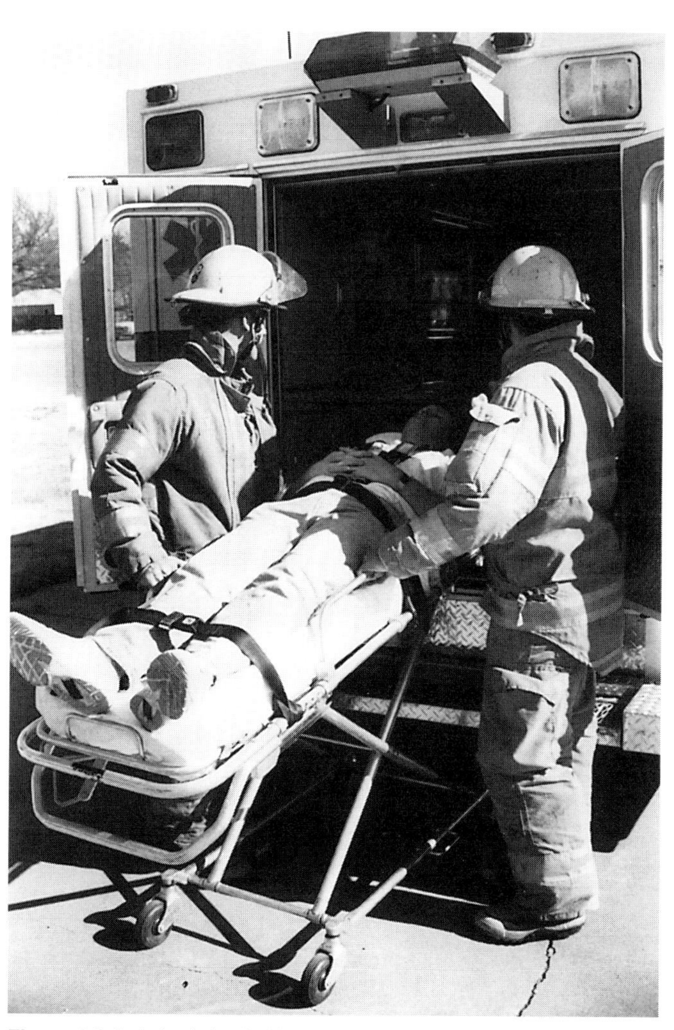

Figure 2.9 A victim is loaded for transport.

- Determine transportation requirements (based on data from the Treatment Group/Sector) and the availability of ambulances and other methods of transportation.
- Report progress and additional resource requirements to the IC.
- Identify ambulance staging and loading areas and determine helicopter landing zones if applicable.
- Verify the victim-handling capabilities of the medical facilities that are to receive the victims.
- Determine the specific entry and exit locations from the triage/treatment area(s).
- Coordinate the order of victim transportation and medical facility allocation with the Treatment Group/Sector.
- Maintain a record of where each victim is taken.
- Establish a means for transporting ambulatory victims.
- Coordinate with other groups/sectors through channels.

Air Transportation

If helicopters are to be used to transport victims from the scene to medical facilities, the Transportation Group/Sector will be responsible for setting up the landing zone. Personnel should locate the largest open area that is as close to the scene as possible. Appropriate sites might include parking lots, open fields, highways, or median strips. The unobstructed open area should be at least 70 x 70 feet (21 m by 21 m), with no more than a 2 percent slope (Figure 2.10). Helicopters rarely land straight down or take off straight up, so the area surrounding the landing zone should be clear of tall obstructions. The landing zone should be well marked so it can be easily seen by the pilot. Any objects that will not be blown about by the downdraft and that contrast starkly with the color of the landing surface may be used for this purpose. Flares may also be used if there is no danger of them starting fires. Hand lights or vehicle headlights may be used at night.

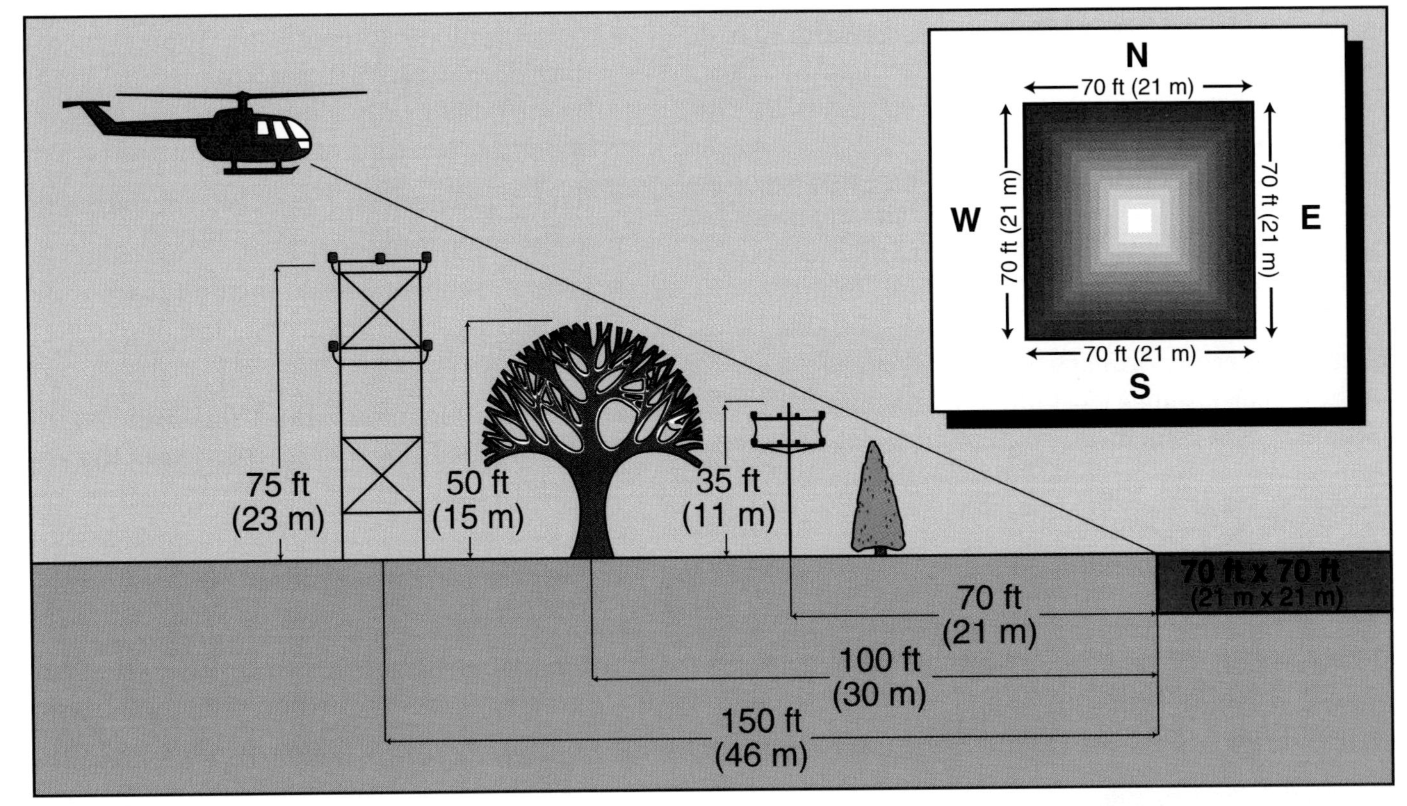

Figure 2.10 Critical clearances for a helicopter landing zone.

CAUTION: During night operations, personnel should *never* shine lights toward an operating helicopter, whether it is aloft or on the ground.

APPARATUS PLACEMENT

Proper placement of apparatus on the emergency scene is an important part of scene management and safety. The goal is to get the vehicles that need to be closest to the operation into that position. Units, such as engine companies, that do not need to be close to the rescue area should leave room for later-arriving EMS and rescue vehicles. Rescue vehicles need to be close enough to the scene to operate effectively, especially if these units are used to supply electrical power, operate hydraulic tools, etc. The following sections give general guidelines for rescue apparatus placement on the emergency scene.

Fire Scene

When operating at a fire scene, it is not as necessary for the rescue vehicle as it is for fire apparatus to be close to the fire building. Rescue vehicle(s) should be positioned where they will not interfere with fire fighting operations but where their equipment will still be readily available. Tools and equipment that might be needed from the rescue vehicle can be carried to the scene. Parking some distance from the fire building also provides a good location for treatment of victims or firefighters and makes it easier to coordinate with ambulances. It would be counterproductive to try to treat people in a rescue vehicle parked in the middle of the smoke.

Rescue Scene

Placement of the rescue vehicle at the nonfire rescue scene depends upon a number of variables. Placement is generally opposite that normally found at a fire scene. At a rescue scene, the rescue vehicle should be positioned nearest the incident. This is necessary because the rescue and extrication equipment may be most important in the situation and should be most readily available. The following are guidelines for the placement of the rescue vehicle(s):

- Place rescue vehicles between the scene and oncoming traffic to protect rescuers if the incident is on the roadway.
- Place the rescue vehicle(s) close enough to the incident to make equipment removal easy and to keep carrying distance to a minimum.

- Do not place the rescue vehicle(s) close enough to be in the way or to expose victims to vehicle exhaust, vibration, or noise.
- Place the rescue vehicle(s) upwind and uphill from the scene whenever possible (Figure 2.11).
- Do not place the rescue vehicle(s) close to downed power lines, damaged transformers, or escaping flammable gas.
- Do not drive heavy vehicles near an open trench because the vibration could cause additional trench wall collapse.
- Do not block the scene. Allow access for ambulances and other emergency vehicles, and allow for the normal flow of traffic if the incident is not in the roadway.

As stated earlier, to provide maximum protection for emergency crews when traffic must be allowed to continue, position emergency vehicles so that they provide a barrier between the traffic and the personnel working on the scene. If it is only possible to close certain lanes of traffic, the lane the accident is in as well as the lane next to it should be closed.

Rescue situations often do not involve vehicles nor are they always located on a street or roadway. They often occur inside buildings or in areas well off the roadway. In these cases, it is important to park rescue and other emergency vehicles in such a way as to not interfere with the normal flow of traffic. Emergency vehicles parked out of traffic lanes should shut down their emergency lights so that

Figure 2.11 Emergency vehicles should park uphill and upwind.

passing motorists will not be attracted by them nor react to them. Shutting down the lights will also reduce the number of spectators drawn to the scene.

CONTROL ZONES

Proper scene management reduces congestion and confusion around the rescue area by reducing the number of personnel within the "hot zone." The *hot zone* is the area where the actual rescue is being performed. The most common method of organizing a rescue scene is to establish three operating zones, commonly labeled "hot," "warm," and "cold" (Figure 2.12). These zones can be described as follows:

- ***Restricted (Hot) Zone*** — the area where the rescue is taking place. Only personnel who are dealing directly with treating or freeing the victim(s) are allowed. This limits crowding and confusion at the scene. The size of this zone may vary greatly depending upon the nature and extent of the problem.
- ***Limited Access (Warm) Zone*** — immediately outside of the hot zone, the area for personnel who are directly aiding rescuers in the hot zone. The warm zone is where those who are handling hydraulic tool power plants, those who are providing emergency lighting, and firefighters who are on standby hoselines would be. Access to this zone should be limited to personnel who are not needed in the hot zone but who are supporting the work being performed there.
- ***Support (Cold) Zone*** — surrounding the previously described zones, this area may include the command post (CP), the public information officer's (PIO) location, and staging areas for personnel and portable equipment. Rescue personnel should remain in this zone until needed in the warm or hot zones. The outer boundary of this area should be cordoned off from the public.

Cordoned Area

In smaller incidents where no evacuation is necessary, cordoning off the area will keep bystanders a safe distance from the scene and out of the way of emergency personnel. There is no specific distance or area that should be cordoned off. The zone boundaries should be established taking into account the amount of area needed by emergency personnel to work, the degree of hazard presented by elements involved in the incident, and the general topography of the area. Cordoning can be done with rope or fireline tape tied to signs, utility poles, parking meters, or any other objects readily available (Figure 2.13). Once the area has been cordoned off, the boundary should be monitored to make sure people do not cross the line.

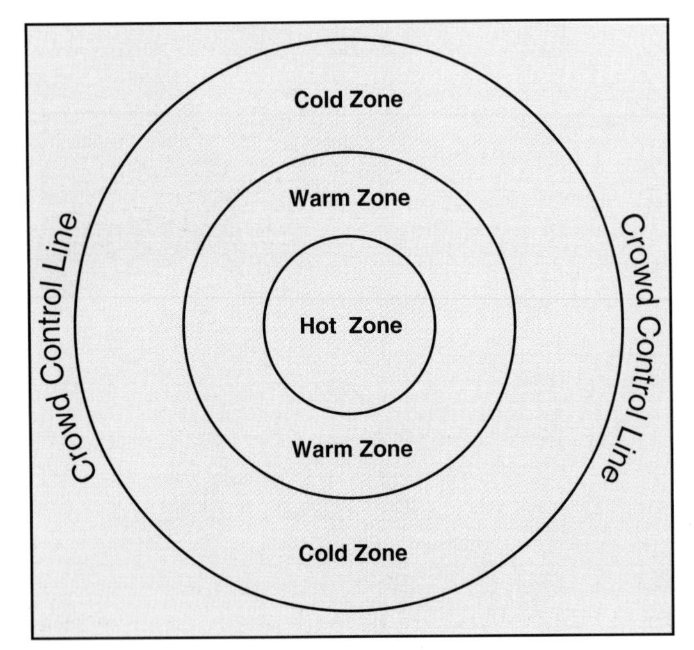

Figure 2.12 Typical zones used to control the scene.

Figure 2.13 The hazardous areas are cordoned off.

Evacuation

Evacuation of a building or neighborhood may be necessary during a rescue operation. Depending on the number of people involved and their condition, evacuation can become a very complex operation. After it has been determined that evacuation is necessary, the first thing the incident commander must decide is what area needs to be cleared. In the case of a structure fire, moving the occupants either out of the involved building or to a safe haven within the building may be all that is needed. However, in the case of a hazardous materials incident, a large number of people may have to be moved a considerable distance to ensure their safety.

The key to a successful evacuation is pre-incident planning. The authorities and the rescue service involved should have an established procedure for various levels of evacuation. Contingency plans should be established for small-, medium-, and large-scale evacuations. A plan of action for notifying people when to evacuate is needed. Arrangements should be made with local television and radio stations to broadcast evacuation orders and information.

Law enforcement and emergency preparedness (Civil Defense) personnel can be extremely helpful in evacuations. Law enforcement personnel can patrol an area and make announcements over public address systems. Both law enforcement and emergency preparedness personnel can institute door-to-door evacuations if necessary (Figure 2.14). Evacuees should be given clear directions as to where they should relocate and approximately how long they will be displaced. Those who cannot evacuate themselves should be assisted in doing so. Some people may refuse to leave their homes or businesses. Depending on the reason for the evacuation and on local protocols, either they may be allowed to stay or they may be placed in protective custody by law enforcement personnel and forced to leave.

Before people are asked to leave, some adequate place to temporarily relocate them must be identified, and security for their unoccupied homes and businesses must also be provided. If an incident involves only a few people, they may go or be taken to the homes of friends or family. Large-scale evacuations may require the use of churches, schools, auditoriums, municipal buildings, or hotels/motels (Figure 2.15). The cooperation of those in charge of these facilities, as well as the Red Cross and similar organizations, should be enlisted during pre-incident planning. Provisions should also be made to feed the anticipated number of people that must be relocated. Evacuees should be checked in when they arrive at a relocation center and checked out when they leave.

Figure 2.14 A police officer assists in an evacuation.

Figure 2.15 Appropriate relocation sites must be established.

In most evacuations, the majority of those displaced will be ambulatory and can provide their own transportation to relocation centers; however, some will be nonambulatory and without personal transportation. Of those who are nonambulatory, most will be in wheelchairs, but some will be completely bedridden. Some will be on full-time oxygen or other life-support systems. These exceptional needs must be anticipated and provisions made for them during pre-incident planning.

Control Of Nonrescue Personnel And Vehicles

Crowd control is essential to managing a well-organized rescue operation. This function is usually the responsibility of the law enforcement agency on the scene, but it may sometimes have to be performed by firefighters or other rescue personnel. It is the responsibility of the IC to ensure that the scene is secured and properly managed.

BYSTANDERS

Even in the most remote locations, bystanders or spectators are often drawn to the scene. Some may be people who were involved in the accident but are not injured. They are often quite curious and try to get as close to the scene as possible. All bystanders should be restrained from getting too close to the incident for their own safety and for that of victims and emergency personnel.

Rescue scenes tend to be emotional situations that should be handled with care. This is particularly true when friends or relatives of the victims are at the scene. These particular bystanders are often difficult to deal with, and rescuers must treat them with sensitivity and understanding. Relatives and friends of victims should be gently but firmly restrained from getting too close, and they should be kept some distance from the actual incident but still within the cordoned area. While they may console each other, they should not be left entirely on their own. A rescuer or other responsible individual should stay with them until the victims have been removed from the scene.

UNINJURED PARTIES

Upon arrival at an incident, rescuers may find people who were involved in the incident but who are uninjured. These uninjured parties need to be handled in an organized manner, and they should not be allowed to wander the scene. Some may be able to assist in handling the incident such as by providing critical information. Others should be escorted to an emergency vehicle or other collection point until it is appropriate to release them. All of these people should be assessed by emergency medical personnel before being released. In the absence of a local protocol to the contrary, anyone who refuses treatment or transportation should be asked to sign a release-of-liability form. Reasons for controlling uninjured parties include the following:

- To keep them from wandering the scene
- To keep the uninjured from getting injured
- To provide a method of accounting for everyone involved in the incident
- To obtain information from those involved in the incident
- To separate witnesses from each other to prevent them from discussing what they saw and perhaps influencing each other to coordinate their stories

The pre-incident plan should identify who is to be responsible for crowd control. These responsibilities are usually assigned to a law enforcement agency; however, if its personnel are busy handling other problems, they may be unable to respond in a timely manner. Uncommitted fire/rescue personnel may be placed in charge of handling uninjured victims, recording necessary information, and coordinating with other agencies such as the Red Cross. The personnel in charge of this operation should maintain contact and coordinate with the IC through the proper channels.

TRAFFIC

A very important function in providing safety around the rescue scene is the control of vehicular traffic. Controlling the flow of traffic makes operations at the scene run smoother and allows for more efficient access and departure of emergency vehicles. Although the law enforcement agency usually handles traffic control, in some cases the rescue organization may have to perform this function. Fire and rescue personnel should be trained in the basics of traffic direction and safety. Some volunteer departments that are located in areas with limited law enforcement agencies have personnel called "fire police." These volunteers are members of the fire department who perform crowd and traffic control, secure the scene, and perform similar functions that would normally be handled by law enforcement personnel.

INCIDENT TERMINATION

The termination phase of a rescue involves such obvious elements as retrieving pieces of equipment used in the operation. But it also involves less obvious elements such as investigating the cause(s) of the incident, releasing the scene to those responsible for it, and conducting critical incident stress debriefings (CISD) with members of the rescue teams.

Equipment Retrieval

Depending on the size, complexity, and length of time involved in the operation, the job of retrieving all of the various pieces of equipment used may be very easy, or it may be very difficult and time-consuming. Under some circumstances, it can also be quite dangerous.

IDENTIFYING/COLLECTING

The process of identifying and collecting pieces of equipment assigned to the various pieces of apparatus on scene is much easier if each piece of equipment is clearly marked (Figure 2.16). However, it may be necessary for the driver/operators of rescue units and other pieces of apparatus on the scene to conduct an inventory of their equipment prior to leaving the scene (Figure 2.17). If the operation was large enough to require the establishment of a Demobilization Unit, it will coordinate the recovery of loaned items, such as portable radios, and the documenting of lost or damaged pieces of apparatus and equipment (Figure 2.18).

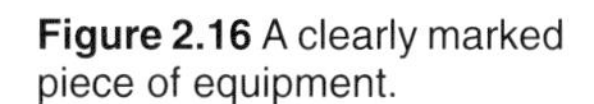

Figure 2.16 A clearly marked piece of equipment.

Figure 2.17 Driver/operators may have to inventory their equipment.

Figure 2.18 A rescue vehicle during demobilization.

ABANDONMENT

In some cases, the environment within the rescue scene may be too hazardous to justify sending rescue personnel back to retrieve pieces of equipment — even expensive ones. Rather than putting rescue personnel at risk, it is sometimes advisable to simply abandon the equipment in place. It may be retrievable after the scene has been restored, or the cost of replacing the abandoned equipment may be recovered from the owner of the property.

Investigation

All rescues should be investigated at some level. At the very minimum, a departmental investigation should be conducted for purposes of reviewing and critiquing the operation. However, if an employee was injured in the incident, the incident will be investigated by the Occupational Safety and Health Administration (OSHA) and perhaps by other entities such as the employer's insurance carrier. Obviously, if the rescue situation was the result of a crime, such as a bombing, law enforcement agencies will also investigate the incident.

Release Of Scene

Once rescuers respond to the rescue scene, they assume control of the scene and the immediate surrounding area. Within certain limits, they can deny access to anyone, including the owner of the property. Legitimate members of the news media have certain constitutionally protected rights of access, but the interpretations of these rights vary from state to state and from country to country. Rescuers should be guided by local protocols.

The process of releasing control of the scene back to the owner or other responsible party is sometimes not as straightforward as it might seem. The owner or responsible party should be escorted on a tour of the scene, or as close to it as possible consistent with safety, and should be given an explanation of any remaining hazards. If the scene is still too hazardous to leave unattended, the owner may be required to post a security guard, erect a security fence around the hazard, or both. Before the scene is released, the department may require the owner to sign a written release that describes the hazards and stipulates the conditions the owner must meet.

Critical Incident Stress Debriefing

Because the injuries suffered by the victims in rescue incidents sometimes can be extremely gruesome and horrific, the members of the rescue teams and any others who had to deal directly with the victims should be *required* to participate in a CISD process (Figure 2.19). Because individuals react to and deal with extreme stress in different ways — some more successfully than others — and because the effects of unresolved stresses tend to accumulate, participation in this type of process should not be optional.

Figure 2.19 Rescue personnel in a CISD session.

The process should actually start *before* rescuers enter the scene if it is known that conditions exist there that are likely to produce psychological or emotional stress for the rescuers involved. This is done through a prebriefing process wherein the rescuers who are about to enter the scene are told what to expect so that they can prepare themselves.

If rescuers will be required to work more than one shift in these conditions, they should go through a minor debriefing, sometimes called "defusing," at the end of each shift. They should also participate in the full debriefing process within 72 hours of completing their work on the incident.

Chapter 2 Review

Directions

The following activities are designed to help you comprehend and apply the information in Chapter 2 of **Fire Service Rescue**, Sixth Edition. To receive the maximum learning experience from these activities, it is recommended that you use the following procedure:

1. Read the chapter, underlining or highlighting important terms, topics, and subject matter. Study the photographs and illustrations, and read the captions with each.
2. Review the list of vocabulary words to ensure that you know the chapter-related meaning of each. If you are unsure of the meaning of a vocabulary word, look up the word in the glossary or a dictionary, and then study its context in the chapter.
3. On a separate sheet of paper, complete all assigned or selected application and review activities before checking your answers.
4. After you have finished, check your answers against those on the pages referenced in parentheses.
5. Correct any incorrect answers, and review material that was answered incorrectly.

Vocabulary

Be sure that you know the chapter-related meanings of the following words:

- ambulatory *(35)*
- cordon *(38)*
- debriefing (42)
- demobilization *(41)*
- havoc *(28)*
- incoherent *(28)*
- mitigate *(30)*
- prebriefing *(42)*
- protracted *(29)*
- rescue vs. recovery *(27 & 29)*
- sector vs. division vs. group *(32-34)*
- triage *(34)*

Application Of Knowledge

1. List your department's standardized procedure (checklist) for taking information from a party reporting a situation requiring a rescue. *(Local protocol)*
2. Read your department's predetermined standardized response for several types of rescue calls. List the resources included for one simple and one complex response. *(Local protocol)*
3. Describe — in outline form — the command/management system used in your jurisdiction. *(Local protocol)*
4. Describe your jurisdiction's critical incident stress debriefing program. *(Local protocol)*

Review Activities

1. Identify each of the following:
   - IMS *(30)*
   - IC *(28)*
   - CP *(29)*
   - CISD *(41)*
   - PIO *(38)*
   - Ops *(33)*
   - OSHA *(41)*
   - Incident Action Plan *(30)*
   - Phoenix Model *(31)*
2. Explain the importance of effectively handling the initial response. *(27)*
3. Describe the primary elements that should be included in a predetermined standardized response for a rescue call. *(27, 28)*
4. State the major functions of the incident commander. *(28)*
5. Explain how fire departments can help their rescue workers develop the confidence and mental toughness needed to command rescue incidents. *(28)*
6. List the initial two critical decisions the rescuer in charge will have to make on arriving at the incident scene. *(29)*
7. List the IC's first priority in regard to safety. *(29)*
8. Explain the factors that determine how the command/management system will be implemented within any given organization. *(29)*
9. Explain when a formal, easily identified command post should be established. *(29)*

10. List the scene evaluation questions that should be asked by the first rescuer arriving on the scene of an emergency. *(30)*
11. Explain what the first-arriving company should do if it is not equipped to begin the rescue. *(30)*
12. Define/describe each of the following terms:
    - chain of command *(30)*
    - unified command *(30)*
    - Incident Action Plan *(30)*
    - Phoenix Model *(31)*
13. List the steps in the decision-making process for a rescue operation. *(30)*
14. State the four main points of the Phoenix decision-making model. *(31)*
15. List the various types of experts that may be needed by an IC. *(31)*
16. State the general rule of thumb employing the different levels of management in the incident command/management system. *(31, 32)*
17. List the two common elements included in every rescue system. *(32)*
18. Explain what is meant by *scene assessment. (32)*
19. List the three most common groups/sectors in an incident command/management system. *(32)*
20. Describe some ways that the IC, the group/sector supervisor, and other position specialists can be easily identified by personnel at the scene. *(34)*
21. List the basic responsibilities of a rescue group/sector. *(34)*
22. List the basic responsibilities of a triage/treatment group/sector. *(34, 35)*
23. List the basic responsibilities of a transportation group/sector. *(35)*
24. Name appropriate landing sites for helicopters used in a rescue operation. *(35)*
25. State the minimum dimensions of a helicopter landing site. *(35)*
26. Compare and contrast apparatus placement at a fire scene and at a nonfire scene. *(36)*
27. List guidelines for placement of rescue vehicles at a rescue scene. *(36-38)*
28. Explain the three operating zones: hot, warm, and cold. *(38)*
29. Briefly explain large incident and small incident evacuation procedures. *(39)*
30. Explain how the rescuer should handle bystanders, relatives, and friends of victims. *(40)*
31. Explain when a Release of Liability form should be used. *(40)*
32. Explain reasons for controlling persons at the scene who were not injured. *(40)*
33. Explain when it is best to abandon rather than retrieve a piece of equipment. *(41)*
34. List the four basic steps in releasing control of the scene back to the owner or other responsible party. *(41, 42)*
35. Discuss critical incident stress, prebriefings, and "defusings." Why should participation in a full critical incident stress debriefing within 72 hours of the incident be mandatory for all involved rescuers? *(42)*

Questions And Notes

3 Rescue Vehicles And Equipment

Chapter 3

Rescue Vehicles And Equipment

INTRODUCTION

Continuing the process of evaluating rescue resources and planning for rescue incidents, this chapter focuses on the special vehicles and equipment needed to perform rescues as quickly and safely as possible. The first section examines different types of rescue vehicles and their advantages and disadvantages. The next section of the chapter discusses the various equipment a rescue unit might need during a rescue incident. The last section examines the personal protective equipment that rescuers should have and wear to safely operate at the emergency scene.

RESCUE VEHICLES

Light Rescue Vehicles

Light rescue vehicles are designed to handle only basic rescue, extrication, and life-support functions; therefore, they carry only basic hand tools and small equipment. Often, a light rescue unit functions as a first responder or rapid intervention vehicle; that is, the unit attempts to handle small incidents and to stabilize larger incidents to keep them from getting worse until heavier equipment arrives. The standard equipment carried on ladder and engine companies may also give them light rescue capabilities.

Light rescue vehicles are generally built on a 1-ton or 1½-ton chassis. The rescue unit's body may resemble a multiple-compartment utility truck. The size of this vehicle limits the amount of equipment it can carry. A light rescue vehicle can carry a variety of small hand tools, such as saws, jacks, and pry bars, as well as smaller hydraulic rescue equipment and small quantities of emergency medical supplies (Figure 3.1). These vehicles are generally capable of transporting two to seven rescue team members. Vehicles carrying more than three people will require either a four-door cab or an enclosed crew compartment in the body of the vehicle.

Figure 3.1 A light rescue vehicle.

Medium Rescue Vehicles

The medium rescue vehicle has more capabilities than the light rescue vehicle (Figure 3.2). In addition to basic hand tools, this vehicle may carry powered hydraulic spreading tools and cutters, air bag lifting systems, power saws, acetylene cutting equipment, and ropes and rigging equipment. Medium rescue units are capable of handling the majority of rescue incidents. They may also carry a

Figure 3.2 A typical medium rescue vehicle.

variety of fire fighting equipment, making them dual-purpose units. These vehicles can carry as many as 8 to 10 rescue team members.

Specialized units are often considered medium rescue vehicles. Specialized units have specific uses, but they may carry generalized equipment that can be used in other incidents. Some types of specialized units are hazardous materials units, water rescue and recovery units, bomb disposal units, mine rescue units, and floodlight/power units (Figure 3.3).

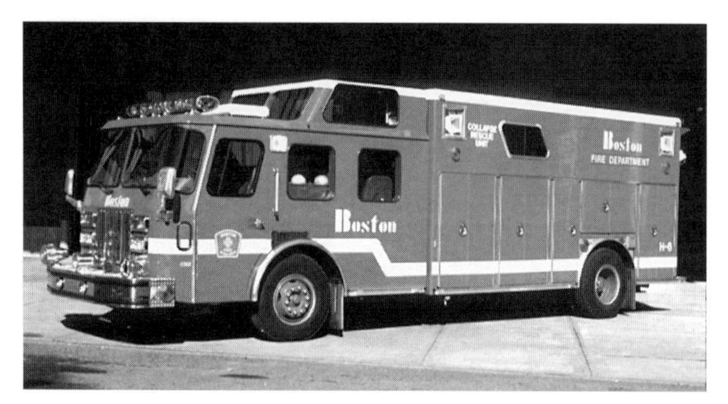

Figure 3.3 A specialized rescue vehicle. *Courtesy of Joel Woods.*

Heavy Rescue Vehicles

Heavy rescue units must be capable of providing the support necessary to rescue victims from almost any entrapment. As their name implies, heavy rescue units have heavier and more specialized equipment than smaller units (Figure 3.4). Additional types of equipment carried by the heavy rescue unit are as follows:

- Booms, A-frames, or gin poles
- Cascade systems
- Larger power plants
- Trenching and shoring equipment
- Small pumps and foam equipment
- Large winches
- Hydraulic booms
- Large quantities of rope and rigging equipment
- Air compressors
- Ladders

Other specialized equipment may be carried according to the responsibilities of the rescue unit

Figure 3.4 A heavy rescue vehicle. *Courtesy of East Greenville (PA) Fire Co. No. 1.*

and the special hazards peculiar to the district. Heavy rescue units are frequently oriented more toward fire fighting than smaller units because they have more space available for fire fighting equipment. They can also carry larger numbers of rescue personnel. Many units have seating for 12 or more people.

Rescue Pumpers

Attempts to streamline operations and combine functions have led the fire service to the use of multipurpose or combination apparatus. One multipurpose vehicle that has gained popularity in recent years is the combination rescue/pumper. This apparatus is designed to carry out the functions of both a structural fire pumper and a rescue vehicle. The result is an apparatus that is useful at almost any type of incident and that has sufficient equipment to handle most rescue and extrication incidents.

Rescue pumpers vary in size. Some departments use minipumpers or midipumpers (initial attack fire apparatus) with light rescue capabilities (Figure 3.5). Other departments use full-sized, custom-designed apparatus that are basically engine companies with extra-large compartments or other modifications for carrying rescue equipment (Figure 3.6). These larger apparatus are equipped with Class A fire pumps and large water tanks.

Standard Engine Company

In some areas, the engine company is expected to provide certain rescue and extrication services (Figure 3.7). Using equipment carried on any standard engine company, personnel should be able to

Figure 3.5 A combination minipumper/rescue vehicle.

Figure 3.6 A rescue/engine combination. *Courtesy of Ron Bogardus.*

Figure 3.7 A standard engine equipped for rescue. *Courtesy of Joel Woods.*

perform light rescue tasks. An early-arriving engine company will sometimes be able to perform a rescue before specialized equipment arrives. If not, the engine company should stand by to provide fire protection and additional personnel for rescue companies that are working the incident.

Ladder Company

In most cases, ladder companies are better equipped to perform rescue operations than the engine companies. Ladder companies usually have a greater variety of equipment (normally used for forcible entry purposes) that can be used for rescue and extrication operations. Ladder companies can provide valuable assistance at large-scale incidents when additional help is needed but no other rescue companies are available.

In areas where budgetary limitations preclude the establishment of a separate rescue unit, ladder companies sometimes carry rescue equipment and routinely do most of the rescue work. Since aerial apparatus typically have a large amount of compartment space, they lend themselves well to carrying additional rescue equipment (Figure 3.8). Personnel who are already trained in ladder company operations are often ideal for rescue operations.

Figure 3.8 Ladder trucks often carry rescue equipment.

TYPES OF RESCUE VEHICLE BODIES

Exclusive Exterior Compartmentation

Vehicles with exterior compartmentation usually do not have a walk-through area or interior storage. Equipment is only accessible from exterior compartments. Exterior compartmentation is advantageous at an emergency scene because the rescuer does not have to enter the vehicle to get needed equipment. The disadvantage is that the number of personnel transported to the scene is limited because of space shortage within the vehicle's cab. Exclusive exterior compartmentation is most commonly found in smaller rescue units, although some larger units are also designed in this manner (Figure 3.9).

Exclusive Interior Compartmentation

Vehicles designed with interior compartmentation exclusively have all their storage in an interior walk-through area. These units are convenient because all the equipment is accessible from the inside of the vehicle. However, interior compartmentation may slow procedures at an emer-

Figure 3.9 A rescue vehicle with exterior compartments only. *Courtesy of Ron Jeffers.*

Figure 3.10 A rescue vehicle with interior compartments only. *Courtesy of Joel Woods.*

gency scene because personnel have to enter the vehicle for equipment. These vehicles can usually carry more people safely inside the vehicle than can vehicles with only exterior compartmentation (Figure 3.10).

Combination Compartmentation

Perhaps the most functional style of rescue body is the combination walk-through body with both exterior *and* interior compartmentation. This type of vehicle offers the advantages of each design. The combination vehicle offers flexibility in setup that makes it convenient to the rescuers (Figure 3.11).

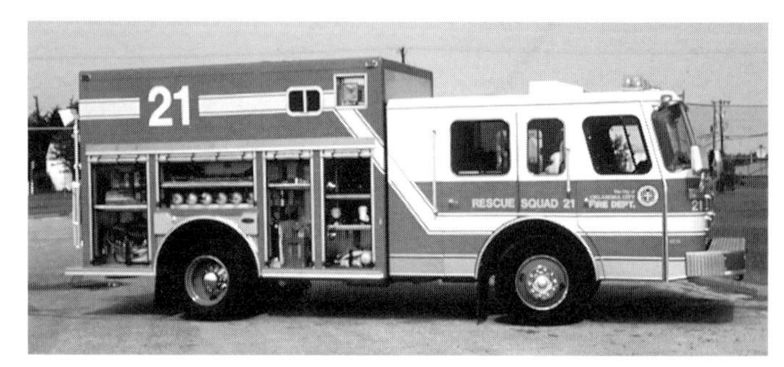

Figure 3.11 Some rescue vehicles have both interior and exterior compartments. *Courtesy of Mike Wieder.*

RESCUE VEHICLE CHASSIS

Commercial Chassis

Commercial chassis are built by commercial truck manufacturers, and they are typically used as commercial service and delivery vehicles. Commercial chassis are also the most commonly used chassis for rescue vehicles. All light- and medium-chassis units and a large percentage of heavy-rescue chassis units have commercial chassis (Figure 3.12). Commercial chassis may be more economical and easier to get serviced than custom-built chassis. The disadvantage is that commercial chassis, particularly smaller ones, may not be designed to withstand the rigorous service that rescue responses require; they may need to be modified.

Figure 3.12 A rescue body on a commercial chassis.

Custom Chassis

Custom chassis are built by manufacturers who specialize in emergency vehicle chassis, so they are designed to withstand the heavy use of the emergency service (Figure 3.13). Custom chassis incor-

Figure 3.13 A custom-built rescue vehicle. *Courtesy of Emmaus (PA) Fire Department.*

porate special features that are requested by the departments purchasing them. Generally, the use of custom chassis is limited to heavy rescue vehicles.

SPECIAL RESCUE VEHICLE EQUIPMENT AND ACCESSORIES

All-Wheel Drive

Rough terrain or severe weather conditions may require the rescue vehicle to be equipped with all-wheel drive (Figure 3.14). All-wheel drive allows safe, reliable vehicle operation under adverse conditions. The type of response area and the type of calls anticipated will determine the need for this capability. Hilly areas and areas where agricultural accidents may occur are good examples of places where off-the-road capability may be necessary. All-wheel drive is also an advantage on vehicles that have to tow or pull other vehicles or objects.

Figure 3.14 An all-wheel-drive rescue vehicle. *Courtesy of Mike Wieder.*

Gin Poles, A-Frames, Or Booms

Gin poles, A-frames, and hydraulic booms are vertical lifting devices that may be attached to rescue apparatus (Figure 3.15). A gin pole consists of a single pole that is supported by guy wires to the vehicle. An A-frame consists of two poles attached several feet apart on the apparatus and whose working ends are connected, forming the shape of the letter "A." A hydraulic boom is usually a telescoping boom that can be rotated 360° to ease the process of locating the working end of the boom directly above the load. All of these lifting devices have a pulley at the working end connected by a cable to a vehicle-mounted winch. Some of these devices have lifting capacities in excess of 3 tons (2 721 kg). Stabilizers should be used to steady the rescue vehicle whenever these devices are used.

Gin poles and A-frames are not designed for lateral (sideways) stress, so care should be taken to avoid such strains. Guide ropes may be used to maintain lateral stability. It is important that vehicles equipped with a gin pole or an A-frame system not be used to lift loads that exceed the rated weight that the apparatus chassis is designed to carry. Exceeding the gross vehicle weight limit may result in damage to the axles, chassis frame, or both.

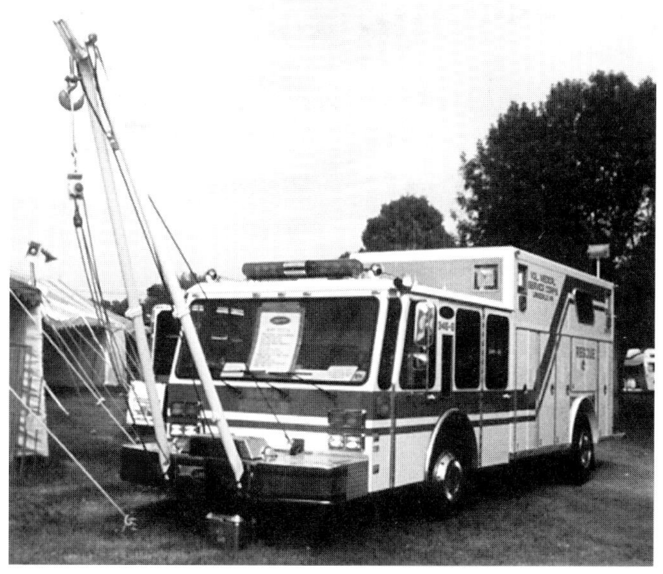

Figure 3.15 A typical A-frame on a rescue vehicle. *Courtesy of Mike Wieder.*

Stabilizers

Stabilizers, also known as *stabilizing jacks* or *outriggers*, are used to stabilize the rescue vehicle when a hydraulic lifting boom, gin pole, or A-frame is in use (Figure 3.16). Stabilizers prevent the vehicle from tipping and reduce strain on the vehicle's suspension system. Stabilizers are generally of two types: remotely controlled hydraulic and manually operated screw-type jacks.

Cascade Systems

Some rescue vehicles are equipped with a series of large-capacity air bottles that are connected by high-pressure tubing. These series of air bottles are called *cascade systems* (Figure 3.17). Their primary use is to refill SCBA cylinders while still at the fire scene. Most cascade systems consist of three to twelve large cylinders that are connected to a

Figure 3.16 Some rescue vehicles must be stabilized. *Courtesy of Joel Woods.*

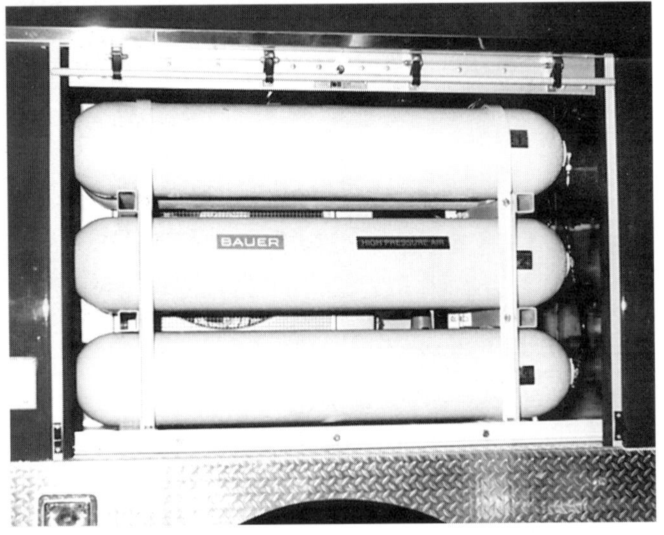

Figure 3.17 Some rescue vehicles have cascade systems.

common manifold. For more information on operating cascade systems, see the IFSTA **Self-Contained Breathing Apparatus** manual.

Air Compressors

BREATHING AIR

Rescue vehicles may be equipped with air compressors that can generate breathing-quality compressed air (Figure 3.18). This air can be used to fill cascade or SCBA cylinders, to support supplied-air breathing equipment in confined spaces, or to purge areas of hazardous, nonflammable, oxygen-depleting gases. These units must be located in a clear atmosphere for proper operation. Generally, they are not operated close to fire scenes because of questionable air quality caused by smoke and other combustive gases in the air. This must be taken into account when positioning the apparatus.

Figure 3.18 A typical rescue air compressor. *Courtesy of Mike Wieder.*

NONBREATHING AIR

Nonbreathing-air compressors supply compressed air for pneumatic (air-powered) tools such as air chisels and air lifting bags. Because of the many types and sizes, as well as their relatively low cost, nonbreathing-air compressors are found on many rescue vehicles (Figure 3.19). Since these compressors are never used to refill SCBA cylinders, they do not require a clean atmosphere for operation.

Figure 3.19 Nonbreathing air is needed to operate tools. *Courtesy of Mike Wieder.*

Power-Generating And Lighting Equipment

Rescue companies are often responsible for providing auxiliary power and lighting at the emergency scene. Electrical power is needed to run electrical equipment, and lighting the scene during nighttime operations is important for safety and efficiency.

POWER PLANTS

Inverters (alternators) are used on rescue vehicles or ambulances when large amounts of power are not necessary and when small electrically operated tools need to be used. The inverter is a step-up transformer that converts the vehicle's 12- or 24-volt DC current into 110- or 220-volt AC current. Advantages of inverters include fuel efficiency and low or nonexistent noise during operation. Disadvantages include limited power supply capability and limited mobility from the vehicle.

Generators can be portable or fixed to the apparatus. They are the most common power source used for emergency services. Portable generators are powered by small gasoline or diesel engines and generally have 110- and/or 220-volt capability (Figure 3.20). Most portable generators are light enough to be carried by two people. They are extremely useful when electrical power is needed in an area that is not accessible to the vehicle-mounted system.

Vehicle-mounted generators usually have a larger power-generating capacity than portable units (Figure 3.21). In addition to providing power for portable equipment, vehicle-mounted generators provide power for the floodlighting system on the vehicle. Vehicle-mounted generators can be powered by gasoline, diesel, or LP-gas engines or by hydraulic or power take-off systems. Fixed floodlights are usually wired directly to the unit through a switch, and outlets are also provided for other equipment. These power plants generally have 110- and 220-volt capabilities with output capacities up to 50 kilowatts—occasionally greater. However, mounted generators with a separate engine are noisy, making communication difficult near them.

LIGHTING EQUIPMENT

Lighting equipment can be divided into two categories: fixed and portable. Portable lights are used in areas that fixed lights are not able to illuminate because of opaque obstructions or when additional lighting is necessary. Portable lights generally range from 300 to 1,000 watts (Figure 3.22). They may be supplied with power by a cord from either a vehicle-mounted power plant or from a self-contained portable power unit. The lights usually have handles for ease of carrying and large bases for stability. Some portable lights are mounted on telescoping stands, which allow them to be directed more effectively (Figure 3.23).

Figure 3.20 A typical portable generator.

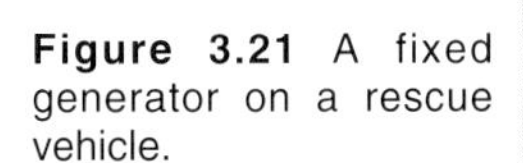

Figure 3.21 A fixed generator on a rescue vehicle.

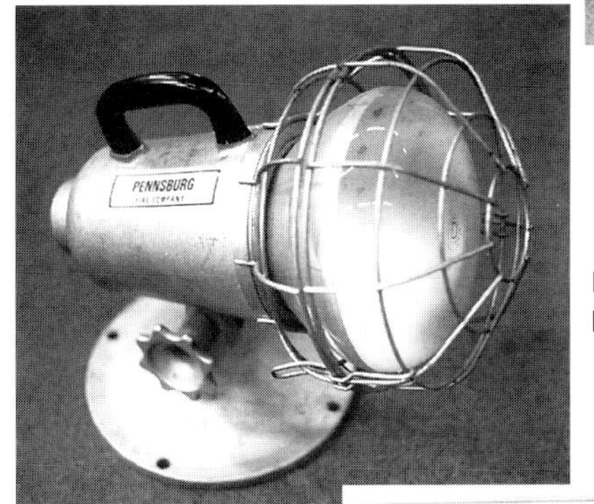

Figure 3.22 A typical portable light.

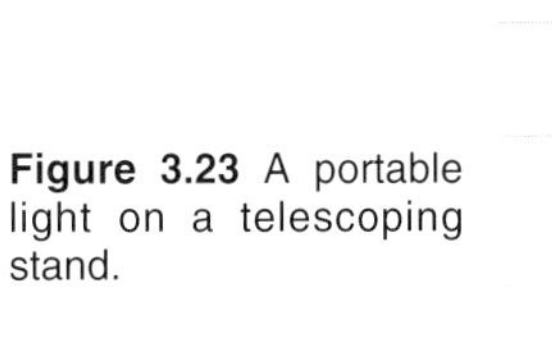

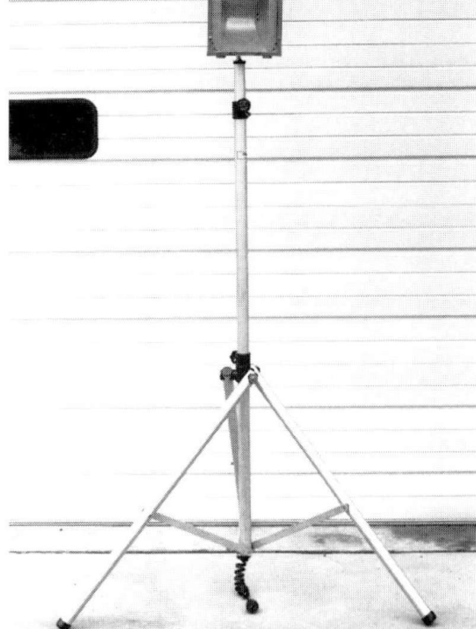

Figure 3.23 A portable light on a telescoping stand.

Fixed lights are mounted to the vehicle, and their main function is to provide overall lighting of the emergency scene. Fixed lights are usually mounted so that they can be raised, lowered, or turned to provide the best possible lighting. Often, these lights are mounted on telescoping poles that allow both vertical and rotational movement (Figure 3.24). More elaborate designs include hydraulically operated booms with a bank of lights (Figure 3.25). These banks of lights generally have a capacity of 500 to 1,500 watts per light. The amount of lighting should be carefully matched with the amount of power available from the power plant. Overtaxing the power plant will give poor lighting, may damage the power-generating unit or the lights, and will restrict the operation of other electrical tools using the same power supply.

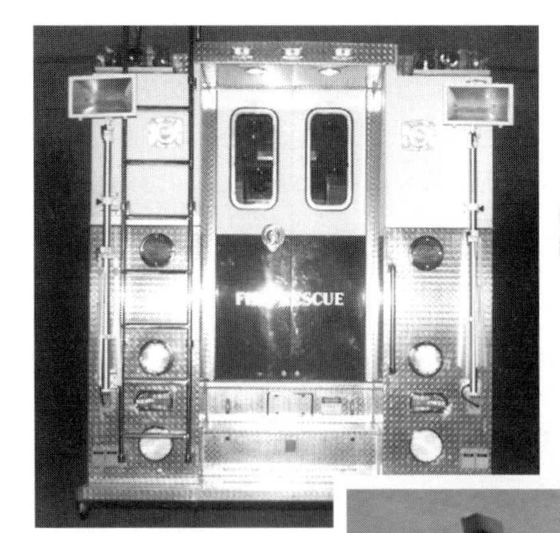

Figure 3.24 Telescoping lights on a rescue vehicle.

Figure 3.25 A typical lighting unit. *Courtesy of Mike Wieder.*

AUXILIARY ELECTRICAL EQUIPMENT

A variety of other equipment may be used in conjunction with power plants and lighting equipment. Electrical cables or extension cords are necessary to conduct electric power to portable equipment. The most common size cable is a 12-gauge, 3-wire type. Cords may be stored in coils, on portable cord reels, or on fixed automatic rewind reels (Figure 3.26). Twist-lock receptacles provide secure, safe connections (Figure 3.27). Electrical cable should be waterproof and explosion-proof and have adequate insulation with no exposed wires. Junction boxes may be used when multiple connections are needed (Figure 3.28). The junction box has several outlets and is supplied through one inlet from the power plant. All outlets should be equipped with ground-fault circuit interrupters and should conform to NFPA 70E, *Standard for Electrical Safety Requirements for Employee Workplaces.*

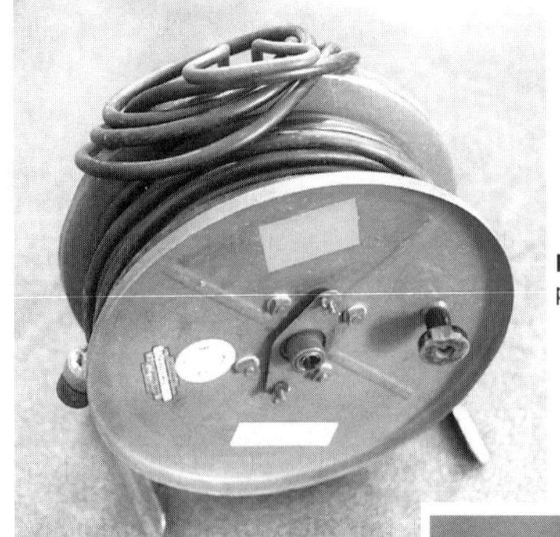

Figure 3.26 Typical power cord reel.

In situations where mutual aid departments frequently work together and have either different sizes or different types of receptacles (for example, one has two prongs, the other has three), adapters should be carried so that equipment can be interchanged (Figure 3.29). Adapters should also be carried to allow rescuers to plug their equipment into standard electrical outlets.

Figure 3.27 A twist-lock adapter.

RESCUE EQUIPMENT AND TOOLS

Because it is difficult to anticipate every aspect of every rescue situation, rescue personnel must be

Figure 3.28 A typical junction box.

Figure 3.29 A variety of electrical adapters.

proficient in the use of all their equipment. Rescue personnel must train to the point that using their equipment is second nature to them. Rescuers must know how and why a tool works, when to use it, and what it will and will not do. Complete knowledge of tools and equipment — and of appropriate procedures — allows rescuers to use or to rapidly devise a safe method for completing almost any rescue. The skills and techniques required for rescue work can be learned only by thorough training supplemented by experience.

The equipment procured for the rescue unit should be determined by the members of the rescue unit considering the needs of the area and the type and size of the rescue vehicle. If rescue needs are properly assessed, the purchase of unnecessary equipment can be avoided.

The following sections give a general guide for possible choices of equipment. Personnel should be familiar with all tools so they can use them safely and efficiently and can keep them in good working order.

NOTE: Rescuers should wear the appropriate protective clothing at all times when using any of this equipment.

HAND TOOLS

Striking Tools

The most common and basic hand tools are striking tools. Most striking tools are characterized by large, weighted heads on long handles. This category of tools includes axes, battering rams, ram bars, punches, mallets, hammers, sledgehammers or mauls, picks, chisels, and spring-loaded center punches (Figure 3.30).

Striking tools can be dangerous and may crush or sever fingers, feet, or other parts of the body when used improperly. Striking tools should be used with short, quick strokes; long, sweeping strokes are difficult to control. Striking tools can also send chips and splinters into the air, piercing skin and eyes. Because of this danger, it is imperative that proper protective clothing be worn.

All tools should be properly maintained, with handles kept solid and well set in the head. Handles can be protected by wrapping inner tubing or tape around the handle near the head of the tool to reduce damage when blows are not made solidly (Figure 3.31). The striking surface should be routinely serviced. Axes or pointed tools should be kept sharp, and blunt striking surfaces should be kept free of chips or cracks (Figure 3.32).

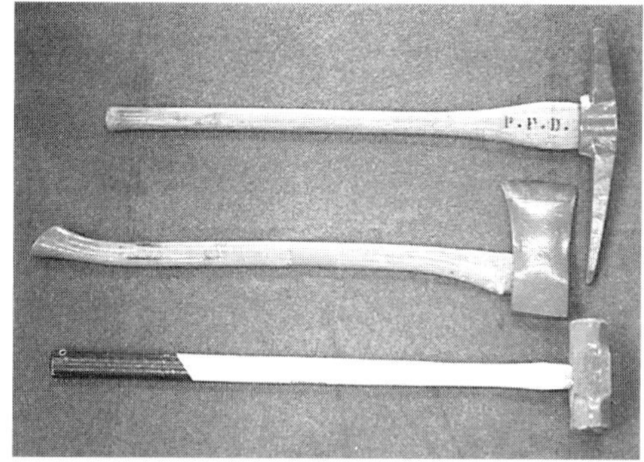

Figure 3.30 Typical striking tools used in rescue.

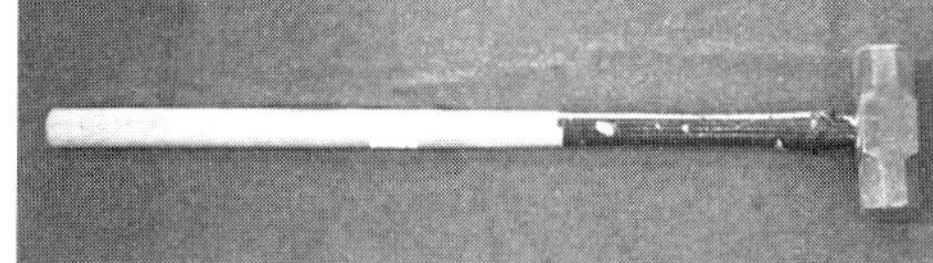

Figure 3.31 Tape can be used to protect tool handles.

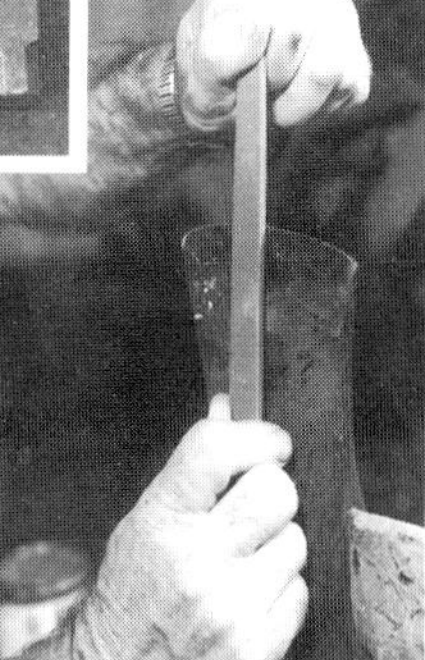

Figure 3.32 Cutting tools must be kept sharp.

Prying Tools

Prying hand tools use leverage to provide a mechanical advantage. This means that the force applied to one end of the tool is multiplied at the other end because of leverage. Prying tools can be used to lift heavy objects. The pry-axe, Halligan (Hooligan) tool, crowbar, claw tool, pry bar, Kelly tool, spanner wrench, and Quick-Bar® are examples of manual prying tools (Figure 3.33). Crowbars and other prying tools are excellent for widening a small opening, making it possible then to use larger power tools.

When used correctly, prying tools are safer than striking tools. As with other tools, using prying tools incorrectly can be hazardous. For example, it is unsafe to use a "cheater" (Figure 3.34). A *cheater* is a piece of pipe slipped over the end of the tool handle to lengthen it and thus provide additional leverage. Using a cheater can put forces on the tool greater than it was designed to withstand. This can damage the tool and injure the operator.

The correct type and size of prying tool for the job should be used. If a job cannot be done with one tool, another should be used. A prying tool should never be used as a striking tool unless it was designed for that purpose.

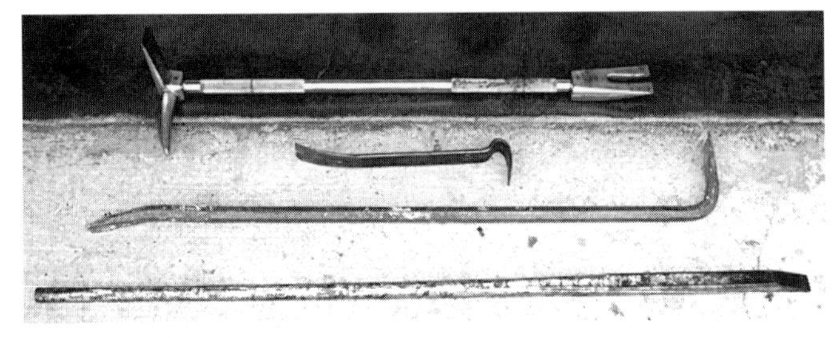

Figure 3.33 Typical prying tools used in rescue.

Figure 3.34 A "cheater" should never be used on a tool handle.

Cutting Tools

Cutting tools are the most diversified of the tool groups. However, cutting tools are designed to cut only specific types of materials. Cutting tools should not be used to cut material that it was not designed to cut. This can destroy the tool and endanger the operator. Manual cutting tools can be divided into four distinct groups:

- Chopping tools
- Scissors or snipping-type tools
- Saws
- Knives

CHOPPING TOOLS

Chopping tools are characterized by a heavy metal head with a cutting edge that is attached to a long handle. They include flat-head axes, pick-head axes, pry-axes, adzes, and various other types of picks (Figure 3.35).

To ensure maximum efficiency of these tools, they must be properly maintained. The cutting head should be unpainted and should be covered with a thin coating of light-grade oil (such as a silicone lubricant or a light machine oil). The cutting head should maintain a slightly sharp edge — but not so sharp that the edges will chip off when the tool is used. Tool handles should be checked regularly for looseness, cracks, or warping.

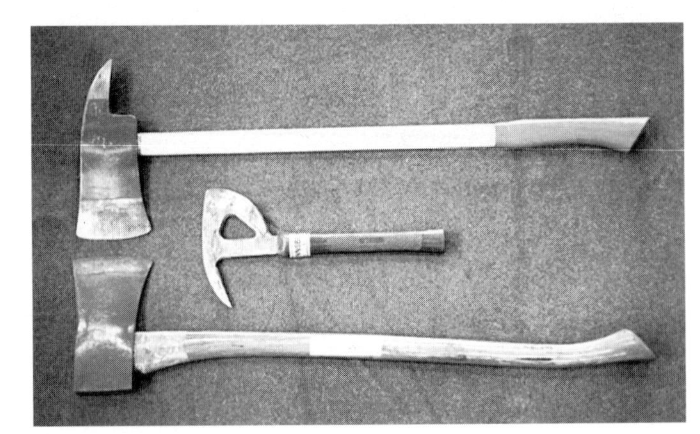

Figure 3.35 Typical chopping tools.

SNIPPING-TYPE TOOLS

Cutting or snipping tools are used in situations where the material can be cut in a controlled fashion. They are most effective on small objects that can easily fit within the jaws of the tool. Using snipping-type tools is generally safer than using other types of cutting devices when working close to a victim. Tools that fall into this category are various kinds of rescue scissors or shears, tin snips, bolt cutters, and wire cutters.

Cutting tools that are often misused are the opposing-jaw metal cutters. Common types of opposing-jaw metal cutters are wire cutters, bolt cutters, and hot-wire cutters (Figure 3.36). These tools have sometimes been used to cut energized electrical wires, but this is not a safe practice and is not recommended (see Chapter 11, Special Rescues).

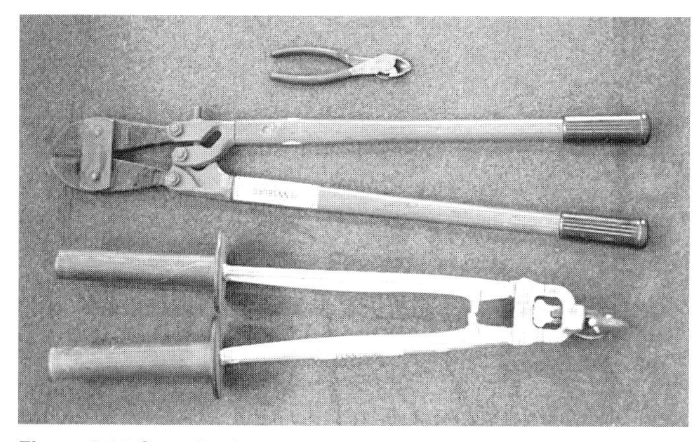

Figure 3.36 Opposing-jaw cutting tools.

WARNING

Standard fire service tools are not dielectric equipment and should never be used to cut or handle energized electrical wires.

SAWS

Handsaws are useful on objects that require a controlled cut but are too big to fit within the jaws of an opposing-jaws cutter. Handsaws are often more time-consuming to use than are powered saws or shears. However, they are also safer to use when working close to the victim or when working in a hazardous atmosphere because they do not create sparks. Handsaws commonly used for rescue include carpenter's saws, hacksaws, coping saws, and keyhole saws (Figure 3.37). All saw blades should be kept sharp, clean, and lightly oiled (Figure 3.38).

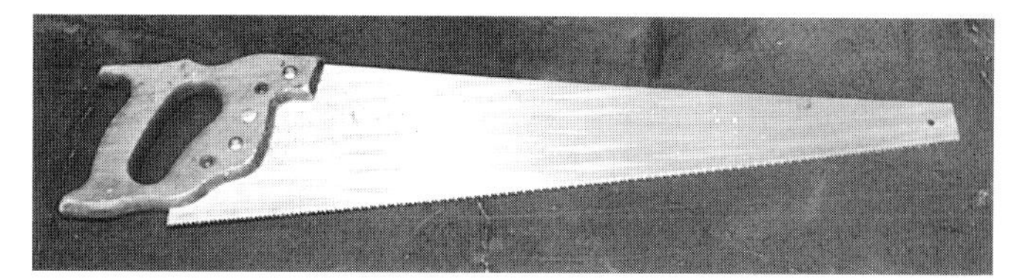

Figure 3.37 A typical handsaw used in rescue.

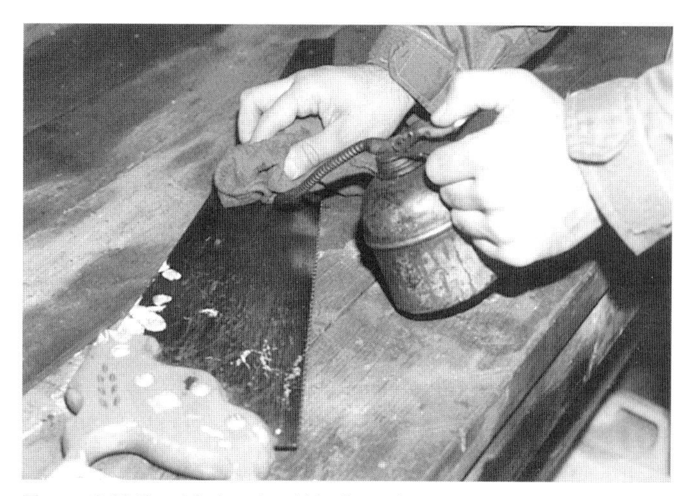

Figure 3.38 Saw blades should be kept clean and lightly oiled.

KNIVES

Various types of knives may be useful to rescuers. A good, sharp pocketknife will probably meet most rescuers' needs. Special knives that may also be carried by rescue companies include V-blade (seat belt) knives, linoleum knives, and razor knives (Figure 3.39). Knife blades should be either sharpened or replaced after each use to ensure maximum readiness for the next use.

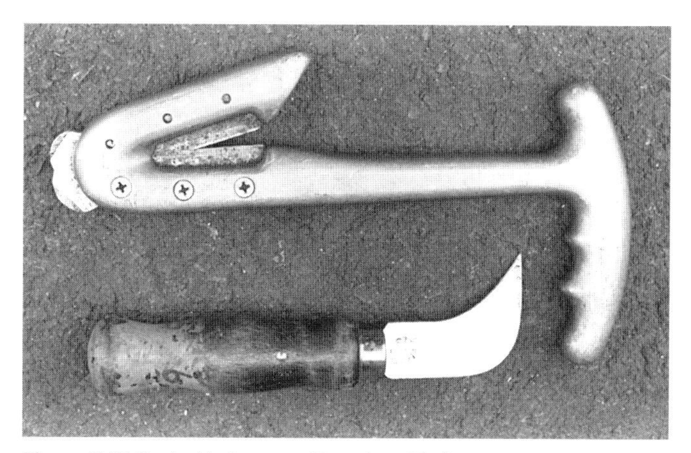

Figure 3.39 Typical knives used to cut seat belts.

Nonhydraulic Jacks

There are several kinds of jacks that can be considered hand tools because they do not operate with hydraulic power. Although these tools are effective for their designed purposes, they do not have the same amount of power as hydraulic jacks. The following sections describe several of the nonhydraulic types of jacks. See the Hydraulic Jack section for safety guidelines when using any type of jack.

SCREW JACKS

Screw jacks can be extended or retracted by turning the shaft. So that they are always in a state of readiness, jacks should be checked for wear after each use. They should also be kept clean and lightly lubricated, with particular attention paid to the screw thread. Foot plates should also be checked for wear or damage. Foot plates make contact with whatever is being stabilized by the jacks.

The two types of screw jacks are the bar screw jack and the trench screw jack. Both jacks have a male-threaded core similar to a bolt and a means to turn the core.

Bar screw jacks. Bar screw jacks are excellent for supporting collapsed structural members (Figure 3.40). These jacks are normally not used for lifting; their primary use is to hold an object in place, not to move it. The jacks are extended or retracted as the shaft is rotated in the base. The shaft is turned by pushing a long bar that is inserted through a hole in the top of the shaft.

Trench screw jacks. Because of their ease of application, durability, and relatively low cost, trench screw jacks are sometimes used to replace wooden cross braces in trench rescue applications. These devices consist of a swivel footplate with a stem that is inserted into one end of a length of 2-inch (50 mm) steel pipe (not to exceed six feet [1.8 m] in length) and a swivel footplate with a threaded stem that is inserted into the other end of the pipe (Figure 3.41). An adjusting nut on the threaded stem is turned to vary the length of the jack and to tighten it between opposing members in a shoring system.

RATCHET-LEVER JACK

Also known as "high-lift" jacks, these medium-duty jacks consist of a rigid I-beam with perforations in the web and a jacking carriage with two ratchets on the geared side that fits around the I-beam (Figure 3.42). One ratchet holds the carriage underneath. The second ratchet is combined with a lever that is pushed down to force the carriage upward. The ratchets can be reversed to move the carriage down.

Ratchet jacks can be dangerous because they are the least stable of all the various types of jacks. If the load being lifted shifts, ratchet-lever jacks may simply fall over allowing the load to suddenly drop to its original position. Also, the ratchets can fail under a heavy load.

WARNING

Rescuers should never work under a load supported only by a jack. If the jack fails or the load shifts, severe injury or death may result. The load should also be supported by properly placed cribbing.

Figure 3.40 A typical bar screw jack.

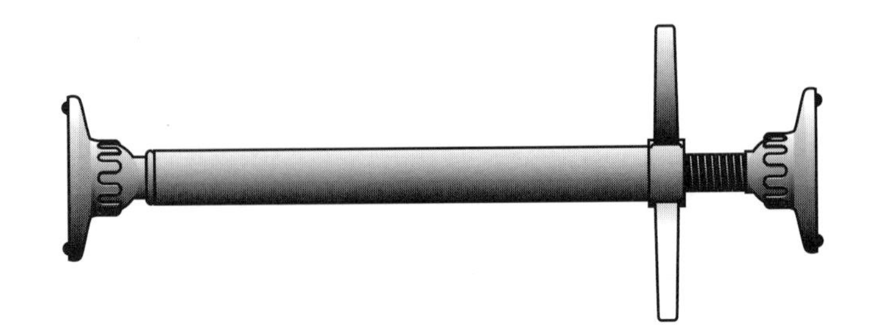

Figure 3.41 A typical trench screw jack.

Figure 3.42 A ratchet-lever jack.

HYDRAULIC TOOLS

Powered Hydraulic Tools

The development of powered hydraulic rescue tools has revolutionized the process of removing victims from various types of entrapments. The wide range of uses, speed, and superior power of these tools has made them the primary tools used in many rescue situations. These tools receive their power from hydraulic fluid pumped through special high-pressure hoses. Although there are a few pumps that are operated by compressed air, most are powered by electric motors or by two- or four-cycle gasoline engines. These units may be portable and carried with the tool, or they may be mounted on the vehicle and may supply power to the tool through a hose reel line (Figure 3.43). Manually operated pumps are also available in case of a power unit failure (Figure 3.44). Four basic types of powered hydraulic tools are used in rescue incidents: spreaders, shears, combination spreader/shears, and extension rams.

SPREADERS

Powered hydraulic spreaders were the first powered hydraulic tools to become available to the fire/rescue service (Figure 3.45). They are capable of either pushing or pulling. Depending on the brand, these tools can produce up to 22,000 psi (154 000 kPa) of force at the tips of the tool. The tips of the tool may spread as much as 32 inches (813 mm) apart.

SHEARS

Hydraulic shear tools are capable of cutting almost any metal object that can fit between their blades, although some models cannot cut case-hardened steel (Figure 3.46). The shears may also be used to cut other materials such as plastics or wood. Shears are typically capable of developing about 30,000 psi (206 850 kPa) of cutting force and have an opening spread of approximately 7 inches (175 mm).

Figure 3.43 A typical hydraulic rescue tool power unit.

Figure 3.44 A manually operated hydraulic pump.

Figure 3.45 Typical hydraulic spreaders.

Figure 3.46 Typical hydraulic shears.

COMBINATION SPREADER/SHEARS

Most manufacturers of powered hydraulic rescue equipment offer a combination spreader/shears tool (Figure 3.47). This tool consists of two arms equipped with spreader tips that can be used for pulling or pushing. The inside edges of the arms are equipped with cutting shears similar to those described in the previous paragraph. This combination tool is excellent for a small rapid-intervention vehicle or for departments where limited resources prevent the purchase of larger and more expensive individual spreader and cutting tools. However, the combination tool's spreading and cutting capabilities are somewhat less than those of the individual units.

Figure 3.47 Combination spreader/shears.

EXTENSION RAMS

Extension rams are designed primarily for straight pushing operations, although they are effective at pulling as well. These tools are especially useful when it is necessary to push objects farther than the maximum opening distance of the hydraulic spreaders (Figure 3.48). The largest of these extension rams can extend from a closed length of 36 inches (914 mm) to an extended length of nearly 63 inches (1 600 mm). They open with a pushing force of about 15,000 psi (103 425 kPa). The closing force is about one-half that of the opening force.

Figure 3.48 A hydraulic ram.

Manual Hydraulic Tools

Manual hydraulic tools operate on the same principles as powered hydraulic tools except that the hydraulic pump is manually powered by a rescuer operating a pump lever. The primary disadvantage of manual hydraulic tools is that they operate slower than powered hydraulic tools, and they are labor-intensive. Two manual hydraulic tools are used frequently in extrication work: the porta-power system and the hydraulic jack.

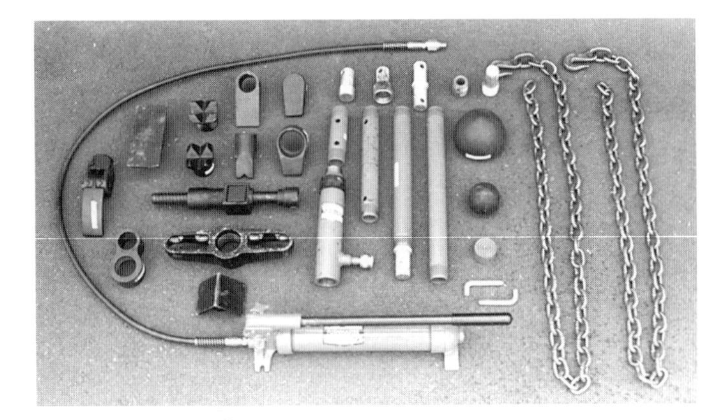

Figure 3.49 A typical porta-power set.

PORTA-POWER TOOL SYSTEM

The porta-power tool system is basically an auto body shop tool that has been adopted by the fire/rescue service (Figure 3.49). It is operated by transmitting pressure from a hand-pumped compressor through a hydraulic hose to a tool assembly. A number of different tool accessories give the porta-power tool a variety of applications.

The primary advantage of the porta-power tool over the hydraulic jack is that the porta-power has accessories that allow it to be operated in narrow places in which the jack either will not fit or cannot be operated. The primary disadvantage of the porta-power is that assembling complex combinations of accessories and actual operation of the tool is time-consuming.

HYDRAULIC JACK

The hydraulic jack is designed for heavy-lifting applications (Figure 3.50). It is also an excellent compression device for shoring or stabilizing operations. Most hydraulic jacks have lifting capacities up to 20 tons (18 144 kg), but units with a higher capacity are available.

Any kind of jack, hydraulic or otherwise, should have flat, level footing, and should be used in conjunction with cribbing (Figure 3.51). On a soft surface, a flat board or steel plate with wood on top should be put under the jack to distribute the force placed on the jack.

Figure 3.50 One type of hydraulic jack.

PNEUMATIC (AIR-POWERED) TOOLS

Pneumatic tools use compressed air for power. The air can be supplied by vehicle-mounted air compressors, apparatus brake system compressors, SCBA cylinders, or cascade system cylinders.

> **WARNING**
>
> **Never use compressed *oxygen* to power pneumatic tools. Mixing pure *oxygen* with grease and oils found on the tools will result in fire or violent explosion.**

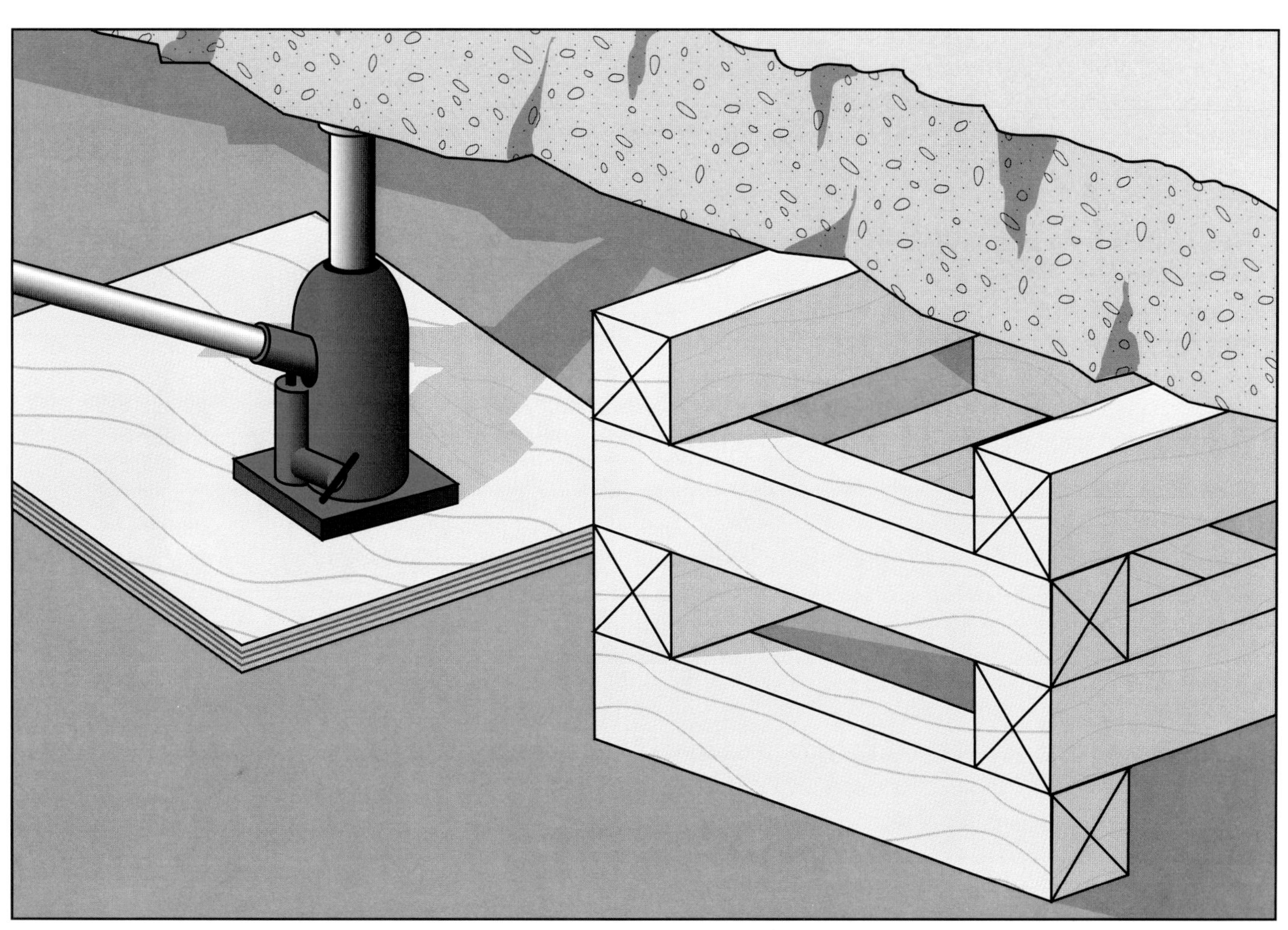

Figure 3.51 Cribbing is used in conjunction with a jack.

Air Chisels

Pneumatic-powered chisels (also called *air chisels*, *pneumatic hammers*, or *impact hammers*) are useful for rescue and extrication work. Most air chisels operate at air pressures between 100 and 150 psi (700 kPa and 1 050 kPa). They will normally use about 4 to 5 cubic feet (113 L to 142 L) of air per minute. These tools come with a variety of interchangeable bits to fit the needs of almost any situation (Figure 3.52). In addition to cutting bits, special bits for such operations as breaking locks or driving in plugs are also available.

Often used in vehicle extrication situations, these tools are good for cutting medium- to heavy-gauge sheet metal and for popping rivets and bolts. Cutting heavier-gauge metal will require more air at higher pressures.

CAUTION: The sparks produced while cutting metal with pneumatic chisels may provide an ignition source for flammable vapors.

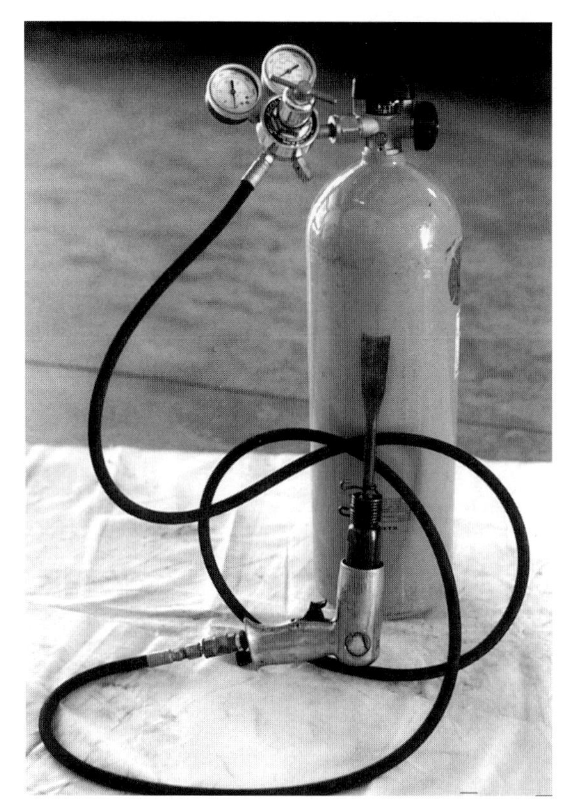

Figure 3.52 A typical air chisel. *Courtesy of Vespra (ONT) Fire Department.*

Pneumatic Nailers

Air-operated nailers can be used to drive nails into wood or masonry. They are especially useful for nailing wedges and other wooden components of shoring systems into place (Figure 3.53).

Figure 3.53
A pneumatic nailer.

OTHER TOOLS AND EQUIPMENT

Power Saws

Power saws are available in various designs, depending upon the purpose for which they are intended. It is important that the operator know the limitations of each type of power saw. When a saw (or any tool) is pushed beyond the limits of its design and purpose, two things may occur: tool failure (including breakage) and/or injury to the operator. Operators should wear full protective clothing with face and eye protection when using power saws.

WHIZZER SAW

The Whizzer saw is an air-driven cutting device with several advantages over other types of power saws (Figure 3.54). At about 2 pounds (0.9 kg), the Whizzer weighs about one-tenth as much as a circular saw, so it is much more maneuverable. Operating at 20,000 rpm, its 3-inch Carborundum™ blade will cut case-hardened locks and steel up to

Figure 3.54 A pneumatic rotary saw. *Courtesy of New York (NY) Fire Department.*

¾ inch in thickness, and produces fewer sparks than metal blades. The tool has a clear Lexan® blade guard to protect the operator and the victim from flying debris. Driven by compressed air at 90 psi from an SCBA cylinder with a regulator, the Whizzer operates much quieter than other power saws and will run for approximately three minutes from a full cylinder.

CIRCULAR SAW

A circular saw, also called a *rotary rescue saw*, is used for cutting a variety of materials because it can be equipped with different blades for specific materials (Figure 3.55). A large-toothed blade produces a faster, less precise cut than does a fine-toothed blade. Carbide-tipped blades are superior to standard blades because they are less prone to dulling with heavy use.

Figure 3.55 A rotary rescue saw.

Circular saws often produce flying debris and sparks when cutting, so victims and rescue personnel in close proximity to the cutting operation should be shielded from the sparks and debris. A firefighter with a charged hoseline or fire extinguisher should also be standing by. During cutting operations, the blade may be cooled with a fine water mist from a hoseline. However, the water must be applied to the blade *before* starting to cut, and the application should be continued throughout the cutting operation. Blade disintegration may result from putting water on a blade that is already hot. Blades from different manufacturers may look alike, but they should not be used interchangeably. Circular saw blades should be stored in a clean, dry environment free of hydrocarbon fumes; these fumes can cause invisible deterioration of composite blades.

CAUTION: In addition to the possible hazards already mentioned, operators need to be aware of the rotational torque developed by circular saws when run at high speeds. This torque causes the saw to twist in the operator's hand.

RECIPROCATING SAW

Reciprocating saws are highly controllable saws that are well suited for cutting wood or metal (Figure 3.56). They have a short, straight blade that moves rapidly forward and backward with an action similar to that of a handsaw. Reciprocating saws are usually more controllable and safer to use than circular saws, and they have less of a tendency to produce sparks and flying debris.

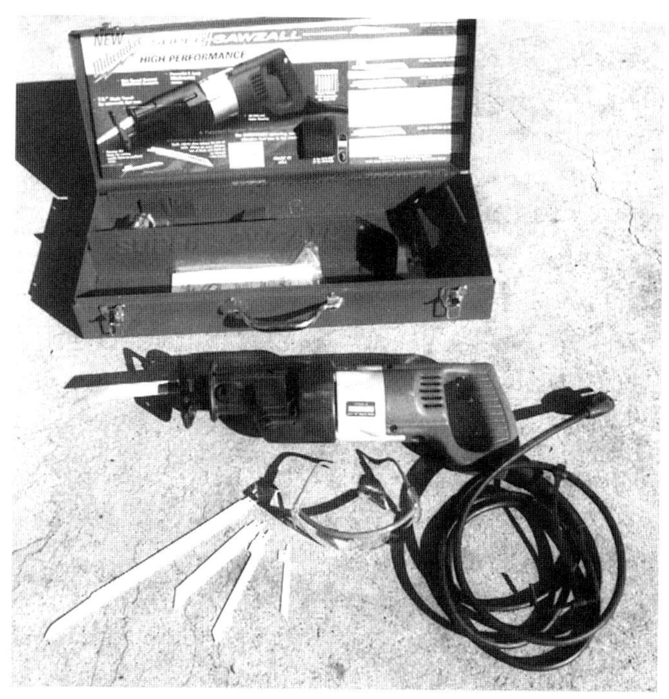

Figure 3.56 A reciprocating saw. *Courtesy of Keith Flood.*

CHAIN SAW

Chain saws are commonly used for rescue, forcible entry, and ventilation work. The most common types are powered by gasoline engines, but electric chain saws are also available (Figure 3.57). The best type of chain saw is one powerful enough to penetrate dense material yet lightweight enough to be easily handled in awkward positions. Chain saws equipped with carbide-tipped chains are capable of penetrating a large variety of materials, including light sheet metal. Although carbide-tipped chains cost almost four times as much as standard chains, they last twelve times longer.

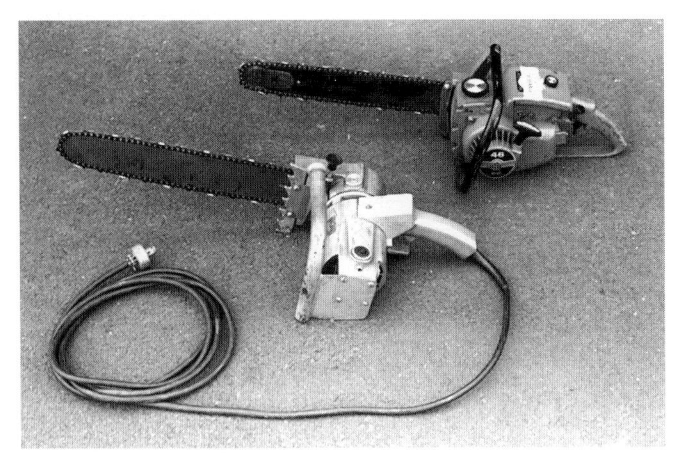

Figure 3.57 Typical rescue chain saws.

CAUTION: Chain saws, especially those not equipped with a chain brake, can jerk violently if the chain binds in the material being cut.

POWER SAW SAFETY RULES

Following a few simple safety rules when using power saws will prevent most typical accidents:

- Operators should always wear proper protective equipment, including gloves and ear and eye protection.
- The type of saw should match the task and the material to be cut.
- Saws should be operated within their design limitations.
- Blades should never be interchanged between saws from different manufacturers or with different operating speeds.
- Blade guards should not be removed or any other safety feature disabled during operation.
- Power saws should not be used in a flammable atmosphere.
- Nonessential personnel should be kept out of the work area.
- Manufacturer's operating guidelines should be followed.
- Blades and chains should be kept well-sharpened.
- Victims and/or rescuers should be protected from sparks and debris.
- An appropriate means of fire extinguishment should be close at hand.
- Saws or other power equipment should not be operated by anyone who is excessively fatigued.

Lifting/Pulling Tools

Rescuers must sometimes lift or pull an object to free a victim. Several rescue tools have been developed to assist in this task. These include tripods, winches, come-alongs, chains, and air bags.

TRIPODS

Rescue tripods are needed to create an anchor point above a manhole or other opening. This allows rescuers to be safely lowered into confined spaces and rescuers and victims to be raised out of them (Figure 3.58).

WINCHES

Vehicle-mounted winches are excellent pulling tools. They can usually be deployed faster than other lifting/pulling devices, generally have a greater travel or pulling distance, and are much stronger. Winches are usually mounted on the front bumper, but some are located at the rear of the vehicle (Figure 3.59). The three most common drives for winches are electric, hydraulic, and power take-off. Either chain or steel cables are used for pulling.

Winches should be equipped with hand-held, remote-control operating devices (Figure 3.60). These devices allow the operator to get a better view of the operation and to stand away from the winch since being near the winch can be dangerous if the cable breaks. Rescuers should position the winch as close to the object being pulled as possible so that if the cable breaks, there will be less cable to suddenly recoil and less chance of injury.

CAUTION: Whenever possible, a winch operator should stay farther away from the winch than the length of the cable from the winch to the load (Figure 3.61).

Winches should be inspected periodically and any needed repairs made. A schedule of preventive maintenance should be followed to ensure that they are in proper working condition.

If no other pulling or lifting devices are available, winches on tow trucks may sometimes be used

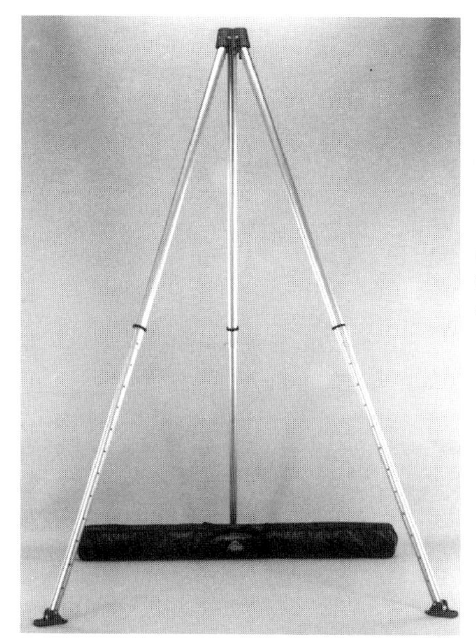

Figure 3.58 A rescue tripod. *Courtesy of SKEDCO, Inc.*

Figure 3.59 A vehicle-mounted winch.

Figure 3.60 The winch operator uses the remote control.

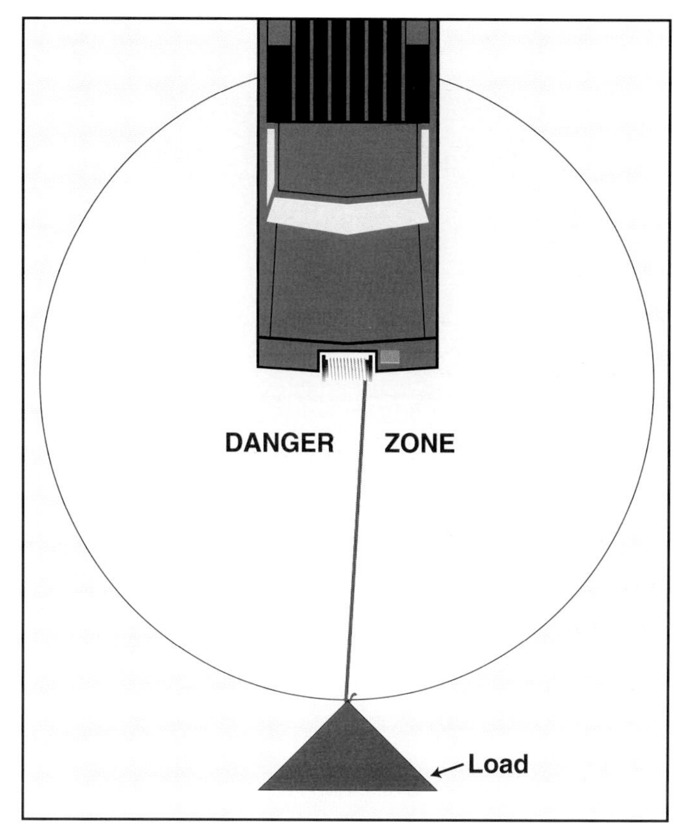

Figure 3.61 The danger zone in a winch operation.

in rescue operations. Tow trucks are often on hand at vehicle accidents, and some fire departments have their own units (Figure 3.62).

COME-ALONG

Another lifting/pulling tool used in rescue is the come-along (Figure 3.63). It is a portable cable winch operated by a manual ratchet. In use, the come-along is attached to a secure anchor point, and the cable is run out to the object to be moved. Once both ends are attached, the lever is operated to rewind the cable which pulls the movable object toward the anchor point. The most common sizes or ratings of come-alongs are from 1 to 10 tons (907 kg to 9 072 kg).

Figure 3.62 A typical tow truck with a winch.

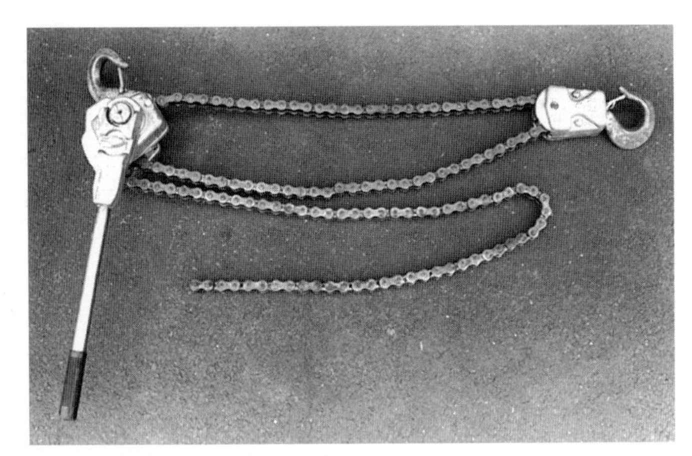

Figure 3.63 One type of come-along.

CHAINS

Winches and come-alongs may use chains as part of a lifting/pulling system. Only alloy steel chains of the correct size should be used in rescue work (Figure 3.64). Alloy steel chains are highly resistant to abrasion, making them ideal for rescue and extrication work. Special alloys are available that are resistant to corrosive or hazardous atmospheres. Proof coil chain, also called *common* or *hardware chain*, is not suitable for emergency situations.

The minimum chain size generally used for rescue operations is ¾ inch (10 mm). For any lifting/pulling operation, it is important to match the rated strength of the chain to the tools being used and the job being done. Table 3.1 lists the safe working loads for various sizes of chain. Additionally, hooks and attachments should have at least the same strength rating as the rest of the chain, and they should be made of the same alloy material.

Chain inspections. Chain should be inspected regularly because failures can occur if the chain was abused or neglected during use or storage. Improper treatment of chain components can lead to metal fatigue and chain failure. Chain should be regularly inspected link by link for signs of cracks, nicks, gouges, bent links, corrosion, elongation, or any other defects. Defective chain should be removed from service. Table 3.2 lists the maximum allowable wear at any point of a link before the chain should be removed from service.

Figure 3.64 Typical rescue chains.

TABLE 3.1
Working Load Limits, Proof Test Loads, And Minimum Breaking Loads For Alloy Steel Chain

Nominal Size Of Chain (in.)	Working Load Limit (lb)	Proof Test (lb)	Minimum Break (lb)
¼	3,250	6,500	10,000
⅜	6,600	13,200	19,000
½	11,250	22,500	32,500
⅝	16,500	33,000	50,000
¾	23,000	46,000	69,500
⅞	28,750	57,500	93,500
1	38,750	77,500	122,000
1⅛	44,500	89,000	143,000
1¼	57,500	115,000	180,000
1⅜	67,000	134,000	207,000
1½	80,000	160,000	244,000
1¾	100,000	200,000	325,000

Source: *Specification for Alloy Chain*, American Society for Testing and Materials, A-391-65. *Alloy Steel Chain Specifications*, No. 3001, National Association of Chain Manufacturers. Reprinted with permission.

Chain safety rules. The following safety rules should be applied when using chain:

- All chains should have an attached tag that has the safe load weight stamped or printed on it.
- A load should not be dragged horizontally across a hard surface if a chain is between the load and the surface over which the load is being pulled.
- Chain should never be tied in a knot to shorten it.
- The listed safe working load of the chain should never be exceeded.

TABLE 3.2 (U.S.)
Maximum Allowable Wear At Any Point Of Link

Chain Size (in.)	Maximum Allowable Wear (in.)
1/4	3/64
3/8	5/64
1/2	7/64
5/8	9/64
3/4	5/32
7/8	11/64
1	3/16
1 1/8	7/32
1 1/4	1/4
1 3/8	9/32
1 1/2	5/16
1 3/4	11/32

Reprinted with permission from the National Safety Council: *Accident Prevention Manual for Industrial Operations: Engineering and Technology*, 9th edition. Chicago: National Safety Council, 1988.

TABLE 3.2 (METRIC)
Maximum Allowable Wear At Any Point Of Link

Chain Size (mm)	Maximum Allowable Wear (mm)
6	1.2
10	2
13	2.8
16	3.6
20	4
22	4.4
25	4.8
29	5.6
32	6
35	7.1
38	7.9
45	8.7

- Damaged or worn-out chains should be destroyed.
- Chains should never be shock-loaded.
- Chain hooks should not be attached directly to loads, only to the chain.
- Broken links on alloy chain should not be rewelded.
- All appliances (hooks, pins, links, etc.) should be strong enough for the load being handled.
- Alloy chain should not be heat-treated or exposed to high heat levels.
- Chain should never be spliced by bolting two links together.
- A load should never be applied to a kinked chain.

AIR BAGS

Air bags give rescuers the ability to lift or displace objects that cannot be lifted with other rescue equipment. There are three basic types of lifting bags: high pressure, medium pressure, and low pressure. A fourth type of bag is used for sealing leaks but has little if any rescue application.

High-pressure bag. High-pressure bags consist of a tough neoprene rubber exterior reinforced with steel wire or Kevlar® aramid fibers. Deflated, the bags lie completely flat and are about 1 inch (25 mm) thick (Figure 3.65). They come in various sizes that range in surface area from 6 x 6 inches (150 mm by 150 mm) to 36 x 36 inches (914 mm by 914 mm). The inflation pressure of the bags is about 135 psi (931 kPa). Depending on the size of the bags, they may inflate to a height of 20 inches (500 mm). The largest bags can lift approxi-

Figure 3.65 A deflated air bag.

mately 75 tons (68 040 kg); however, an air bag's weight-lifting capacity decreases as the height of the lift increases. For example, a bag rated at 10 tons (9 072 kg) will only lift 5 tons (4 536 kg) to 8 inches (203 mm); one rated at 75 tons (67.5 t) will only lift 37 tons (33.3 t) to 20 inches (508 mm).

Low- and medium-pressure bags. Low- and medium-pressure bags are considerably larger than high-pressure bags and are most commonly used to lift or stabilize large vehicles or objects (Figure 3.66). They are often used to right overturned vehicles after all occupants are out of the vehicle. Their primary advantage over high-pressure air bags is that they have a much greater lifting distance. Depending on the manufacturer, a lifting bag may be capable of lifting an object 6 feet (2 m) above its original position. Low-pressure bags generally operate on 7 to 10 psi (49 kPa to 70 kPa), while medium-pressure bags use 12 to 15 psi (84 kPa to 105 kPa), depending on the manufacturer.

Air bag safety rules. Operators should follow these safety rules when using air bags:

- The lifting operation should be planned before starting.
- Operators should be thoroughly familiar with the equipment — its operating principles and methods — and its limitations.
- Operators should follow the manufacturer's recommendations for the specific system used.
- All components should be kept in good operating condition with all safety seals in place.
- Operators should have available an adequate air supply and sufficient cribbing before beginning operations.
- The bags should be positioned on or against a solid surface.
- The bags should never be inflated against sharp objects.
- The bags should be inflated slowly and monitored continually for any shifting.
- Rescuers should never work under a load supported only by air bags.
- The load should be continuously shored up with enough cribbing to adequately support the load in case of bag failure.
- When box cribbing is used to support an air bag, the top layer should be solid; leaving a hole in the center may cause shifting and collapse (Figure 3.67).
- Bags should not be allowed to contact materials hotter than 220°F (104°C).
- Bags should never be stacked more than two high; with the smaller bag on top, the bottom bag should be inflated first (Figure 3.68). A single multicell bag is preferred.

CAUTION: Air bags should be inspected regularly and should be removed from service if any evidence of damage or deterioration is found.

Figure 3.66 Low-pressure bags in operation. *Courtesy of Joel Woods.*

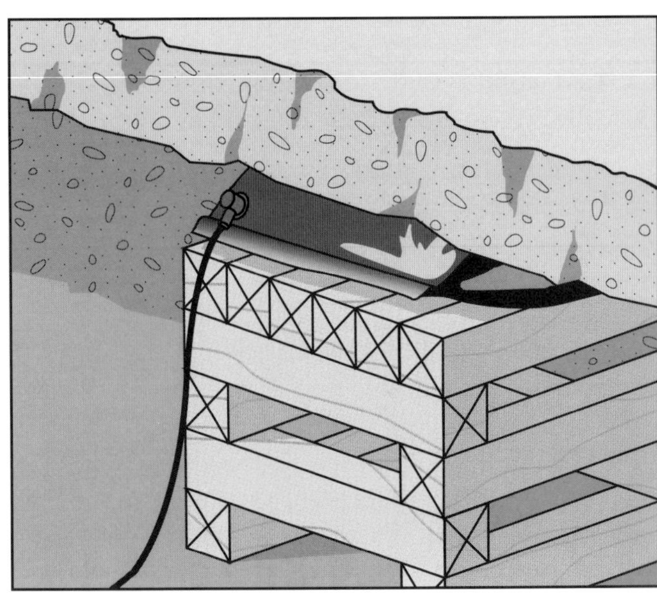

Figure 3.67 Air bags can be supported by cribbing.

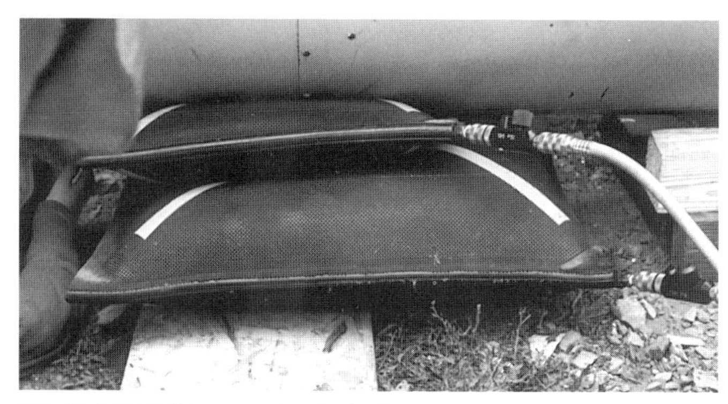

Figure 3.68 Air bags may be stacked.

Cutting And Burning Equipment

To free trapped victims, rescuers must sometimes cut through structural members that are too dense to be cut with power saws. In these cases, some form of gas or arc cutting or burning device may have to be used. While a variety of devices, such as burning bars and plasma cutters, are carried on some rescue units, the most common exothermic cutting device used by firefighters is the oxyacetylene cutting torch.

OXYACETYLENE CUTTING TORCH

Oxyacetylene cutting torches cut by burning (Figure 3.69). They may be used to cut through heavy metal enclosures that are resistant to other rescue equipment. These torches preheat metal to its ignition temperature and then burn a path in the metal with an extremely hot cone of flame created by introducing pure oxygen into the flame. For preheating metal, the flame temperature is approximately 4,200°F (2 316°C). When pure oxygen is added, a flame of over 5,700°F (3 149°C) is created. This flame is hot enough to burn through iron and steel with relative ease.

Figure 3.69 A rescuer uses a cutting torch.

Acetylene is an unstable gas with a wide flammability range (2.5 to 81.0 percent) and is both pressure- and shock-sensitive. However, acetylene storage cylinders are designed to keep the gas stable and safe to use. The cylinders contain a porous filler of calcium silicate, which prevents accumulations of free acetylene within the cylinder. They also contain liquid acetone in which the acetylene is dissolved. When the cylinder valve is opened, the acetylene gas is liberated from the acetone and the gas flows through the hose to the torch assembly.

WARNING

Acetylene cylinders must be kept in an upright position to prevent the loss of acetone which could cause the cylinder to explode.

The following safety rules should be observed when using oxyacetylene cutting equipment:

- Acetylene cylinders must be stored and used in an upright position to prevent a loss of acetone (Figure 3.70). Even when an acetylene cylinder is empty of acetylene, it is still full of acetone.
- Acetylene cylinders must be handled carefully to prevent damage to the cylinder or the filler. A dent in the cylinder may indicate that the filler is damaged. If the filler is damaged, voids may be created where free acetylene can pool and decompose, making it dangerously unstable. Dropping a cylinder may loosen the fuse plug and allow acetylene gas to leak out.
- Acetylene cylinders must not be stored or used in an ambient air temperature exceeding 130°F (54°C).
- Acetylene cylinders should not be stored on wet or damp surfaces. Cylinders can rust at the bottom if protective paint is worn away.

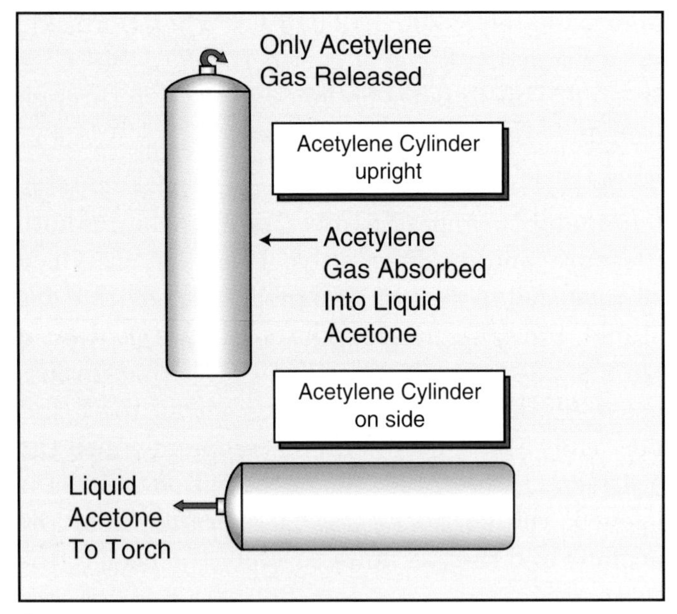

Figure 3.70 Acetylene cylinders must kept upright.

- Acetylene cylinders should not be stored near cylinders of oxygen or other oxidizing gases. Full acetylene cylinders should not be stored with empty or partially full cylinders. Cylinder storage areas should be designed to prevent cylinders from falling over if they are bumped.
- A solution of soap and water should be applied to the regulator, torch, hose, and cylinder connections to detect gas leaks. Even slow leaks may allow dangerous concentrations of acetylene to accumulate in confined spaces. Leaking cylinders should be moved to an open area immediately.
- Acetylene cylinder valves should be opened no more than a three-quarter turn. Inoperative valves should not be forced open. The cylinder should be taken out of service and returned to the supplier for service.
- Acetylene operating pressure should be set at no more than 15 psi (103 kPa). Acetylene decomposes at high pressures and may explode in the process.
- Acetylene withdrawal rate should not exceed one-seventh of the cylinder capacity per hour.
- Cylinder valves should be kept closed when not in use and when the cylinder is empty. After cylinder valves are closed, the pressure in the regulator and torch assembly should be bled off. When not in use, whether full or empty, cylinders should be capped to prevent damage to fittings.

In some cases, acetylene has been replaced by methylacetylene-propadiene, stabilized (MPS), marketed as MAPP GAS® and APACHE GAS®. MPS is less pressure-sensitive than acetylene and is not prone to explosive decomposition. The cutting equipment used with acetylene is easily adapted for use with MPS.

BURNING BARS

Burning bars are ultrahigh-temperature cutting devices capable of cutting through virtually any metallic, nonmetallic, or composite material. They will cut through materials that cannot be cut with an oxyacteylene torch, such as concrete or brick, and they will cut through metals much faster. The torch feeds oxygen and up to 200 amperes of electrical power to an exothermic cutting bar that produces temperatures in excess of 10,000°F (5 538°C). The cutting bars or rods range in size from ¼ to ¾ inch (6 mm to 10 mm) in diameter and from 22 to 36 inches (550 mm to 900 mm) in length.

PLASMA CUTTERS

Plasma arc cutters are also ultrahigh-temperature metal-cutting devices. As the arc rod melts the metal being cut, a jet of extremely hot gas (usually air, nitrogen, or a mixture of argon and hydrogen or argon and helium) is used to blow the molten metal from the cutting area.

Ropes And Related Equipment

Many of the lifting and pulling tools previously described in this chapter must have either rope or webbing attached between the objects and the tool. It is vital that rescuers know the uses and limitations of ropes and webbing. Using them beyond their limitations can result in their failure. Except where noted, webbing (of the appropriate size and strength) and rope may be used interchangeably.

ROPE

Rope is one of the most versatile and valuable pieces of equipment used by fire departments and rescue squads. Rope can be used for anything from a lifeline to hoisting, lowering, anchoring, rigging, or even for crowd control. When arranged with pulleys in mechanical advantage systems, rope can greatly assist rescuers in lifting. One of the most common uses for rope in rescue situations is to secure and stabilize vehicles that are in precarious positions.

The rope must be of high quality to withstand the stresses exerted on it. It is important that rescuers know how rope is made as well as how to use it. Any rope or rope equipment that is used in a life safety system for any reason should conform to the standards set forth in NFPA 1983, *Standard on Fire Service Life Safety Rope, Harness, and Hardware*. Life safety rope should be clearly marked as such and should be stored separately from utility rope so that one cannot easily be mistaken for the other. For more information on ropes and rope systems, see Chapter 4, Rope Rescue.

BLOCK AND TACKLE

Because of their mechanical advantage in converting a given amount of pull to a working force greater than the pull, blocks and tackle are useful for lifting or pulling heavy loads. A *block* is a wooden or metal frame containing one or more pulleys called *sheaves*. *Tackle* is the assembly of ropes and blocks through which the line passes to multiply the pulling force (Figure 3.71).

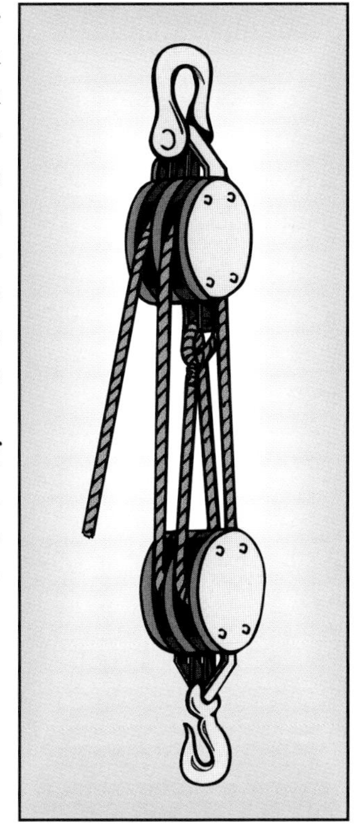

Figure 3.71 A double block and tackle.

> **WARNING**
>
> **Block and tackle systems are not life safety devices and should only be used for lifting or stabilizing objects and NOT for lifting people.**

Operators should observe the following safety rules when using block and tackle:

- The rope must be the right size for the weight being lifted and the blocks being used.
- Those pulling on the fall line should exert a steady, simultaneous pull and hold on to the gain.
- The anchor to which the standing block is attached must be strong enough to hold the load and the pull.
- The pull should be in a direct line with the sheaves.
- The pull should be downhill whenever possible.
- No one should stand under the load in case the system fails.
- Suspended loads should be lowered gradually, without jerking.
- Open hooks should always be moused to prevent slings or ropes from slipping off (Figure 3.72).

Figure 3.72 Open hooks should always be moused.

WEBBING

Conventional webbing is made from the same materials used in synthetic ropes, so the same precautions and maintenance procedures apply (see Chapter 4, Rope Rescue). The size of webbing will vary with the intended use, but most webbing used for lifting and pulling operations starts at about 2 inches (50 mm) in width (Figure 3.73). The minimum strength for webbing should be the same as for chain used in a similar situation. A major

Figure 3.73 A vehicle stabilized with webbing.

Figure 3.74 A trench rescue trailer. *Courtesy of Fire Wagons, Inc.*

disadvantage of webbing is that it is more susceptible to abrasion and chemical degradation than chain. This makes webbing impractical for use in some rescue applications. If the way webbing must be used subjects it to abrasion, it should be protected with a salvage cover or similar material.

There are two main types of webbing construction. One has a solid, flat design and the other a tubular design. Both look the same unless viewed cross-sectionally at the ends. The tubular weave also has two designs: One has a spiral structure while the other has a chain structure. Overall, the spiral structure is stronger and more resistant to abrasion than the chain structure.

Trench And Shoring Equipment

To be accomplished safely, trench rescue operations require special equipment. The use of special equipment is important because both the victim and the rescuer are in a very dangerous environment. Some trench rescue equipment may be too large or too infrequently used to be carried regularly on the rescue vehicle, but provisions should be made for this equipment to be available without delay (Figure 3.74).

CRIBBING

Rescue vehicles should carry an adequate amount of appropriately sized cribbing. Cribbing is essential in many rescue operations. It is most commonly used to stabilize objects but also has many other uses. Large wedges may be used to shim up loose cribbing (Figure 3.75). The wedges may be driven in with a mallet or a piece of cribbing.

Wood selected for cribbing should be solid, straight, and free of such major flaws as large knots

Figure 3.75 Wedges can be used to shim up cribbing.

or splits. Various sizes of wood can be used, but the most popular are 2- x 4-inch and 4- x 4-inch hardwood lumber. The length of the pieces may vary, but 16 to 18 inches (400 mm to 450 mm) is standard. The ends of the blocks may be painted different colors for easy identification by length. Other surfaces of the cribbing should be free of paint or any other finish because they can make the wood slippery, especially when it is wet. Individual pieces of cribbing may have a hole drilled through the end with a loop of rope or webbing tied through the hole for easy carrying (Figure 3.76).

Cribbing can be stored in numerous ways. It can be stacked in a compartment with the grab handles facing out for easy access (Figure 3.77). It can also be placed on end in a storage crate (Figure 3.78).

Figure 3.76 Cribbing with rope handles attached.

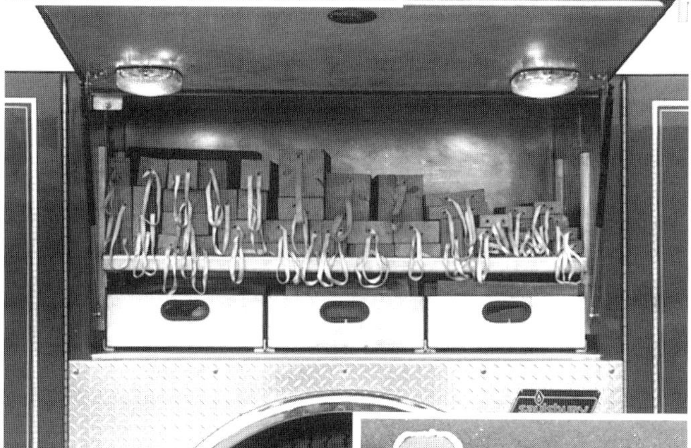

Figure 3.77 Cribbing stored in a compartment.

Figure 3.78 Cribbing stored in crates is easy to move.

Figure 3.79 Aluminum hydraulic shores.

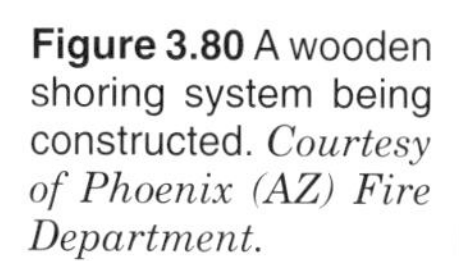

Figure 3.80 A wooden shoring system being constructed. *Courtesy of Phoenix (AZ) Fire Department.*

TRENCH TOOLS

Due to space constraints in a trench, special scaled-down tools are sometimes needed to perform routine tasks such as removing dirt and debris. Ideally, spades and picks carried on rescue units should have shortened handles; the working ends of these tools may be scaled down as well.

It may be useful to have a supply of buckets in order to haul dirt out of a hole. Ropes may be attached to the handles for hauling purposes, so only sturdy buckets with secure handles should be used.

While screw jacks or pneumatic or aluminum hydraulic shores (Figure 3.79) may be available on scene, rescuers should be prepared to supply their own. Rescuers may also have to construct a wooden shoring system (Figure 3.80). For more information, see Chapter 10, Trench Rescue.

SHEETING

Sheeting is material that has a large surface that is placed directly against the wall of a trench. Sheeting must be sturdy and be able to withstand the pressures exerted by the wall of the trench or the cross-bracing. The two principle types of sheeting are planks and plywood. Planks are usually 2 x 8 inches or 4 x 8 inches. Plywood is at least 1¼-inch (30 mm) CDX or ¾-inch (18 mm) Fin-form white birch plywood (Figure 3.81). Due to its bulkiness, it may not be practical to carry sheeting on the rescue vehicle. Contingency plans with local lumberyards, mills, or building supply stores may be necessary if these materials are to be readily available when needed.

Figure 3.81 Fin-form plywood.

Special Elevator-Rescue Equipment

Elevator rescues require specialized equipment that is not used in other types of rescue incidents. Two common types of specialized equipment used in elevator rescue are formed emergency keys (lunar keys) and interlock release tools (IRT).

FORMED EMERGENCY KEYS (LUNAR KEYS)

Emergency keys are used to open hoistway doors that are equipped with unlocking devices. These keys are molded into different shapes and sizes depending on the manufacturer of the elevator equipment. The three most common types are the T-shaped, moon-shaped, and drop keys (Figure 3.82).

It is usually impractical for the rescue company to carry every type of key necessary as standard equipment. Therefore, the key to the elevators of a particular occupancy should be kept in a lockbox or other emergency enclosure somewhere on the premises — generally, in the main elevator lobby.

INTERLOCK RELEASE TOOL (IRT)

An interlock release tool (IRT) is designed to slide through the crack of a closed door and release the interlock that holds the door shut (Figure 3.83). There are no tools of this type on the market; however, they can easily be formed from a thin strip of steel. The IRT tool resembles the tools that are made to open locked car doors, but it is somewhat longer.

Cots, Baskets, And Stretchers

There are several methods used to move victims during rescue operations. While some victims may be ambulatory, others will need to be carried

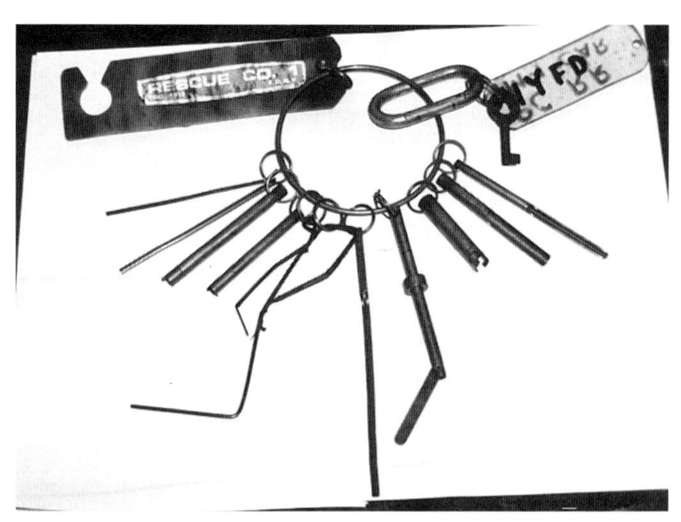

Figure 3.82 Typical elevator door keys. *Courtesy of New York (NY) Fire Department.*

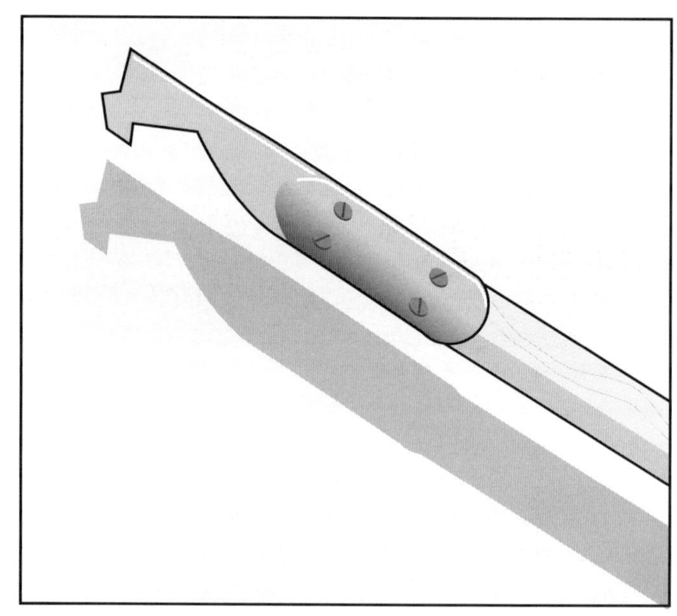

Figure 3.83 A typical IRT.

from the accident site to the medical treatment or transportation area. Under the general term "litters," there are three different types of devices used to carry injured victims: cots, baskets, and stretchers.

AMBULANCE COT (GURNEY)

An *ambulance cot* (gurney) is a pad on a collapsible frame with four wheels. The frame on most gurneys can be raised from its normal, collapsed position to waist height. These features make it highly transportable and easy to use for patient loading (Figure 3.84). Some ambulance cots allow either the head or the feet to be raised to facilitate patient treatment (Figure 3.85).

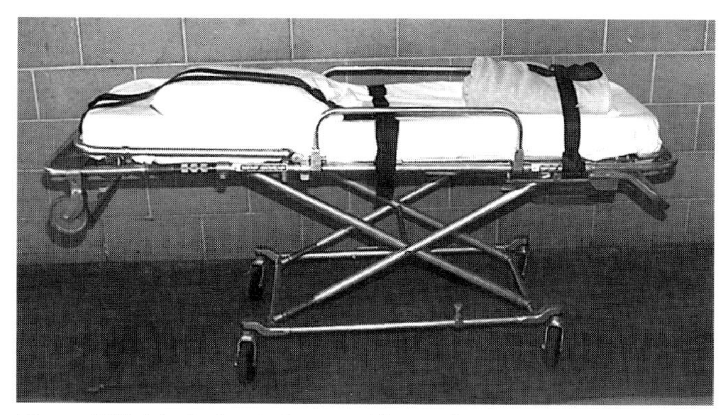

Figure 3.84 A typical ambulance cot (gurney).

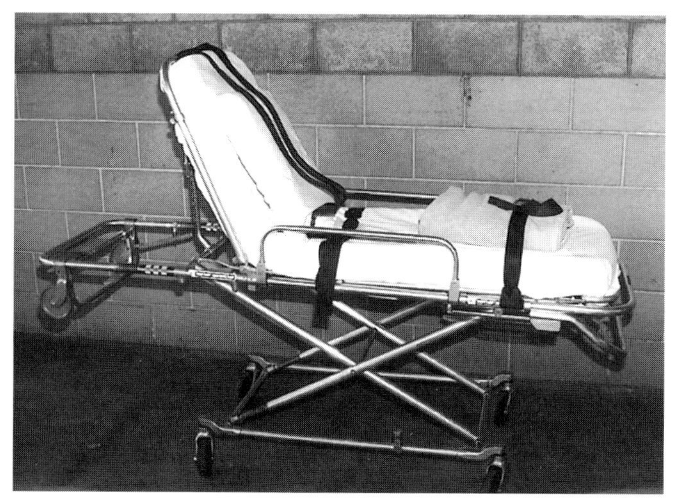

Figure 3.85 Some can be raised to a sitting position.

Figure 3.86 Modern and traditional basket litters.

Figure 3.87 Basket litters may be equipped with floats. *Courtesy of SKEDCO, Inc.*

BASKETS

The original Stokes basket consisted of a steel frame covered with woven wire mesh contoured to fit the human form. While some baskets are still produced that closely follow that original design, most modern baskets are made of rigid plastic or fiberglass, and they will accommodate a backboard (Figure 3.86). Some baskets are designed to be separated into two pieces to allow easier access into confined spaces. Baskets are stronger, are easier to handle, and offer greater protection and comfort for the patient than other types of litters. The basket is especially useful for moving patients from heights or over rough terrain or rubble. Baskets can be used in water rescue situations with the addition of flotation devices to provide buoyancy (Figure 3.87). Special kits are available to upgrade existing litters.

STRETCHERS

There are several different types of stretchers in use for rescue purposes. The most common types are the collapsible, scoop, and D-ring. Some of these stretchers are more versatile than others, but in general, they are better in confined spaces than the rigid baskets. The most practical type of stretchers for confined space rescue are those made of a flexible heavy-duty plastic that wraps snugly around the victim (Figure 3.88).

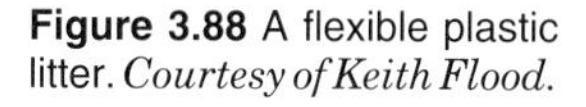

Figure 3.88 A flexible plastic litter. *Courtesy of Keith Flood.*

Monitoring/Detection Equipment

Rescuers may have to use a variety of atmospheric and scene monitoring and/or hazard detection equipment. The most common types allow rescuers to assess and monitor the atmosphere in and around the rescue scene and to detect downed wires or other electrical equipment that are still energized.

ATMOSPHERIC MONITORING EQUIPMENT

One of the most important pieces of information rescuers need is an accurate assessment of the atmosphere in which the rescue will have to be performed. Rescuers need to know if the atmosphere is deficient in oxygen and/or if it contains combustible or toxic gases. Direct-reading, broad-spectrum gas analyzers should be used to sample the atmosphere in any confined space where rescuers will have to work (Figure 3.89).

AC POWER DETECTION EQUIPMENT

Rescuers often need to survey the scene for the presence of electricity in order to identify and isolate electrical hazards. The AC power locator is an excellent tool for determining the presence of electrical hazards (Figure 3.90).

Figure 3.89 A modern broad-spectrum gas analyzer. *Courtesy of AIM Safety USA, Inc.*

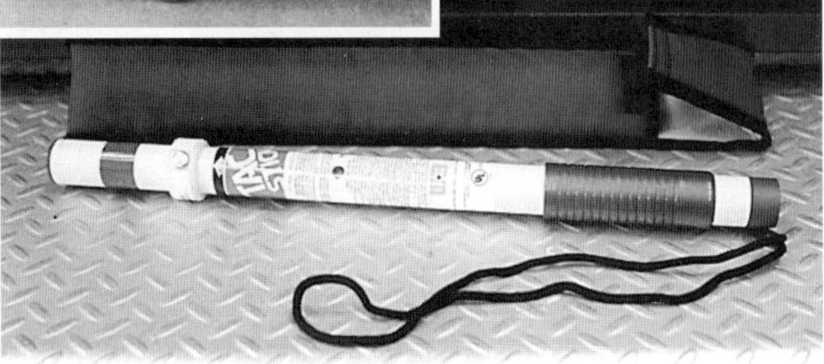

Figure 3.90 An AC power locator. *Courtesy of Keith Flood.*

Search Tools/Equipment

Tools and equipment needed for a thorough search include various types of hand lights and a means for marking searched areas. While there may be regional variations, the most common types in use are discussed in the following sections.

HAND LIGHTS

Rescuers assigned to search the dark interior of buildings, both standing and collapsed, and/or voids or other confined spaces need effective and reliable hand lights. Because rescuers may find themselves in or around potentially flammable atmospheres, the hand lights should be of the explosion-proof type.

MARKING DEVICES

As rescuers search buildings, or the rubble of collapsed buildings, they need to be able to mark their progress. A number of different marking systems have been developed, and rescuers need to have whatever is needed to use the system that is most appropriate for the situation. Common marking devices are markers, latch straps, and spray paint.

Markers. As rescuers progress from room to room or floor to floor in a fire building, they may use felt-tip markers, lumber chalk, or lumber crayons to mark those areas that have been searched or are being searched (Figure 3.91).

Latch straps. Many fire/rescue units use latch straps to mark the progress of search operations and to prevent doors from latching and/or locking at an inopportune time (Figure 3.92).

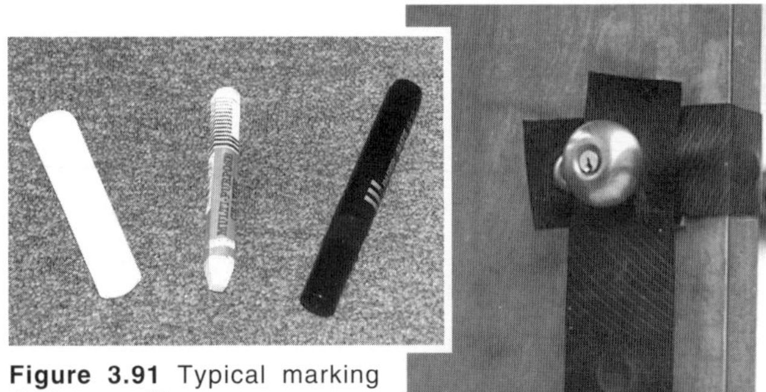

Figure 3.91 Typical marking devices used in rescue.

Figure 3.92 Typical latch straps.

Spray paint. A uniform system for marking collapsed buildings as described in Chapter 6, Structural Collapse Rescue, involves the use of spray paint. International Orange is the color most often used (Figure 3.93).

Figure 3.93 Typical search marking on a collapsed building.

PERSONAL PROTECTIVE EQUIPMENT (PPE)

The environment in which rescuers must work requires that they be provided with the best personal protective equipment (PPE) available. Special operations, such as water and ice rescues, rope rescues, rescues from confined spaces, or mine rescues, may require specialized personal protective equipment. Standard structural fire fighting turnout gear may be needed in some rescue operations; in others, it will not be appropriate. The following sections describe the basic types of protective clothing, as well as other equipment that may be required by rescuers. Those wearing the protective clothing should be aware of the dangers of wearing loose clothing, long hair, and jewelry during rescue operations.

Head And Face Protection

In all rescue situations, personnel should wear the appropriate protective headgear. Special head protection may be required for confined space, high angle, or rope rescue situations. Helmets protect personnel from falling objects, from bumps against protruding objects, and from head injuries due to slips or falls (Figure 3.94). All helmets used in rescue incidents should conform to NFPA 1972, *Standard on Helmets for Structural Fire Fighting*, or ANSI Z87.1, *Occupational and Educational Eye and Face Protection*.

Figure 3.94 Typical head protection.

In addition to wearing helmets to protect their heads, it is essential that rescue personnel use proper face and eye protection. The face and eyes must be protected from flying objects and spraying liquids that may cause severe injury. For rescuers, the main type of protection for the face is approved safety goggles or safety glasses (Figure 3.95).

Although they may not need them in most rescue operations, all rescue personnel should be equipped with fire-resistant, protective hoods (Figure 3.96). These hoods provide excellent heat protection for the neck, ears, head, and sides of the face. In addition to providing protection from fire, hoods can provide warmth during cold weather. They may also prevent some cuts and scratches caused by brushing against sharp edges or objects.

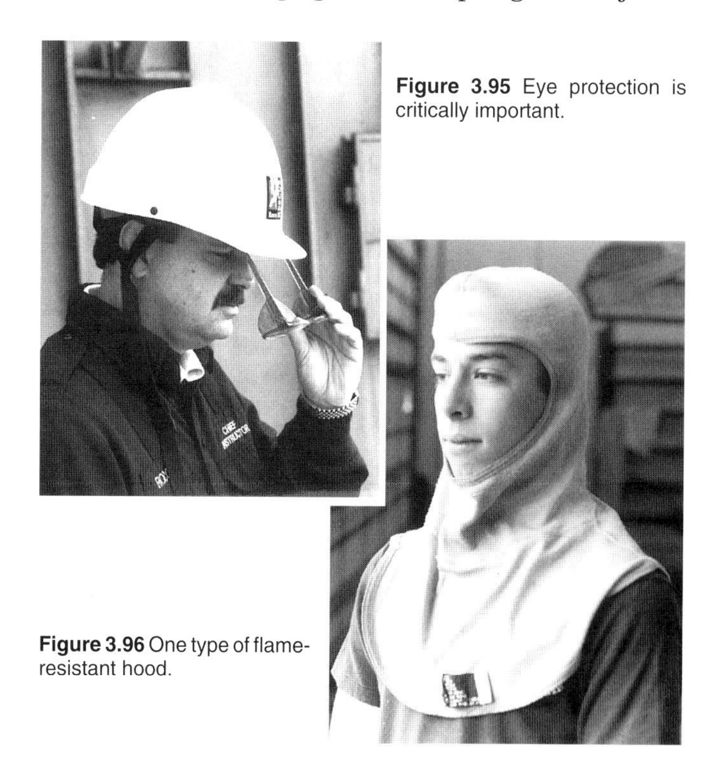

Figure 3.95 Eye protection is critically important.

Figure 3.96 One type of flame-resistant hood.

Body Protection

Appropriate protective clothing should be worn at all rescue incidents. Rescue company members should be in complete gear upon arrival at the scene, but the nature of the incident may dictate that other clothing be worn. Structural turnout gear, in addition to providing protection against fire, protects the rescuer from sharp or protruding objects, flying objects, and many other hazards that may be found on the rescue scene. All structural turnout gear should conform to the standards set forth in NFPA 1971, *Standard on Protective Clothing for Structural Fire Fighting* (Figure 3.97). However, structural turnout gear may not be appropriate for all rescue situations; other clothing, such as close-fitting jumpsuits or dry suits for protection from the cold, may be required (Figure 3.98). Some mountain rescue situations make shorts and tee shirts most practical.

It is imperative that all parts of the appropriate protective ensemble be worn while a rescuer is engaged in the operation. Personnel not wearing protective equipment appropriate for the duties to which they are assigned should not be permitted within the controlled zones.

Figure 3.97 A rescuer in structural turnout gear.

Figure 3.98 The most appropriate clothing should be worn.

For their own safety, it is important that rescue personnel be highly visible on the emergency scene, especially if they are exposed to vehicular traffic (Figure 3.99). Therefore, its visibility should be considered when protective clothing is chosen. NFPA 1971 lists specific requirements for the amount of striping necessary on structural turnouts. Reflective materials can also be applied to helmets.

Figure 3.99 Rescue gear should be highly visible.

Foot Protection

Hazards to the feet are numerous in fires and other rescue situations, so proper foot protection is essential. When selecting foot protection, the hazards that will be encountered should be considered: potential injuries from heat, punctures, and impact. The type of footwear needed will vary with the rescue situation, but all footwear should meet the applicable ANSI and/or NFPA standards (Figure 3.100). Proper fit is important for adequate utility, for reducing fatigue, and for preventing blisters and irritation.

Figure 3.100 A typical rescue boot. *Courtesy of Warrington Group, Ltd.*

Hand Protection

Gloves are an important part of personal protective equipment. The type of gloves worn by rescuers will also vary with the job they are doing and the type of protection required. If rescue personnel are engaged in fire fighting, they should wear gloves that meet the standards set forth by NFPA 1973, *Standard on Gloves for Structural Fire Fighting* (Figure 3.101). However, during rescue incidents where fire is not involved, this type of glove may be too bulky and may restrict manual dexterity.

When performing rescue work, rescuers need gloves that protect their hands from sharp objects, splinters, and rope burns. Rescuers should wear comfortable, close-fitting leather gloves that allow dexterity but are sturdy enough to protect the hands (Figure 3.102). When performing emergency medical aspects of rescue, latex exam gloves should always be worn.

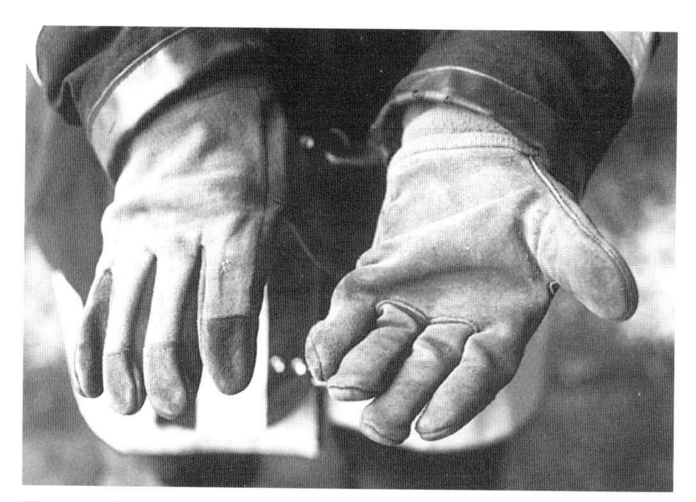

Figure 3.101 Typical gloves used for structural fire fighting.

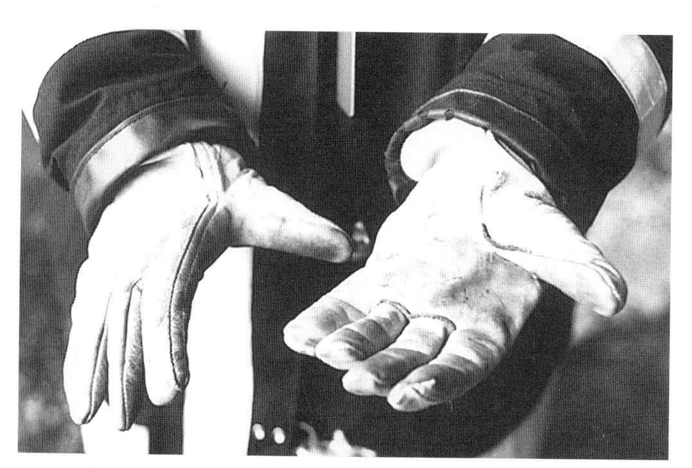

Figure 3.102 Typical leather rescue gloves.

Respiratory Protection

Rescuers often work in atmospheres that require the use of protective breathing apparatus, so rescue personnel should be thoroughly trained in its operation, use, and limitations. Hazardous vapors, fumes, smoke, dust, heat, or lack of available oxygen are examples of conditions that require this type of equipment. In hazardous atmospheres or oxygen-deficient atmospheres, such as those found in some confined space incidents, both rescuers and victims may need respiratory protection (Figure 3.103).

WARNING

Rescuers must never compromise their protective breathing system to share their air supply with another. Only approved devices should be used for "buddy breathing."

Figure 3.103 An example of "buddy breathing."

Except for water rescues, rescue personnel generally use one of four types of breathing equipment. These are the positive-pressure self-contained breathing apparatus (SCBA), supplied-air respirators (both of which are open-circuit types), rebreather units of the closed-circuit type, and filter masks. Open-circuit units draw breathing air from the source of supply and discharge the exhaust air into the atmosphere. Closed-circuit units draw air from the source, contain the exhaust air within the unit, filter out the carbon dioxide (CO_2), reoxygenate the air, and resupply it to the wearer. SCBA is not intended to be used underwater. Those involved in underwater rescues or body recoveries must use equipment designed and approved for that purpose. Filter masks are sometimes needed in dust-laden atmospheres such as in collapsed buildings and trenches (Figure 3.104).

Figure 3.104 Rescuers wearing filter masks. *Courtesy of Phoenix (AZ) Fire Department.*

SELF-CONTAINED BREATHING APPARATUS (SCBA)

The main feature of SCBA is that the wearer is independent of the surrounding atmosphere as long as there is a supply of air in the air cylinder. All SCBA must be of the positive-pressure type and should meet the requirements of NFPA 1981, *Standard on Open-Circuit Self-Contained Breathing Apparatus for Fire Fighters* (Figure 3.105).

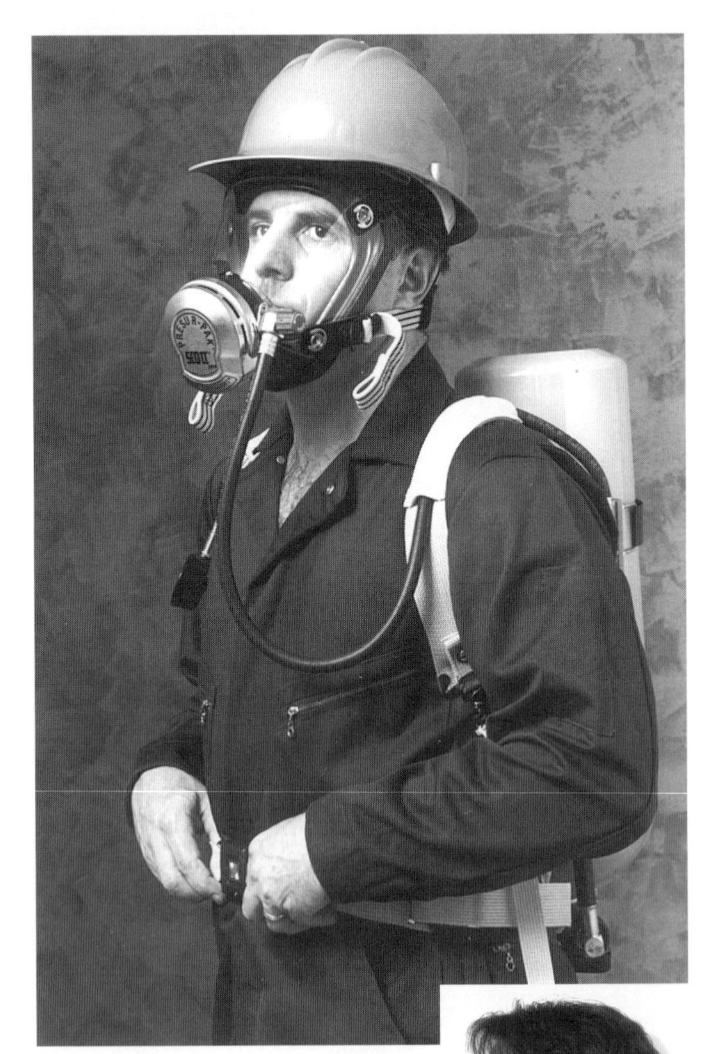

Figure 3.105 A rescuer in typical SCBA. *Courtesy of Scott Aviation.*

The positive-pressure regulator is designed to provide a constant pressure within the facepiece that is slightly above atmospheric pressure. This positive pressure prevents the contaminated atmosphere from entering the facepiece in case of a leak in the facepiece.

SCBA facepieces are available in different sizes in order to accommodate a variety of facial contours. With an ill-fitting facepiece, a gap between the wearer's face and the facepiece may exist that cannot be closed even by fully tightening the straps. Fit testing is critically important and should be done when a rescuer is issued a facepiece. Facial hair, usually in the form of beards or long sideburns, is unacceptable when wearing any type of respiratory protection because it may interfere with the seal of the facepiece (Figure 3.106). Anyone required to wear SCBA should *not* be allowed to have facial hair within the seal area of the facepiece. While an improperly sealed facepiece on positive-pressure SCBA will not allow contaminants into

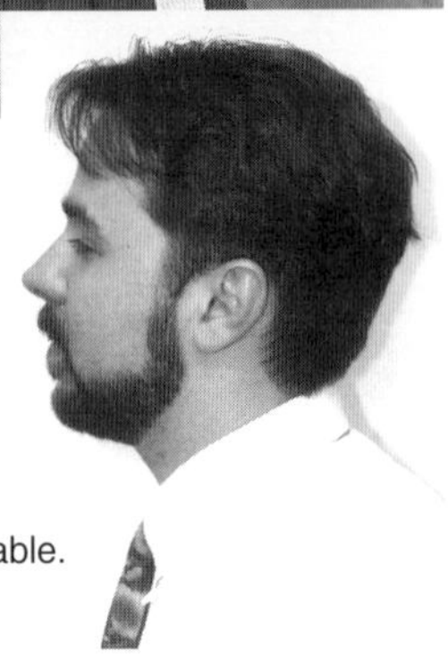

Figure 3.106 Some facial hair is unacceptable.

the facepiece, it will allow breathing air to escape into the atmosphere and thereby deplete the air supply faster.

The most common type of SCBA in use by fire departments today employs a cylinder of compressed air with a 30- or 45-minute rated service time (Figure 3.107). However, since several variables can influence how long a given cylinder of breathing air will last, it must be understood that these service time ratings represent the *maximum* amount of working time to be expected from a full cylinder. Because these units have a relatively short duration per cylinder, they may not be the best choice for extended rescue operations in contaminated or oxygen-deficient atmospheres.

For extended rescue operations, SCBA with a greater service time will allow rescuers to work more efficiently. The two most common types of long-duration SCBA are supplied-air respirators and rebreather units. Further information on self-contained breathing apparatus is available in the IFSTA **Self-Contained Breathing Apparatus** manual.

Figure 3.107 A typical SCBA unit.

SUPPLIED-AIR RESPIRATORS

Supplied-air respirators consist of a harness, harness-mounted regulator, a low-pressure hose connected to a conventional facepiece, and a small, 5-minute escape cylinder of compressed air. The regulator is coupled to a larger air supply (compressor or cascade cylinders) by a high-pressure hose of any length and a low-pressure hose up to 300 feet (90 m) long (Figure 3.108).

Figure 3.108 A typical airline breathing apparatus. *Courtesy of ISI.*

Breathing-air systems on elevated platforms can also be adapted to supply the air to one of these systems by adding the necessary extension hose and fittings. The same can also be done with cascade systems on rescue or other apparatus.

SELF-CONTAINED UNDERWATER BREATHING APPARATUS (SCUBA)

As mentioned earlier, conventional fire service SCBA is not designed for use underwater, and anyone involved in underwater rescues and/or body recoveries should be equipped with self-contained underwater breathing apparatus (SCUBA) (Figure 3.109). The safe use of SCUBA in underwater operations is a highly specialized skill. Only those personnel who are certified rescue divers, properly equipped for this type of work, should be allowed to attempt it.

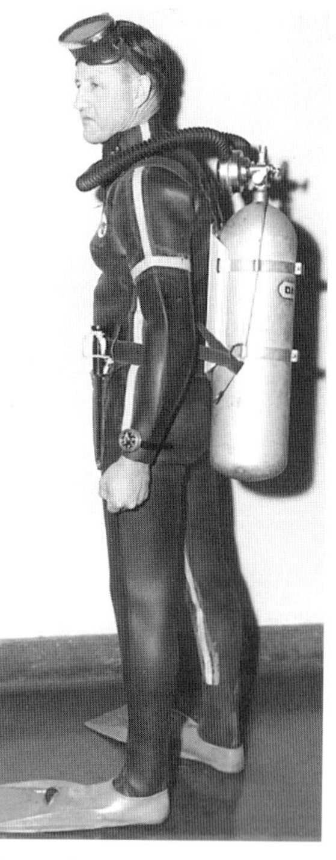
Figure 3.109 SCUBA gear is used by dive rescue teams.

Special Protective Equipment

In some situations, special protective equipment may be required. Rescuers may have to rely on specially trained and equipped hazardous materials technicians or underwater specialists to perform rescues. Only those personnel who are specifically trained and equipped to safely operate in these environments should be allowed to use this equipment. Special equipment may include hazardous materials suits, proximity or entry suits, and wet suits (Figure 3.110).

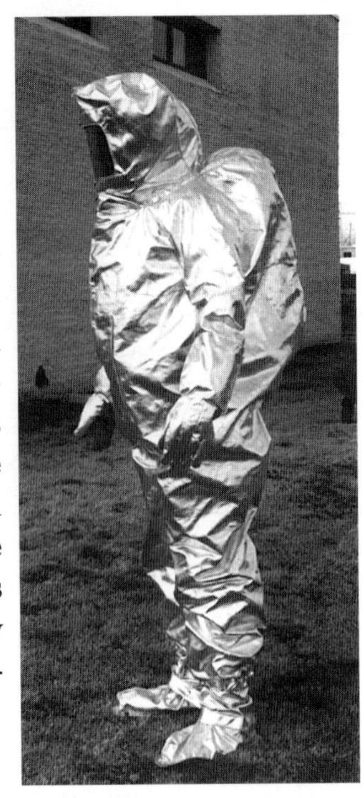

Figure 3.110 Rescuers sometimes need special protective clothing.

Care Of Personal Protective Equipment

Whether the tools used by the rescue company are homemade or the latest equipment available commercially, the importance of the rescuer's thorough knowledge of and ability to use the tools cannot be overemphasized. None of the techniques covered in this manual can be done safely and efficiently if the rescuers are not able to operate and care for their equipment properly.

Proper care and maintenance of personal protective equipment is vital to maintaining its intended level of protection. Each time rescue personnel use a piece of equipment, they should clean it and inspect it for defects that might pose a hazard or reduce its effectiveness the next time it is used. All equipment — from turnout gear to breathing apparatus — should be cared for according to the manufacturer's recommendations. Breathing apparatus should be tested regularly. For more information on personal protective equipment, see the IFSTA **Fire Department Occupational Safety** and **Self-Contained Breathing Apparatus** manuals.

Chapter 3 Review

Directions

The following activities are designed to help you comprehend and apply the information in Chapter 3 of **Fire Service Rescue**, Sixth Edition. To receive the maximum learning experience from these activities, it is recommended that you use the following procedure:

1. Read the chapter, underlining or highlighting important terms, topics, and subject matter. Study the photographs and illustrations, and read the captions with each.
2. Review the list of vocabulary words to ensure that you know the chapter-related meaning of each. If you are unsure of the meaning of a vocabulary word, look up the word in the glossary or a dictionary, and then study its context in the chapter.
3. On a separate sheet of paper, complete all assigned or selected application and review activities before checking your answers.
4. After you have finished, check your answers against those on the pages referenced in parentheses.
5. Correct any incorrect answers, and review material that was answered incorrectly.

Vocabulary

Be sure that you know the chapter-related meanings of the following words:

- alloy *(66)*
- ambient *(69)*
- aramid *(67)*
- corrosive *(66)*
- cribbing *(58)*
- degradation *(71)*
- dexterity *(79)*
- ensemble *(78)*
- exothermic *(69, 70)*
- hydraulic *(59)*
- lateral *(51)*
- litter *(74)*
- metal fatigue *(66)*
- mouse *(71)*
- neoprene *(67)*
- opaque *(53)*
- pneumatic *(61)*
- precarious *(71)*
- procure *(55)*
- purge *(52)*
- ratchet *(58)*
- sheave *(71)*
- shim *(72)*
- shoring *(72)*
- trenching *(72)*
- unstable gas *(69)*
- versatile *(70)*
- void *(69 & 76)*

Application Of Knowledge

1. Classify your department's rescue vehicle(s) as light rescue, medium rescue, or heavy rescue. *(Local protocol)*
2. Classify your department's rescue vehicle compartmentation as exclusive exterior, exclusive interior, or combination. *(Local protocol)*
3. Choose from the following tool and equipment procedures those appropriate to your department and equipment, or ask your training officer to choose appropriate procedures. Mentally rehearse the procedures, or practice the chosen procedures under your training officer's supervision.
   - Set up gin poles and A-frames. *(51)*
   - Deploy vehicle stabilizers. *(51)*
   - Recharge an SCBA cylinder from a cascade system. *(51, 52)*
   - Recharge an SCBA cylinder from a breathing-air compressor. *(52)*
   - Set up and operate fixed and portable lighting equipment. *(53, 54)*
   - Choose and use appropriate striking, prying, and cutting hand tools for given tasks. *(55, 56)*
   - Use bar and trench screw jacks to support collapsed structural members. *(58)*
   - Use cribbing and ratchet-lever jacks. *(58)*
   - Operate a hydraulic jack. *(59)*
   - Operate a hydraulic spreader. *(59)*
   - Operate hydraulic shears. *(59)*
   - Operate combination spreader/shears. *(60)*
   - Operate a hydraulic extension ram. *(60)*
   - Operate a porta-power tool system. *(60)*
   - Operate a pneumatic nailer. *(62)*
   - Operate a pneumatic chisel/hammer. *(62)*
   - Operate a Whizzer saw. *(62, 63)*
   - Operate a circular saw. *(63)*
   - Operate a reciprocating saw. *(63)*
   - Operate a chain saw. *(63, 64)*
   - Operate a truck-mounted winch. *(64)*
   - Use a come-along. *(65)*
   - Operate a nonbreathing-air compressor to inflate air lifting bags. *(52)*
   - Cut metal with an oxyacetylene cutting torch. *(69)*
   - Cut metal with a burning bar. *(71)*
   - Cut metal with a plasma cutter. *(71)*

- Use a block and tackle. *(71)*
- Use trench tools and sheeting. *(73)*
- Use elevator lunar keys and interlock release tool. *(74)*
- Place a victim on a litter (gurney, basket, and stretcher). *(74, 75)*
- Don and doff articles of personal protective clothing. *(77-80)*
- Don SCBA and start the airflow. *(80, 81)*

Review Activities

1. Identify the following abbreviations and acronyms:
   - IRT *(74)*
   - mps *(70)*
   - PPE *(77)*
   - rpm *(62)*
   - SCBA *(80)*
   - SCUBA *(81)*
2. Make a table in which you complete columns headed Type, Functions, Tools Carried, Rescue Capability, Design Specs, and Personnel Capability for each of the following rescue vehicles:
   - Light rescue vehicle *(47)*
   - Medium rescue vehicle *(47, 48)*
   - Heavy rescue vehicles *(48)*
   - Rescue pumper *(48)*
   - Standard engine company *(48, 49)*
   - Ladder company *(49)*
3. List the advantages and disadvantages of each of the following types of rescue vehicle body designs:
   - Exclusive exterior compartmentation *(49)*
   - Exclusive interior compartmentation *(49, 50)*
   - Combination compartmentation *(50)*
4. Compare and contrast commercial and custom rescue vehicle chassis. *(50, 51)*
5. Briefly describe the functions of each of the following special rescue vehicle equipment and accessories:
   - All-wheel drive *(51)*
   - Gin poles and A-frames *(51)*
   - Stabilizers *(51)*
   - Cascade systems *(51, 52)*
   - Breathing-air compressors *(52)*
   - Nonbreathing-air compressors *(52)*
   - Inverters *(52, 53)*
   - Vehicle-mounted generators *(53)*
   - Fixed lighting equipment *(53, 54)*
   - Portable lighting equipment *(53, 54)*
   - Electrical cables and extension cords *(54)*
   - Ground-fault circuit interrupters (GFCIs) *(54)*
   - Twist-lock receptacles and receptacle adapters *(54)*
6. List 12 types of hand tools used in the fire service for striking. *(55)*
7. State two safety guidelines for using manual striking tools. *(55)*
8. List three maintenance procedures for manual striking tools. *(55)*
9. List eight types of hand tools used in the fire service for prying. *(56)*
10. State two safety guidelines for using manual prying tools. *(56)*
11. Provide examples of each of the following types of manual cutting tools:
    - Chopping tools *(56)*
    - Scissors or snipping tools *(56, 57)*
    - Saws *(57)*
    - Knives *(57)*
12. Describe proper maintenance procedures for chopping tools. *(56)*
13. List seven types of fire service snipping tools. *(57)*
14. Explain why standard fire service snipping tools *should not* be used to cut energized electrical wires. *(57)*
15. Explain the advantages of using handsaws in some rescue operations. *(57)*
16. Name four handsaws commonly used for rescue. *(57)*
17. Name three types of knives that may be carried by rescue companies. *(57)*
18. Make a table in which you complete columns headed Type, Design, and Primary Use(s) for each of the following types of nonhydraulic jacks:
    - Bar screw jack *(58)*
    - Trench screw jack *(58)*
    - Ratchet-lever jack *(58)*
19. Describe the uses and design capabilities of each of the following hydraulic tools:
    - Spreaders *(59)*
    - Shears *(59)*
    - Combination spreaders/shears *(60)*
    - Extension ram *(60)*
    - Jack *(61)*
20. Explain the advantages and disadvantages of the porta-power tool. *(60)*
21. List types of power sources available to the rescue team for pneumatic tools. *(61)*
22. Explain why compressed oxygen must never be used to power pneumatic tools. *(61)*

23. Describe the rescue uses and design capabilities of the following pneumatic tools:
    - Air chisel *(62)*
    - Pneumatic nailer *(62)*
24. Compare and contrast the following power saws:
    - Whizzer saw *(62, 63)*
    - Circular saw *(63)*
    - Reciprocating saw *(63)*
    - Chain saw *(63, 64)*
25. List power saw safety rules. *(64)*
26. List five examples of lifting/pulling tools used for rescue. *(64)*
27. Describe the rescue uses of the tools listed in Activity 26. *(64-66)*
28. Describe chain inspection procedures. *(66)*
29. List safety rules for using chain in rescue procedures. *(66, 67)*
30. Compare the rescue uses of high-pressure and low-pressure air lifting bags. *(67, 68)*
31. List air bag safety rules. *(68)*
32. Explain how acetylene is stabilized in an acetylene cylinder. *(69)*
33. Explain how an oxyacetylene cutting torch operates. *(69)*
34. List safety rules for using oxyacetylene cutting equipment. *(69, 70)*
35. Compare and contrast the operation and uses of burning bars and plasma cutters. *(70)*
36. List safety rules for using a block and tackle. *(71)*
37. Explain the advantages and disadvantages of using webbing in rescue operations. *(71, 72)*
38. Briefly identify and list the rescue use(s) for each of the following tools and pieces of equipment:
    - Cribbing *(72)*
    - Sheeting *(73)*
    - Lunar key *(74)*
    - Interlock release tool *(74)*
    - Gurney *(74)*
    - Stokes basket *(75)*
    - Stretcher *(75)*
    - Spectrum gas analyzer *(76)*
    - AC power locator *(76)*
    - Markers and spray paint *(77, 78)*
    - Latch strap *(76)*
39. Describe the personal protective equipment available to protect the following parts of the rescuer's body:
    - Head and face *(77)*
    - Body *(78)*
    - Feet *(78)*
    - Hands *(79)*
    - Respiratory system *(79, 80)*
40. Explain the differences among the following types of respiratory protective gear:
    - Self-contained breathing apparatus *(80, 81)*
    - Supplied-air respirators *(81)*
    - Self-contained underwater breathing apparatus *(81)*

Questions And Notes

4

Rope Rescue

Chapter 4
Rope Rescue

INTRODUCTION

When victims are located above or below grade in rescue situations, the most efficient and sometimes the only means of reaching them and getting them to ground level may be by the use of ropes, knots, and rope systems. It may be necessary to lower rescuers into a confined space and to hoist a victim out with a mechanical advantage system made of rescue rope. Victims stranded on a rock ledge or on an upper floor of a partially collapsed building may have to be lowered to the ground with rescue ropes.

Rescue rope, webbing, and appropriate hardware are used for a variety of purposes. Rescue rope and harness are used to protect rescuers and victims as they move and work in elevated locations where a fall could cause injury or death. Ropes in combination with webbing are the primary tools for raising and lowering rescuers, equipment, and victims. Rope and appropriate hardware are used to create a variety of mechanical advantage and safety systems.

This chapter reviews the equipment and systems used in rope rescue. The construction, care, and maintenance of rescue rope and webbing are discussed. Also covered are the various pieces of hardware used with rescue rope and the mechanical advantage and safety systems they are used to create. Finally, the tactical considerations involved in rope rescue situations are reviewed.

BACKGROUND

Regulations

NFPA 1983, *Standard on Fire Service Life Safety Rope, Harness, and Hardware*, is the primary standard covering the types of equipment discussed in this chapter. The standard provides minimum performance requirements for the life safety rope, harness, and hardware that rescuers use to support themselves and victims during actual or simulated rope rescue operations. NFPA 1983 covers this topic in greater depth and greater detail than is appropriate or necessary in this manual; however, rescuers are encouraged to familiarize themselves with the content of the standard.

Also applicable in some rope rescue incidents are the Occupational Safety and Health Administration (OSHA) regulations contained in Title 29 of the Code of Federal Regulations (CFR). In the U.S., work of any kind, including rescues, within confined spaces is regulated under 29 CFR 1910.146.

EQUIPMENT

Personal Protective Equipment

The structural fire fighting turnout gear worn by most firefighters is often not appropriate for rescue situations in general or for rope rescues in particular — unless there is a fire in close proximity or an imminent threat of one starting. The type of fire fighting personal protective equipment (PPE) that is most appropriate for rope rescues is similar to that worn by wildland firefighters. In the majority of rope rescue operations, rescuers wear the following PPE.

HELMETS

The type of helmets most often worn in rope rescue operations are similar to those worn by wildland firefighters, but they have chin straps and are designed specifically for rescue work (Figure 4.1). These helmets are lighter in weight than structural fire fighting helmets, they allow better peripheral vision, and they are less cumbersome in tight places.

EYE PROTECTION

Eye protection is recommended in some rope rescue applications. In these operations there are numerous opportunities for airborne particulates and flying debris to enter rescuers' eyes. Just keeping dust out of rescuers' eyes is reason enough to wear goggles or safety glasses (Figure 4.2).

Figure 4.1 Typical helmets used in rope rescue. *Courtesy of Laura Mauri.*

OUTERWEAR

Except where protection from actual or potential fires is needed or where extremely wet or cold environmental conditions warrant, structural turnout gear is usually not appropriate for rope rescues. Clothing suitable for the rescue environment should be worn (Figure 4.3). The clothing selected should fit snugly enough to avoid becoming entangled in the rope system but loose enough to allow freedom of movement.

Figure 4.2 Eye protection is very important. *Courtesy of Laura Mauri.*

Figure 4.3 Clothing that is appropriate for rope rescue.

FOOTWEAR

Unless the operation must be carried out in standing water or other liquids, rubber boots are usually not needed and are not the best choice for rope rescue. Leather boots often provide a good combination of function and protection. However, the footwear selected should be lightweight, have nonslip tread, provide ankle support, and be appropriate for the particular environment in which the rescue is to be performed (Figure 4.4).

Figure 4.4 Typical boots used in rope rescue.

GLOVES

Medium-weight leather gloves with reinforced palms are the most appropriate for rope rescues (Figure 4.5). They allow good freedom of movement and tactile dexterity while providing an acceptable level of protection.

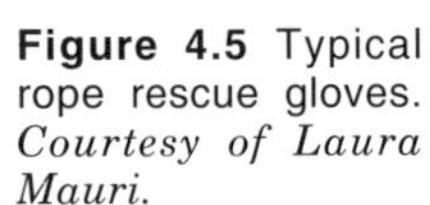

Figure 4.5 Typical rope rescue gloves. *Courtesy of Laura Mauri.*

Software

In this context, the term *software* refers to rope, webbing, accessory cord, and harness. These items can be used alone or in combination to protect rescuers and victims in situations involving significant elevation differences.

ROPE

Rescue rope falls into two classifications: life safety rope and utility rope. Life safety rope is used to support rescuers and/or victims during actual incidents or training. Utility rope is used for hoisting equipment and/or securing objects in place, not for supporting rescuers or victims. All life safety rope must conform to the requirements of NFPA 1983, which defines *life safety rope* as "rope dedicated solely for the purpose of supporting people during rescue, fire fighting, or other emergency operations, or during training evolutions." Only rope of block creel construction using continuous-filament virgin fiber for load-bearing elements is suitable for life safety applications.

According to NFPA 1983, the rope manufacturer must supply the purchaser with information regarding use criteria, inspection procedures, maintenance procedures, and criteria for retiring the rope from service. The manufacturer is further required to supply criteria to be considered before life safety rope is reused. Included in these criteria are the following conditions that must be met before a life safety rope is considered for reuse:

(a) Rope has not been visibly damaged.

(b) Rope has not been exposed to heat, direct flame impingement, or abrasion.

(c) Rope has not been subjected to any impact load.

(d) Rope has not been exposed to liquids, solids, gases, mists, or vapors of any chemical or other material that can deteriorate rope.

(e) Rope passes inspection when inspected by a qualified person following the manufacturer's inspection procedures both before and after each use.

The manufacturer must provide the user with information about not using a piece of life safety rope and removing it from service if it does not meet all of the previously stated conditions, if it does not pass inspection, or if there is any reason to doubt its safety or serviceability. Any life safety rope that fails to pass inspection or has been impact-loaded should be destroyed immediately. However, in this context, "destroy" means that it is altered in such a manner that it cannot be mistakenly used as a life safety rope. This could include disposing of the rope or removing the manufacturer's label and cutting the rope into shorter lengths to be used as utility rope.

Some departments dye the last few inches of each piece of rope to identify its status. The ends of life safety rope are dyed green, training rope yellow, and utility rope red (Figure 4.6). Once a life safety rope has been retired, it can be cut into shorter lengths and the ends cut off, resealed, and remarked in red for utility use.

Figure 4.6 Rope may be color-coded. *Courtesy of Laura Mauri.*

Life safety rope is usually purchased in 100- to 600-foot (30 m to 186 m) lengths depending upon local needs. It is usually 7⁄16 or ½ inch (11 mm or 13 mm) in diameter and should be of the kernmantle type. *Kernmantle* is a German word meaning core and sheath. The core (kern) is made up of continuous fibers, usually nylon, that run the entire length of the rope. The core provides about 75 percent of the rope's strength and carries the majority of the working load. The core is protected by a sheath (mantle) made of tightly woven nylon or other synthetic fiber that provides the remainder of the rope's strength (Figure 4.7). The sheath protects the core from abrasion, dirt, and the effects of sunlight, which can weaken nylon with prolonged exposure.

NFPA 1983 specifies that the minimum breaking strength for a one-person life safety rope is 4,500 pounds (2 045 kg); for a two-person rope, 9,000 pounds (4 090 kg). The *maximum working load* for life safety rope is defined as the maximum amount of weight that may be supported by the

Figure 4.7 The components of kernmantle rope.

rope in use. The maximum safe working load for a one-person life safety rope is 300 pounds (136 kg); it is 600 pounds (272 kg) for a two-person life safety rope. These figures result from dividing the respective minimum breaking strengths by 15. This provides an adequate margin of safety for both rescuers and victims suspended from ropes.

Traditionally, two types of rope have been used in life safety situations: dynamic rope and static rope. Each type has its advantages and disadvantages because of different design and performance criteria.

Dynamic rope. Dynamic (high-stretch) rope is used when long falls are a possibility, such as in rock climbing. To reduce the shock of impact on both climbers and their anchor systems in falls, dynamic rope is designed to stretch up to 60 percent of its length without breaking. However, this elasticity is a disadvantage when trying to raise or lower heavy loads, so dynamic rope is not considered practical for hauling applications and is usually employed only in mountain rescue work.

Static rope. Static (low-stretch) rope is the rope of choice for most rescue incidents. It is designed to stretch up to 20 percent of its length without breaking, and this makes it better suited for raising and lowering heavy loads (Figure 4.8). It is made with a heavy sheath that protects it from abrasion damage. Static rope is used for hauling, rescue, rappelling, and where no falls or very short falls are expected.

NOTE: The 1995 edition of NFPA 1983 only addresses "life safety rope" and does not distinguish between dynamic and static types. The standard specifies that new life safety rope must be capable of elongating at least 15 percent, but not more than 45 percent, of its length at 75 percent of breaking strength.

Figure 4.8 Static rope is used for hauling and most rescues. *Courtesy of Laura Mauri.*

ACCESSORY CORD

The term *accessory cord* refers to rope that is of a smaller diameter than life safety rope. It is usually synthetic rope of about ¼ or 5⁄16 inch (6 mm or 8 mm) in diameter. Accessory cord is used for lashing litters and for forming Prussik loops.

Prussik loops are formed from a length of accessory cord about 5 feet (1.5 m) long tied with a Double Overhand Knot (Double Fisherman) (Figure 4.9). Prussiks were originally designed to allow someone to climb up a vertical rope to rescue themselves if they fell into a crevasse. However, in modern rescue systems they perform three important functions: pulling, braking, and ratcheting.

A Pulling Prussik seizes the rope and pulls it into motion (Figure 4.10). A Braking or Tandem

Prussik seizes the rope and prevents it from moving (Figure 4.11). A Ratchet Prussik allows mechanical advantage pulley systems to be reset repeatedly for a series of pulls (Figure 4.12).

Tandem Prussik brakes or safeties are versatile, reliable, and easy to set up (Figure 4.13). They are an effective means of securing a rescue load. They are also capable of stopping and holding a rescue load dropped from nominal heights without serious damage to the main rescue line.

Figure 4.9 A typical Prussik loop.

Figure 4.10 A Prussik used for pulling. *Courtesy of Laura Mauri.*

Figure 4.11 Prussiks used for braking. *Courtesy of Laura Mauri.*

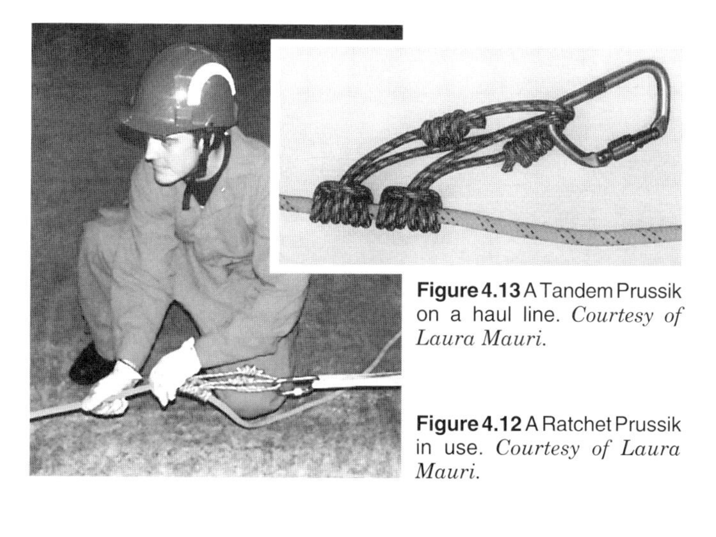

Figure 4.13 A Tandem Prussik on a haul line. *Courtesy of Laura Mauri.*

Figure 4.12 A Ratchet Prussik in use. *Courtesy of Laura Mauri.*

ROPE MAINTENANCE

In order for rescue rope to be ready and safe for use when needed, it must be properly maintained. This maintenance involves proper inspection, cleaning, and storage of the rope.

Inspection. Inspecting kernmantle rope for damage is somewhat difficult because the damage may not be obvious. Rescue rope should be checked both visually and by feel before and after each use. This check can be done before use by putting a slight tension on the rope while feeling for any lumps, depressions, or soft spots in the rope (Figure 4.14). A temporary soft spot resulting from hard knots or sharp bends in the rope may be felt; however, if the rope is undamaged, the fibers within the core may realign themselves over time. The only way to determine whether such a spot is a damaged spot or just temporarily misaligned core fibers is by carefully inspecting the outer sheath. Any damage to the outer sheath indicates probable damage to the core. The core of a kernmantle rope

Figure 4.14 A firefighter inspects a kernmantle rope.

can be damaged without evidence of such damage being visible on the outer sheath. If there is any doubt about the rope's integrity, the rope should be downgraded to utility status.

The rope should also be inspected for irregularities in shape or weave, foul smells, discoloration from chemical contamination, roughness, abrasion, or fuzziness. A certain amount of fuzziness is normal and is not necessarily a cause for concern. If there is a greater amount of fuzziness in one spot or if the overall amount is excessive based upon judgment and experience, the rope should be downgraded. Each use of a rescue rope should be entered in its log.

Rope logs. When a piece of rescue rope is purchased, a log for that piece of rope must be started and must be kept with the rope throughout its working life. The log tracks the use and maintenance of that piece of rope and will help to determine when to retire that piece of rope (Figure 4.15). The log should be kept in a waterproof envelope, usually stored in a pocket sewn on the side of the rope bag (Figure 4.16).

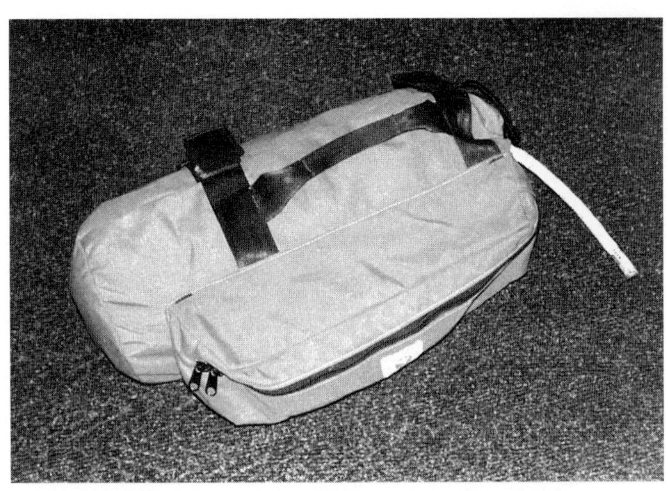

Figure 4.16 The pouch containing the rope log. *Courtesy of Laura Mauri.*

Cleaning. Methods of washing and drying synthetic fiber rope vary with each manufacturer, so it is always advisable to contact them for specific cleaning instructions for the type of rope or ropes in use. Cool water and mild soap are least likely to damage the rope. Any type of bleach or strong cleaners should not be used. Some synthetic rope may feel stiff after washing, but this is not a cause for concern.

Oklahoma State University
Fire Service Training
Rope Log

Rope Type:________ Rope Size:________ Rope#:________
Manufacturer:________ Model:________ Rope Color:________
Purchased From:________ Date:________ Bag Color:________

Date	Sign Out	Use	Possible Damage/Comments	Sign In

Figure 4.15 A typical rope log.

There are three principal ways to clean synthetic rope: by hand, by using a special rope-washing device, or by using a regular clothes-washing machine. Washing by hand consists of wiping the rope with a cloth or scrubbing it with a brush and thoroughly rinsing it with clean water (Figure 4.17).

Commercial rope-washing devices that can be connected to a standard faucet or a garden hose are available (Figure 4.18). Rope is fed manually through the device, and multidirectional streams of water clean all sides of the rope at the same time. These devices do an adequate job of cleaning mud and other surface debris from the rope, but for a more thorough cleaning, the rope should be washed in a clothes-washing machine.

Front-loading washing machines that have a plastic window are not recommended because the plastic can cause enough friction with the rope during the spin cycle to damage the rope. Top-loaders may also damage the rope during agitation. So, front-loaders without a plastic window are the best type to use for washing synthetic rope. The washer should be set on the coolest wash/rinse temperature available, and only a small amount of mild soap, if anything, should be used. The rope can be protected by putting it in a cloth bag before placing it in the washer, or it can be first coiled into a bird's-nest coil (Figure 4.19).

Once the rope has been washed, it may be dried in the same way fire hose is dried. It can be spread out on a hose rack (out of direct sunlight), suspended in a hose tower, or loosely coiled in a hose dryer.

Figure 4.17 Rope may need to be scrubbed.

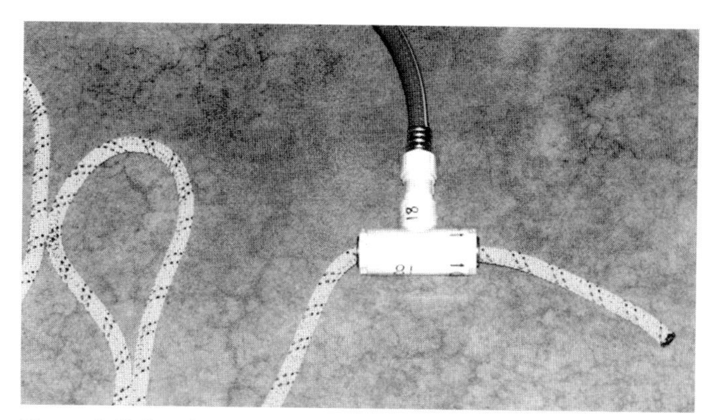

Figure 4.18 A typical hose-bib rope washer. *Courtesy of Laura Mauri.*

Figure 4.19 A rope is loaded into a washing machine.

Storage. Rescue ropes can be stored in various coils, but most rescue units store their ropes in stuff bags (Figure 4.20). Regardless of how rope is stored, where it is stored is of critical importance. Rescue rope should be stored in spaces or compartments that are clean and dry but that have adequate ventilation. The rope should not be exposed to chemical contaminants, such as battery acid or hydrocarbon fuels, or the fumes or vapors of these substances. Rescue rope should not be stored in the same compartments as gasoline-powered rescue tools or the spare fuel for these tools.

Figure 4.20 Rescue rope can be inspected as it is stuffed into a bag.

WEBBING

Webbing is used extensively in rescue to construct anchor systems and harness, to package and secure victims, and to lash rescue components together. Webbing is preferable to rope for constructing a harness because it lies flat against the body and provides better support. Because of its strength and pliability, webbing is an excellent material to use for slings; it loses very little strength when bent around a carabiner. Webbing is relatively inexpensive, lightweight, easy to tie, and can be cut into short lengths for many uses. Although there are different methods by which webbing is made, webbing used in rescue applications is most often spiral weave or shuttle-loom tubular nylon.

Webbing is available in various widths from 1 to 2 inches (25 and 50 mm). The 1-inch size is most commonly used for lashing and has a 4,000-pound (1 818 kg) breaking strength. NFPA 1983 specifies that any load-bearing webbing making body contact be at least 1¾ inches (44 mm) wide and have a 6,000-pound breaking strength. For constructing pelvic and chest harnesses, 2-inch (50 mm) webbing is recommended.

Webbing is available in rolls of up to 1,000 feet (300 m) in length and in a variety of colors. A system of color-coding webbing to identify its length greatly aids in constructing rescue systems. For example, if all pieces of red webbing are 20 feet (6 m) in length, it is easy to select the proper piece of webbing to construct a seat harness or to secure a victim in a litter. The following is a recommended color-coding system for rescue webbing:

Color	*Length*
Green	5 feet (1.5 m)
Yellow	12 feet (3.5 m)
Blue	15 feet (4.5 m)
Red	20 feet (6.0 m)
Black	25 feet (7.5 m)

If the webbing is of one color throughout its length, the ends of each piece should be dyed according to the chart to identify its length. Most webbing is constructed of nylon, but regardless of the constituent material, it is produced in two basic forms: flat and tubular.

Flat webbing. Flat webbing is constructed of a single layer of material and is similar to that used in automobile seat belts. While not as expensive as tubular webbing, flat webbing is stiffer and more difficult to tie into knots. Flat webbing is mainly used in rescue work for straps and harness.

Tubular webbing. Because tubular webbing is more supple and easier to tie, it is the most commonly used webbing in rescue work. Its tubular shape can be seen if it is viewed from the end while the two edges are squeezed toward each other (Figure 4.21).

There are two types of tubular webbing: edge-stitched and spiral weave. Edge-stitched webbing is formed by folding a piece of flat webbing lengthwise and stitching the edges together (Figure 4.22). Because of abrasion, edge-stitched webbing can become unstitched, but edge-stitched webbing that is lock-stitched is not as prone to this occurrence.

Also known as *shuttle-loom construction*, spiral weave tubular webbing is the type most preferred in rescue work. Spiral weave webbing is constructed by weaving the tube as a unit and is so named because the threads spiral around the tube as it is being woven.

Figure 4.21 The ends of tubular webbing. *Courtesy of Laura Mauri.*

Figure 4.22 Edge-stitched tubular webbing. *Courtesy of Laura Mauri.*

HARNESS

There are two general types of life safety harness used in rescue incidents: (1) those identified in NFPA 1983 as Class I, Class II, and Class III and (2) a variety of improvised harness. The more common types of life safety harness are discussed in this section.

NOTE: Ladder belts are no longer classified as rescue harness (Figure 4.23). They are intended only as a positioning device for a person on a ladder.

Class I harness. This type of harness fastens around the waist and around thighs or under buttocks and is intended to be used for emergency escape with one-person loads (Figure 4.24).

Class II harness. This type of harness fastens around the waist and around the thighs or under the buttocks and may be used in two-person rescues. Class I and Class II harness may appear to be identical; the difference between them is in their rated working loads which can only be determined by reading the label on the harness.

Class III harness. This type of harness fastens around the waist, around the thighs or under the buttocks, and over the shoulders (Figure 4.25). It is designed to support two-person loads and to prevent inverting.

Improvised harness. In the absence of a manufactured harness, rescuers may have to construct a life safety harness using the equipment available on scene — rope, webbing, and carabiners. The most common types of improvised harness are the Rescue Knot, Swiss Seat, and Swiss Seat with Shoulder Harness.

Rescue Knot. The Rescue Knot has the advantage of needing only the rescue rope and does not require webbing or hardware. It can be tied onto a conscious or unconscious victim. The rescue knot is tied as follows:

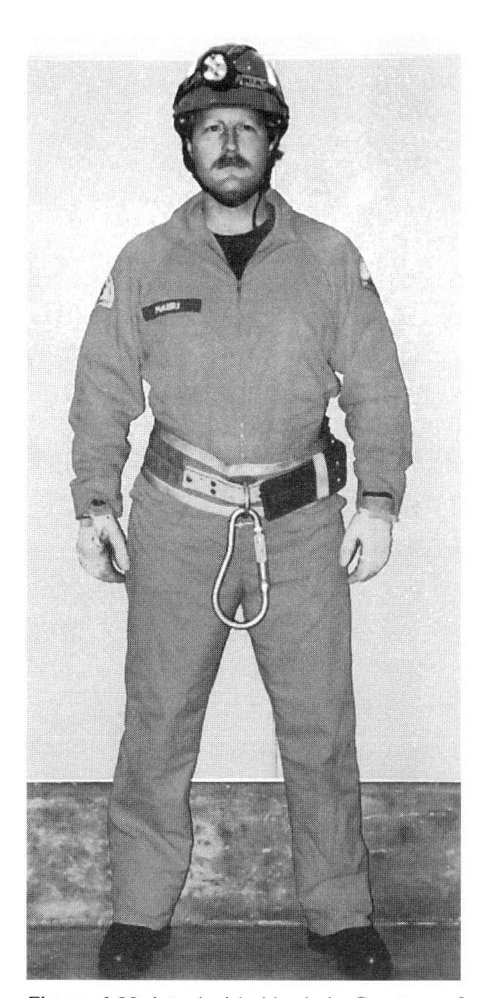

Figure 4.23 A typical ladder belt. *Courtesy of Laura Mauri.*

Figure 4.24 A typical Class I or Class II harness. *Courtesy of Laura Mauri.*

Figure 4.25 A typical Class III harness. *Courtesy of Laura Mauri.*

Step 1: A Double-Loop Figure-Eight or a Bowline on a Bight (see Knots section) is tied in the end of the rope.

Step 2: The wearer's legs slip into the loops formed by the knot, and the loops are pulled up until they are snug just below the wearer's buttocks (Figure 4.26).

Step 3: A Half Hitch is then formed around the wearer's upper torso, just below the armpits, and the standing part of the rope is passed over one shoulder (Figure 4.27).

Step 4: A small bight is formed in the standing part, and it is passed down and behind part of the Half Hitch (Figure 4.28).

Step 5: The bight is then passed up in front of the Half Hitch and through the loop formed when the bight was passed behind the Half Hitch (Figure 4.29).

Step 6: The loose end of the rope is then passed up through the bight, and an upward pull on the standing part of the rope will cause the small bight to snap over and complete the knot (Figure 4.30).

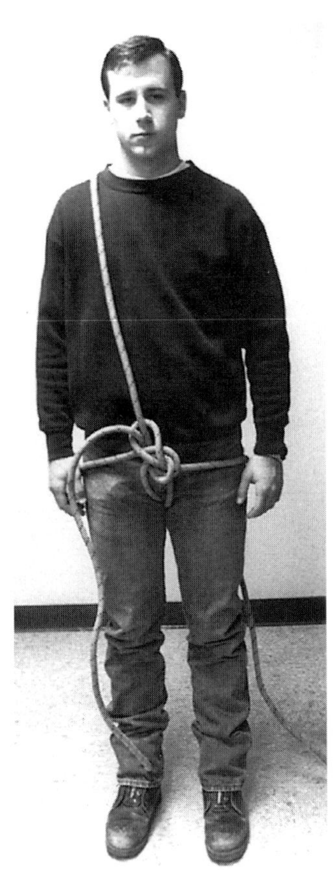

Figure 4.26 Loops around victim's legs.

Figure 4.27 A Half Hitch around the victim's upper torso.

Step 7: The loose end is tied off with an overhand safety (Figure 4.31).

Figure 4.28 The bight passes down behind the Half Hitch.

Figure 4.29 The bight passes up through the bend.

Figure 4.30 The loose end passes up through the bight, and the standing part is pulled upward.

Figure 4.31 The knot is finished when it is secured with an overhand safety.

Swiss Seat. Also called a *hasty seat*, a Swiss Seat is an improvised Class II harness made up of one large carabiner and a piece of webbing about 12 feet (4 m) long and at least 1¾ inches (44 mm) wide. A Swiss Seat is constructed as follows:

Step 1: If the webbing is not in the form of a loop, it must be made into one by buckling the ends together (if so equipped) or by tying the ends together with a Water Knot and overhand safeties (Figure 4.32).

Step 2: The point where the ends of the webbing are joined is placed at the small of the wearer's back (Figure 4.33).

Step 3: The bights formed at each end of the loop are brought around to the front of the wearer and held there (Figure 4.34).

Step 4: Reaching between the wearer's legs, a bight is formed in the middle of the lower part of

Figure 4.32 A loop of webbing is formed.

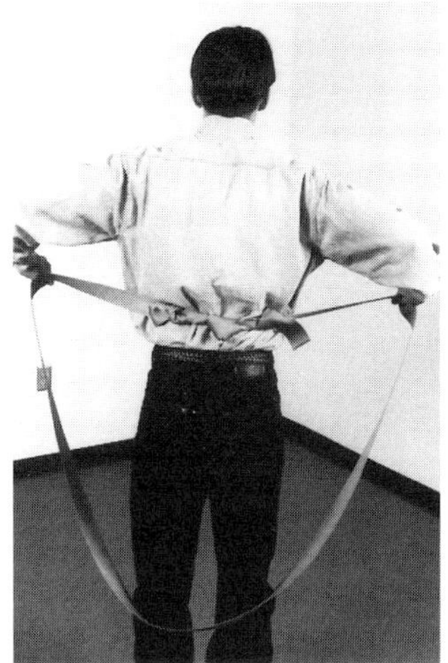

Figure 4.33 The loop across the wearer's back.

Figure 4.34 The ends of the loop are brought to the front.

the webbing, and it is brought forward (Figure 4.35).

Step 5: Reaching through the ends of the loop that came around each side, each leg of the bight coming between the wearer's legs is pulled up through the loop ends (Figure 4.36).

Step 6: The newly-formed bights are pulled up until the harness is snug, and they are snapped into a large carabiner (Figure 4.37).

Swiss Seat with shoulder harness. If there is a possibility of the wearer inverting, an improvised shoulder harness can be constructed and linked to the Swiss Seat. Another carabiner and a slightly shorter loop of webbing are needed to form the shoulder harness as follows:

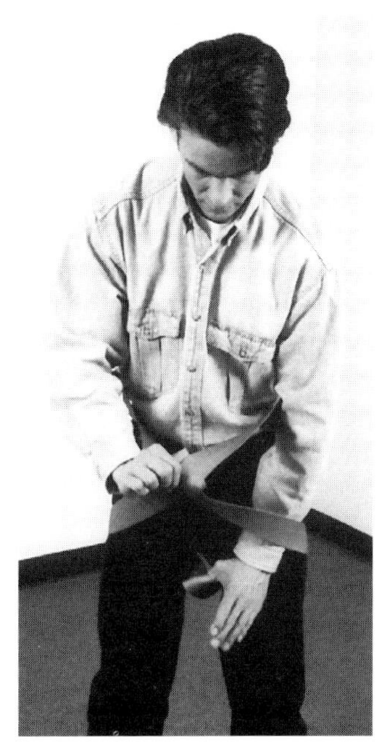

Figure 4.35 A bight is pulled forward.

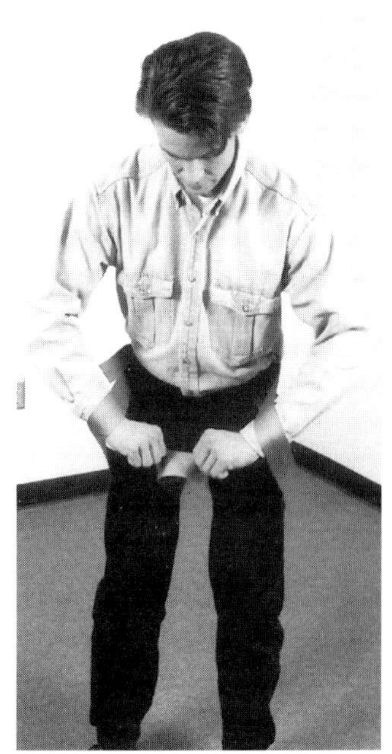

Figure 4.36 The legs of the bight are pulled through the ends of the loop.

Figure 4.37 A carabiner is snapped into the bights.

Step 1: The loop is crossed over itself to form a figure-eight (Figure 4.38).

Step 2: The wearer's arms are slipped through the ends of the figure-eight, with the crossover point in the middle of the wearer's upper back (Figure 4.39).

Step 3: The ends are brought together in front of the wearer's chest and are joined by being snapped into a carabiner (Figure 4.40).

If the carabiner in the seat harness will reach the one in the shoulder harness, they can be snapped together to form an improvised full-body harness (Figure 4.41). If there is not enough slack for the

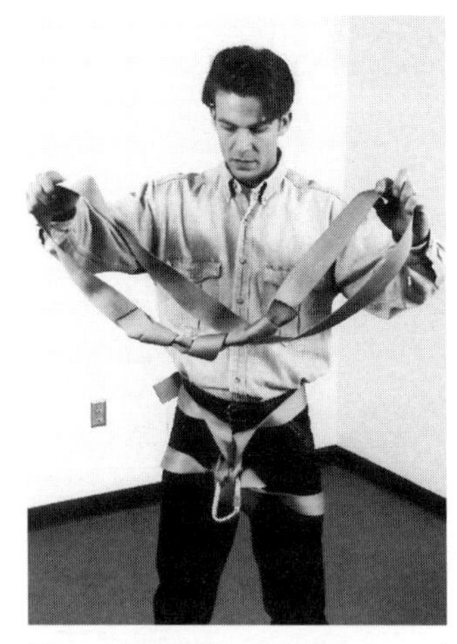

Figure 4.38 A figure-eight is formed in the loop.

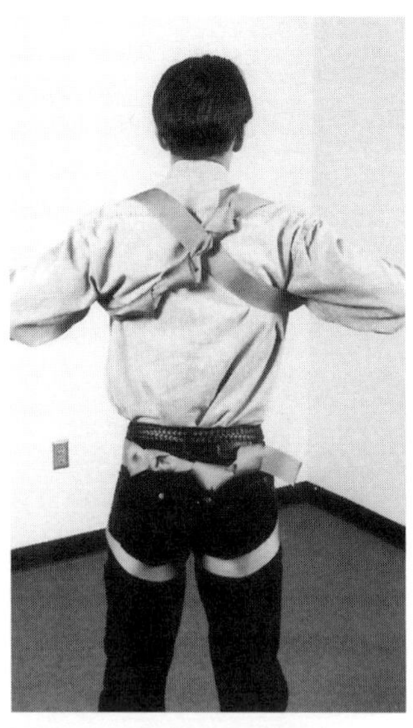

Figure 4.39 The wearer's arms through the figure-eight.

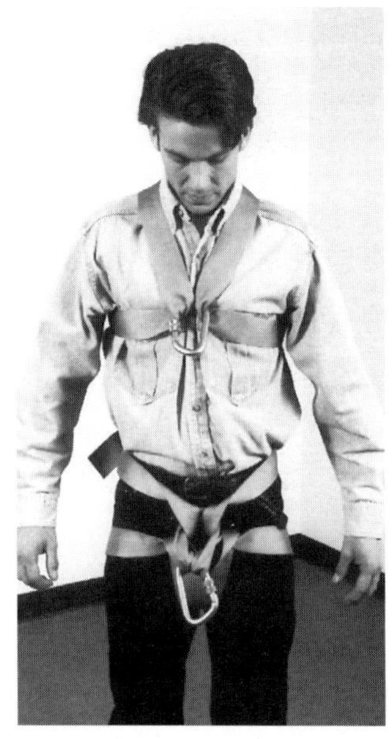

Figure 4.40 A carabiner joins the ends.

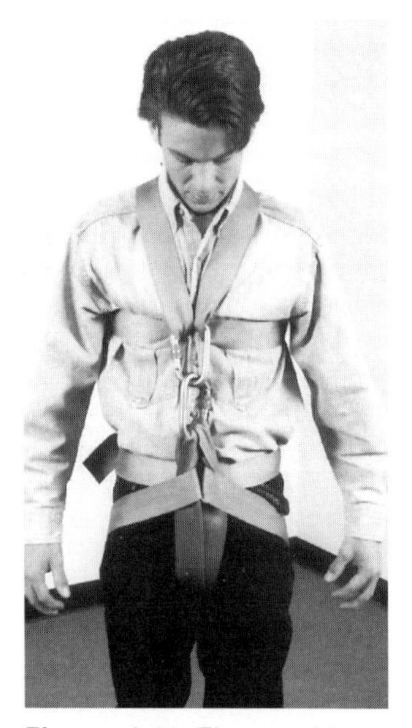

Figure 4.41 The carabiners snapped together.

two carabiners to be snapped together, they can be lashed together with a short piece of webbing tied with a Water Knot.

Hardware

In this context, the term *hardware* refers to the mechanical devices needed to fully and safely utilize rescue rope and to construct mechanical advantage systems with rope. According to NFPA 1983, all load-bearing hardware must withstand both a 1,200-pound (545 kg) tensile test without permanent distortion and a 5,000-pound (2 273 kg) test without failure. The most commonly used pieces of rescue hardware are load-bearing fasteners, ascending and descending devices, and pulleys.

LOAD-BEARING FASTENERS

As the name implies, load-bearing fasteners connect rope and/or webbing together or to other objects and support the weight of rescuers, victims, and the equipment involved. The fasteners most often used in rescue work are carabiners, tri-links, rescue rings, swivels, and anchor plates.

Carabiners. A carabiner consists of an open metal loop with a hinged gate to close the opening (Figure 4.42). Some gates can be locked in the closed position, and others cannot (Figure 4.43). While carabiners are available in other metals, most are made of aluminum or steel. A variety of sizes and shapes are available, but most carabiners are either oval or some form of "D" shape (Figure 4.44). Large, D-shaped, steel carabiners with locking gates are recommended for rescue work.

If nonlocking carabiners must be used in a life safety situation, whether an actual incident or a training exercise, they should be used in pairs and should be snapped on with the gates opposite and opposed (Figure 4.45).

If aluminum carabiners are used, they must be handled with care because they may develop invisible stress fractures if they are subjected to impact. These fractures may later cause the carabiner to fail under load.

NOTE: There is no practical way to field test the reliability of a carabiner after impact. So, a good rule of thumb is that if a carabiner was dropped from waist height onto a hard surface, it should not be used in a life safety application until it has been lab tested.

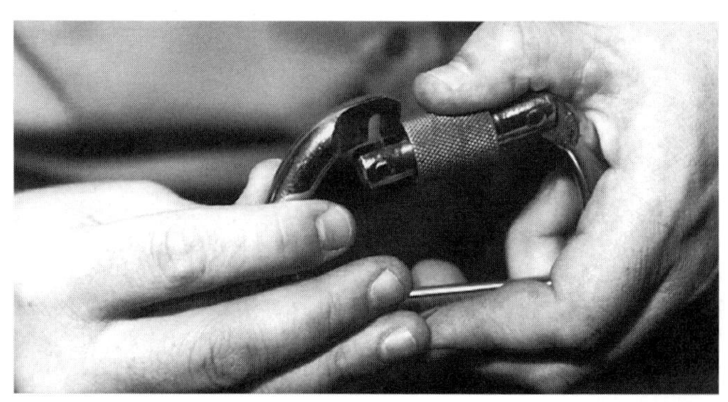

Figure 4.42 A typical locking carabiner.

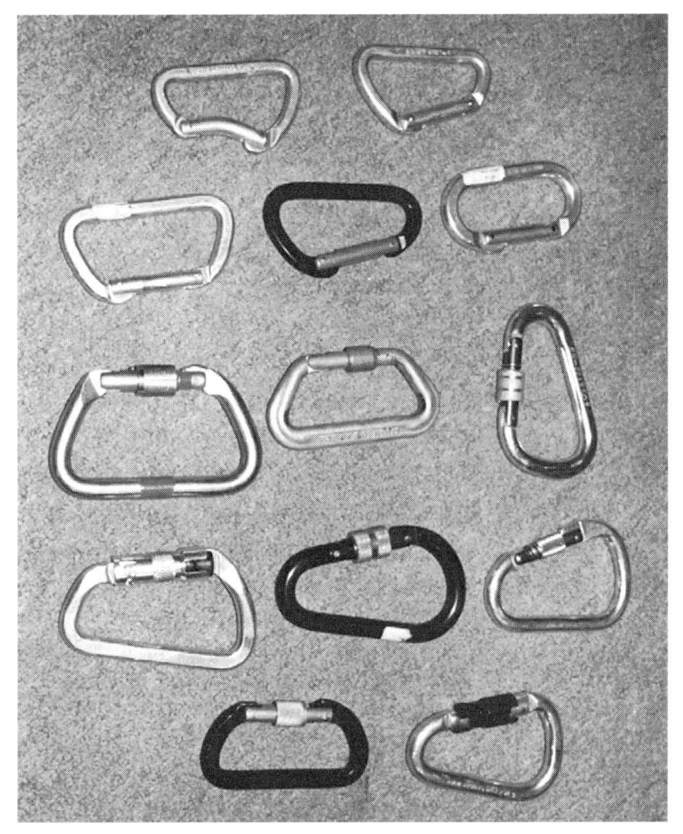

Figure 4.43 A variety of carabiners. *Courtesy of Laura Mauri.*

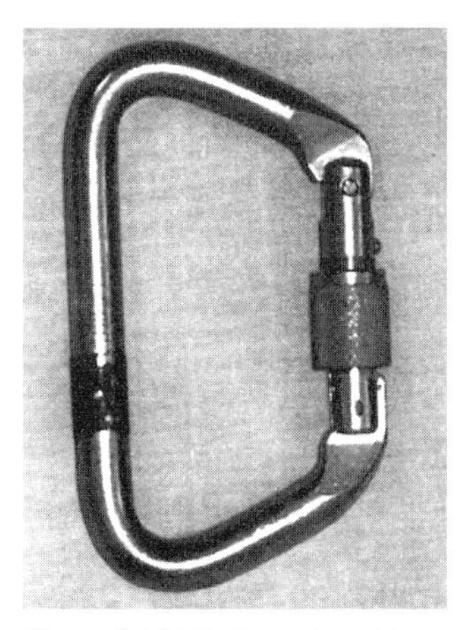

Figure 4.44 A D-shaped carabiner. *Courtesy of Laura Mauri.*

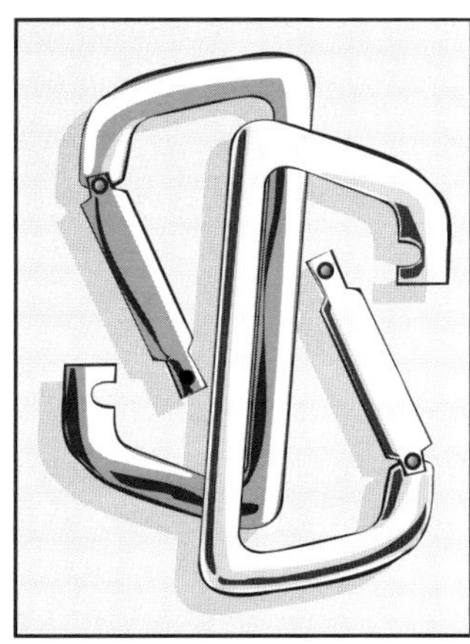

Figure 4.45 Nonlocking carabiners reversed and opposed.

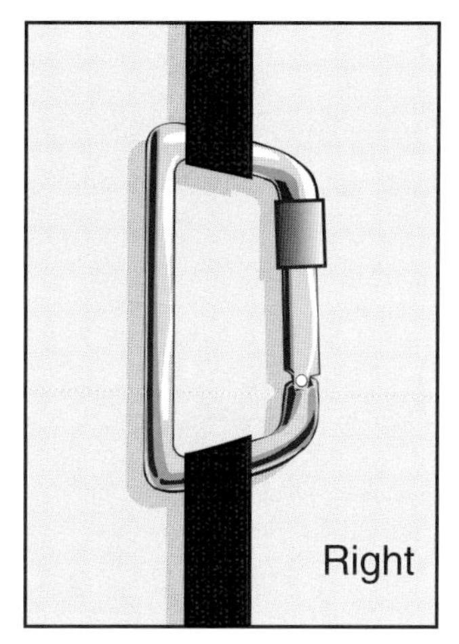

Figure 4.46 Correct and incorrect loading.

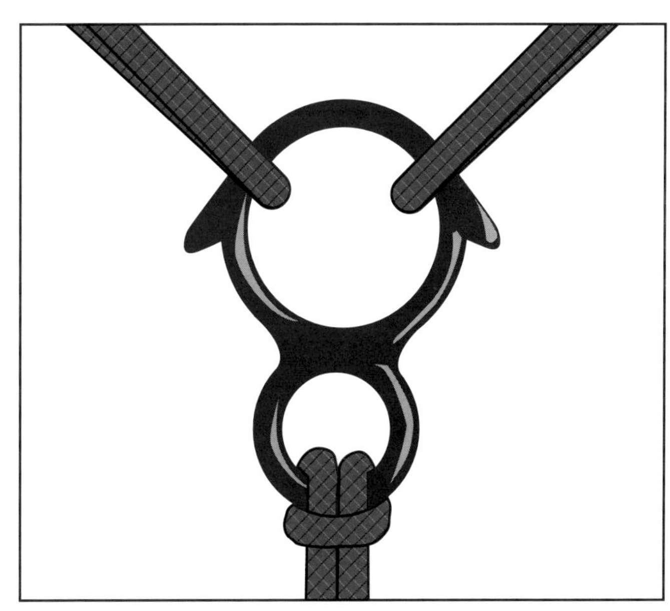

Figure 4.47 Multidirectional loads on a Figure 8 plate.

Carabiners are strongest when loaded vertically near the spine and are weakest when loaded horizontally across the gate (Figure 4.46). They are designed to be loaded in one direction only. If a multidirectional load must be supported, a tri-link, rescue ring, or Figure 8 plate should be used (Figure 4.47).

Tri-links. Tri-links are similar to carabiners, but they have a screw-type locking sleeve to close the opening, and they are usually triangular or semicircular in shape. They are specifically designed to handle multidirectional loads.

Rescue rings. As the name implies, rescue rings are steel rings specifically designed for rescue applications; that is, they meet NFPA 1983 specifications for load-bearing devices. Rescue rings are also designed to be used when a multidirectional load must be supported (Figure 4.48).

Swivels. Applied at the point of attachment to the anchor, these devices prevent the twists that sometimes develop in mechanical advantage systems (Figure 4.49). Without swivels, the twists in the rope can add a significant amount of friction to the system. Only swivels specifically approved for life safety applications should be used in these situations.

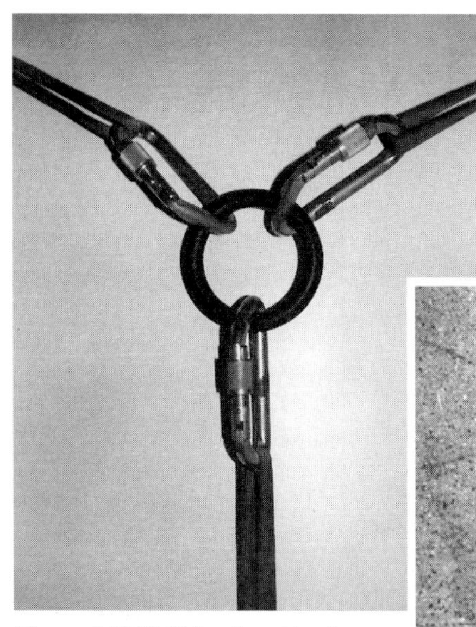

Figure 4.48 Multidirectional loads on a rescue ring. *Courtesy of Laura Mauri.*

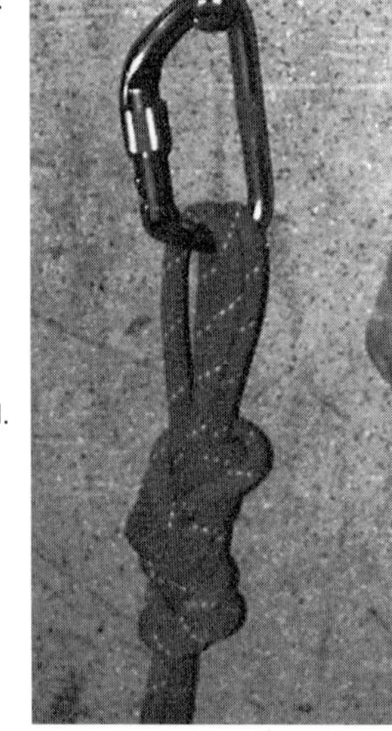

Figure 4.49 A typical rescue swivel. *Courtesy of Laura Mauri.*

Anchor plates. Also called *belay plates*, these are essentially braking devices normally used to stop or slow the descent of a rescuer if the lifeline fails. These devices may also be used for attaching multidirectional loads.

ASCENDING DEVICES

These devices are designed to allow someone to climb (ascend) a fixed vertical rope. However, ascenders are not designed to stop a falling load. In drop tests, these devices caused serious damage to the rope or severed it at approximately one third of the rope's rated strength. When used for their intended purpose, ascenders do little or no damage to the rope on which they are applied. There are two basic types of ascending devices used in rescue work: cams and handled ascenders.

Cams. Often called *Gibbs ascenders*, these devices can be used as ascenders and in hauling systems. A cam device consists of a metal frame that surrounds the vertical rope and is the base for the pivot pin of a lever having a serrated cam on one end and a loop on the other (Figure 4.50). For a cam device to be applied to a rope, the device must be disassembled by removing the pivot pin and cam lever. The rope is then slipped into the frame, and the device is reassembled in the reverse order of disassembly. The load is attached to the device through the loop at the exposed end of the lever. When a load is applied, the lever pivots and forces the cam against the rope within the frame (Figure 4.51).

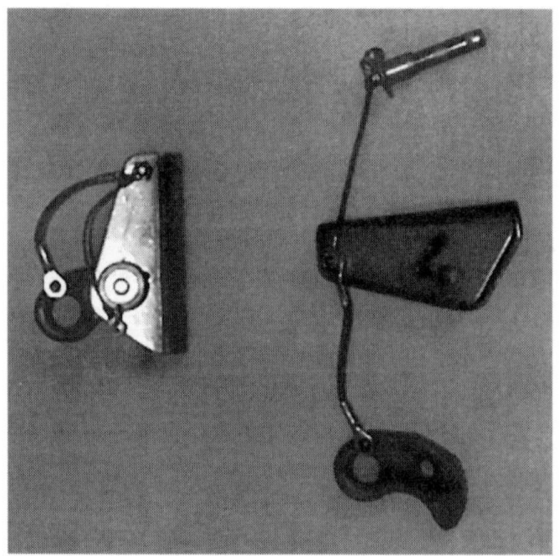

Figure 4.50 A Gibbs ascender. *Courtesy of Laura Mauri.*

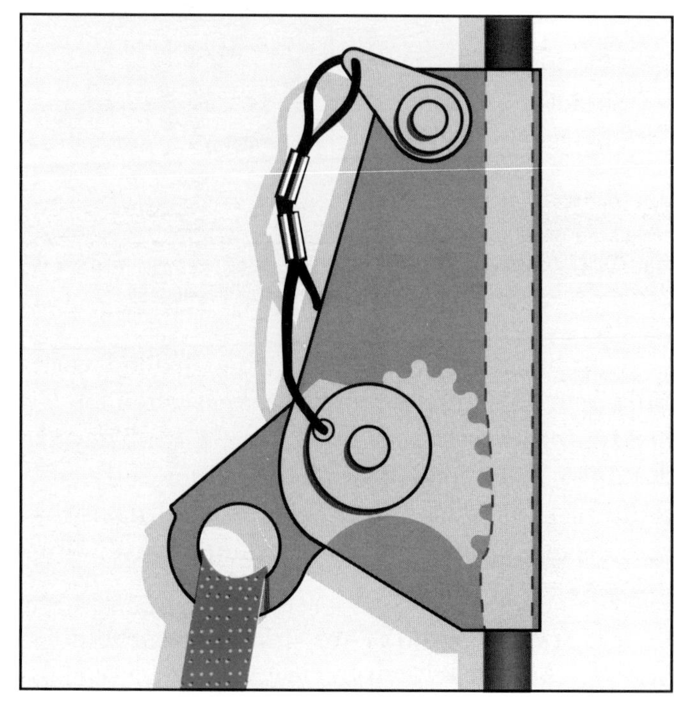

Figure 4.51 How the cam operates.

There are two basic types of cam devices: free-running and spring-loaded. In the free-running type, the cam does not engage the rope unless and until a load is applied to the other end of the lever. In the spring-loaded type, the spring keeps the cam in contact with the rope at all times, regardless of whether or not a load is applied.

Handled ascenders. These devices also operate with a form of cam, but they are significantly different from the cam devices previously described. Handled ascenders are not designed to be used in hauling systems and are intended to support only the weight of one person. Normally used in pairs, handled ascenders consist of either a right-handed or a left-handed metal frame that surrounds the vertical rope and incorporates some form of handle (Figure 4.52). Unlike other cam devices, handled ascenders do not have to be disassembled to be applied to the rope, and they are designed with a cam release that allows them to be applied to the rope with one hand. In operation, with the climber's weight on the foot loop attached to the lower ascender, the top one is pushed up the rope as far as the climber can reach. When the climber puts weight on the foot loop of the upper ascender, the cam engages the rope and allows the climber to ascend. The lower ascender is then pushed up to the top one and the sequence is repeated until the climber reaches the desired elevation.

DESCENDING DEVICES

Sometimes called *rappel devices*, descenders are devices that function by creating friction on the rope being descended. They are designed to allow a person to descend a fixed vertical rope at a safe and controlled rate, and they may also be used to control the rate of descent of litters and other equipment. The most common types of descending devices used in rescue are Figure 8 plates, rappel racks, and belay plates.

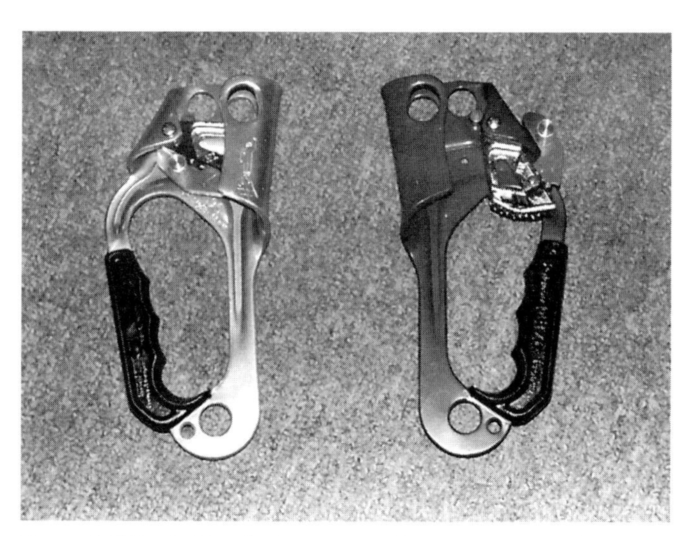

Figure 4.52 Typical handled ascenders. *Courtesy of Laura Mauri.*

Figure 8 plates. These devices consist of a double-ring of steel or anodized aluminum, with one ring larger than the other (Figure 4.53). In operation, the smaller ring is attached to the user's harness by a carabiner, and the rope being descended is looped through the larger ring and around the waist of the Figure 8 (Figure 4.54). On some Figure 8s, often called *Rescue 8s*, appendages called *ears* have been incorporated into the larger ring (Figure 4.55). The ears are intended to prevent the rope from slipping out of place and forming a Girth Hitch at the top of the ring (Figure 4.56). If the rope does slip into a Girth Hitch, the user is

Figure 4.53 A basic Figure 8 plate. *Courtesy of Laura Mauri.*

Figure 4.54 A figure-eight plate rigged for rappelling. *Courtesy of Laura Mauri.*

Figure 4.55 A Figure 8 plate with ears. *Courtesy of Laura Mauri.*

Figure 4.56 Results of the rope slipping out of place. *Courtesy of Laura Mauri.*

essentially stuck at that point. The user must then be able to ascend far enough to relieve the tension on the hitch in order to slip the rope back into position.

Rappel racks. Also called a *brake bar rack*, a rappel rack consists of an elongated U-shaped rod with an eye on one end, a nut (stopper) threaded onto the other, and six friction bars in between (Figure 4.57). The first bar is penetrated by both legs of the rack, and like the other five bars, the first bar can be moved backward and forward on the rack. Unlike the first bar, the others are penetrated by the longer leg of the rack only at one end. Except for the first one, each bar has a groove machined into it near the unattached end so that it can align parallel to the first bar when in place across the rack (Figure 4.58).

Rappel racks are superior to Figure 8 plates for rappelling in several ways. By varying the number and spacing of bars engaged, rappel racks allow the amount of friction applied to the rope to be adjusted during descent (Figure 4.59). This adjustment provides greater control of the rate of descent than is possible with Figure 8s. Because the rope passes straight through a rappel rack, there is less of a tendency for the rope to twist. On long rappels, they generate less heat, which could damage the rope.

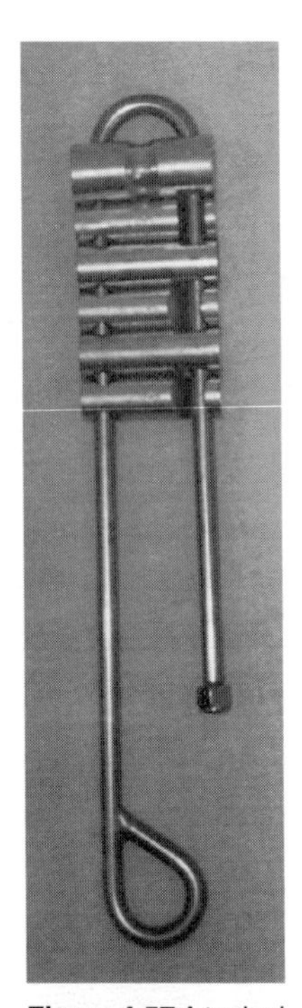

Figure 4.57 A typical rappel rack. *Courtesy of Laura Mauri.*

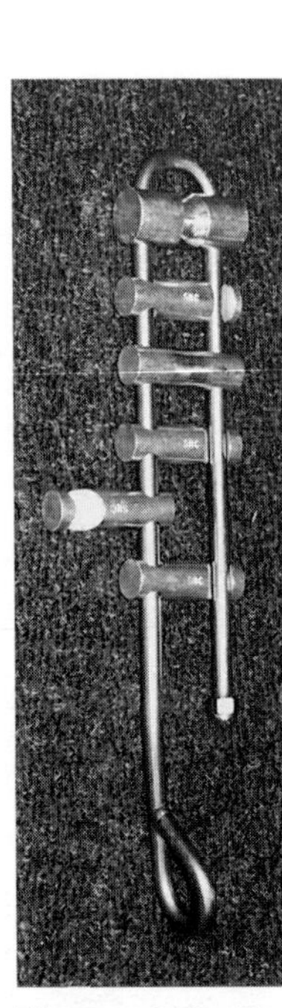

Figure 4.58 The brake bars are grooved at one end. *Courtesy of Laura Mauri.*

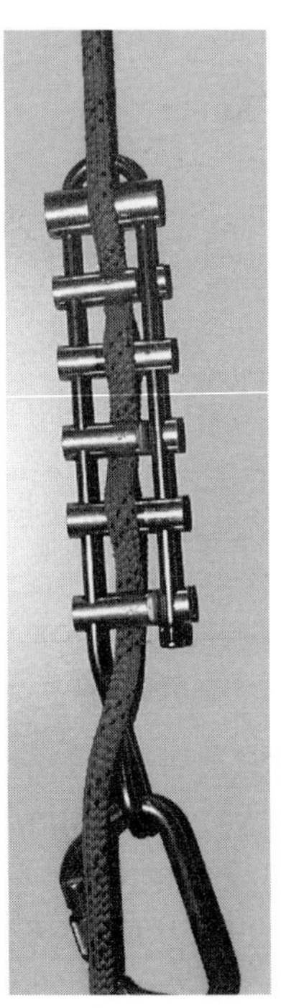

Figure 4.59 A fully reeved rappel rack. *Courtesy of Laura Mauri.*

Belay plates. Frictioning devices of all sorts can be used for belaying (applying a friction brake to) a single-person load; however, belay plates are designed specifically for this purpose and may do it better than devices designed for other uses. Belay plates are intended to allow a rescuer's rate of descent to be controlled and stopped if the rescuer falls or the main line fails. A belay plate consists of a metal plate with two or more holes in it (Figure 4.60). In operation, a bight of the safety line is passed through the belay plate and attached to an anchor with a carabiner. When the running part of the rope is pulled, more friction is added to the rope, and the rescuer's progress can be slowed or stopped (Figure 4.61).

Figure 4.60 Typical belay plates. *Courtesy of Laura Mauri.*

Figure 4.61 A reeved belay plate. *Courtesy of Laura Mauri.*

PULLEYS

Pulleys are used to change the direction and point of application of a rope in a way that protects the rope from damage and keeps friction to a minimum. In rescue, they are sometimes used as a means of hoisting equipment, but they are most often used as part of a mechanical advantage system for raising or lowering rescuers and victims. Rescue pulleys are of all-metal construction and consist of one or more sheaves (wheels), an axle with a bearing or bushing, and two sideplates (Figure 4.62). The types most often used in rescue are single- and double-sheave pulleys. The diameter of the sheaves should be three to four times the diameter of the rope to which it is applied. This prevents putting the rope into too tight a turn, which would increase friction and might damage the rope. While edge rollers are not technically pulleys, they serve the same purposes and share many of the characteristics of pulleys, so they also are discussed in this section.

Single-sheave pulleys. These are the simplest pulleys used with rescue rope. They consist of the components described earlier with only one sheave. While the sideplates are designed to meet, they are not connected; each can pivot in either direction around the axle. Pivoting them in opposite directions opens up the pulley and allows it to be applied to a rope at any point in its length (Figure 4.63). This saves having to thread the pulley onto the rope from one end.

Double-sheave pulleys. Double-sheave pulleys are similar to single-sheave types, but as the name implies, they have two sheaves instead of one. In addition, the sheaves are separated by a plate that also separates the two sideplates (Figure 4.64). Double-sheave pulleys are much more versatile than single-sheave types because they are adaptable to a variety of operational demands such as constructing mechanical advantage systems.

Edge rollers. These devices are also of all-metal construction and are designed to protect rope from damage and to reduce friction when the rope must be pulled over sharp edges such as cornices or parapet walls. While there is a variety of different designs available, they all incorporate one or more rollers, similar to wide sheaves, and they are capable of conforming to the shape of the edge over which the rope must be pulled (Figure 4.65).

KNOTS

To use the equipment just described, rope and webbing must usually be tied into knots. While there is technically a differences between knots,

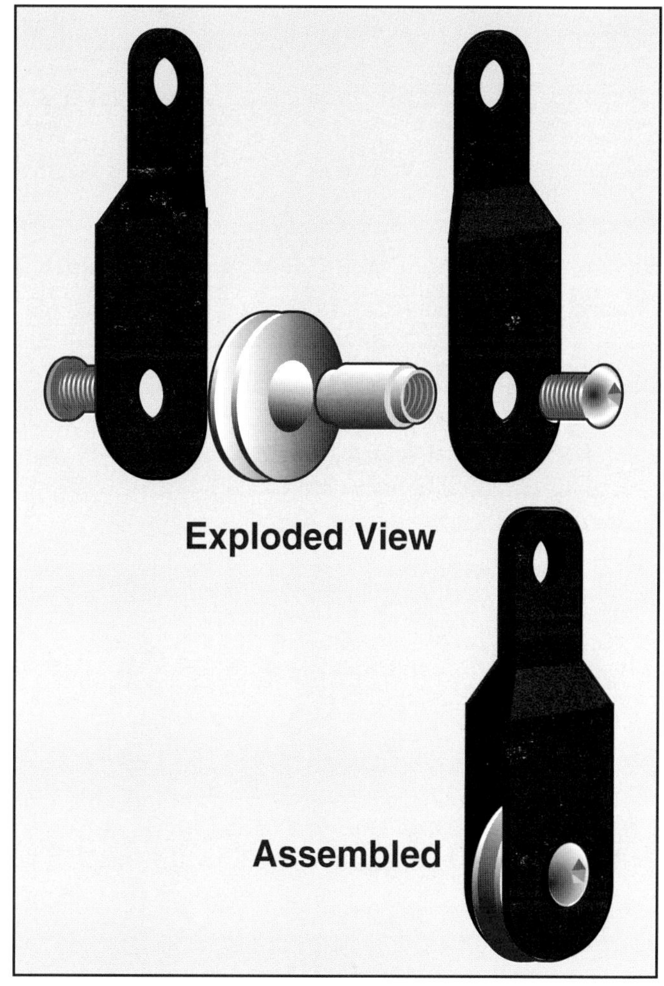

Figure 4.62 Components of a rescue pulley.

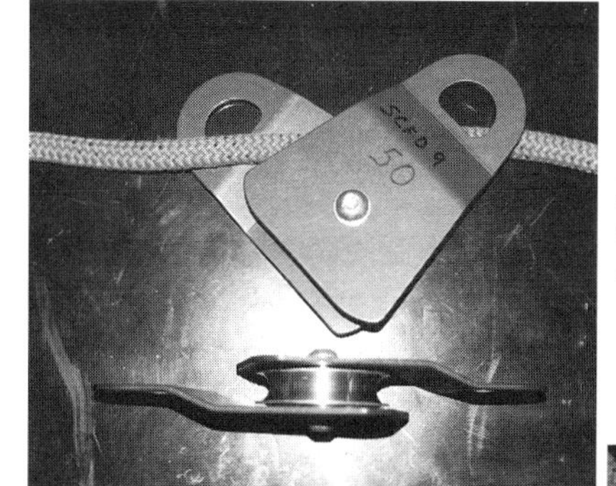

Figure 4.63 A pulley applied in the middle of a rope. *Courtesy of Laura Mauri.*

Figure 4.64 A double-sheave pulley has three plates. *Courtesy of Laura Mauri.*

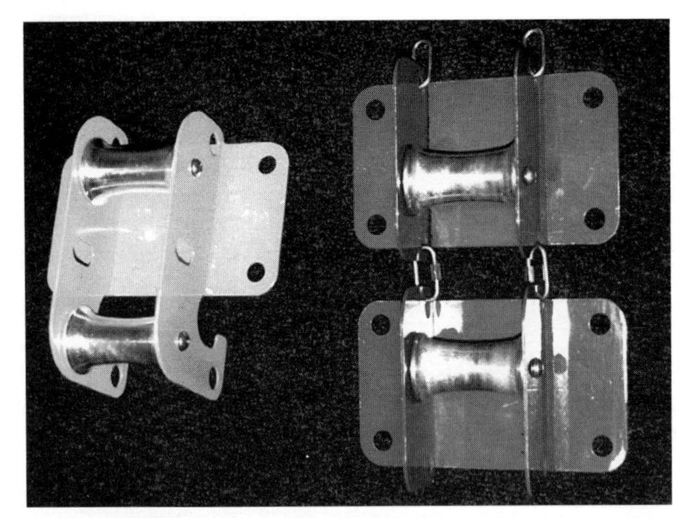

Figure 4.65 Typical edge rollers. *Courtesy of Laura Mauri.*

bends, and hitches, the general term *knots* is used here to denote any of the three. The exception is whenever specific names such as Clove Hitch, Girth Hitch, etc., are used.

To be suitable for use in rescue, a knot must be easy to tie and untie, be secure under load, and reduce the rope's strength as little as possible. A rope's strength is reduced whenever it is bent. The tighter the bend, the more strength is lost. Some knots create tighter bends than others and thereby reduce the rope's strength to a greater degree. All knots should be *dressed* after they are tied; that is, they should be tightened until snug and all slack is removed. Except for the Rescue Knot described earlier, knots are used in rescue to join or connect things or to form loops.

Joining/Connecting Knots

These knots may be used to join two pieces of rope or webbing together or to connect rope or webbing to an object such as an anchor point. The most common joining/connecting knots are the Becket Bend, Double Fisherman (Double Overhand knot), Figure-Eight Follow Through, Water Knot, and the hitches (Clove, Girth, and Prussik).

BECKET BEND

Also called a *Sheet Bend*, the Becket Bend is a very quick and easy knot for connecting two ropes of unequal size or connecting a rope and a piece of webbing. A bight is formed in the end of one piece, and the end of the other is fed up through the bight, around the two standing parts, and under its own standing part (Figure 4.66).

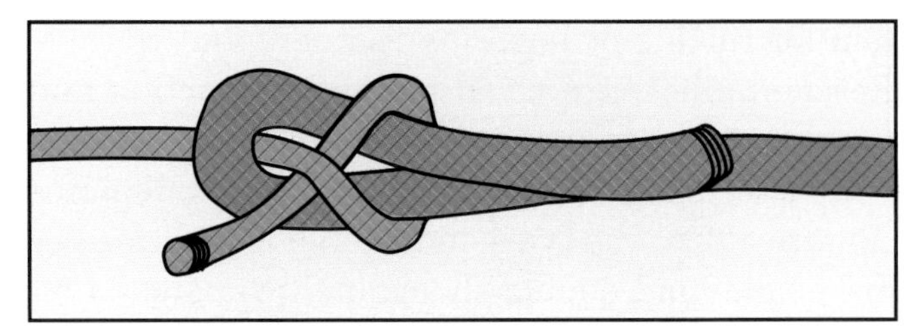

Figure 4.66 A Becket Bend (Sheet Bend).

DOUBLE FISHERMAN (DOUBLE OVERHAND)

This knot makes an extremely secure connection between two pieces of rope, or it can be used to join the ends of the same piece of rope to form a loop. It consists of a single knot tied around the standing part of the other piece of rope, and the two knots are pulled together to dress the finished knot (Figure 4.67).

FIGURE-EIGHT FOLLOW THROUGH

This knot is one of the most secure and one of the easiest to untie and is formed by tying a loose Figure-Eight in the end of the rope and then threading the end of the second rope through the loosely tied Figure-Eight in the first rope (Figure 4.68).

WATER KNOT

This knot is the preferred knot for joining two pieces of webbing or the ends of the same piece if a loop is needed. Similar to the Figure-Eight Follow Through, the Water Knot is formed by tying a simple overhand knot in the end of one piece and following it through in the reverse direction with the end of the other piece (Figure 4.69). It is often secured with an Overhand Safety on each side of the knot.

CLOVE HITCH

This knot is used to attach a rope or piece of webbing to an object. If the knot will be subjected to repeated loading and unloading, it should be backed up with an Overhand Safety. The Clove Hitch is formed by dropping the end of the rope over the object and pulling the end under the object and up and across itself before dropping it again on the other side of the object. The end is once again brought up on the near side of the object and is inserted under itself where it crosses the top of the object (Figure 4.70).

GIRTH HITCH

This knot is also used to attach rope or webbing to an anchor point or other object. It is formed by

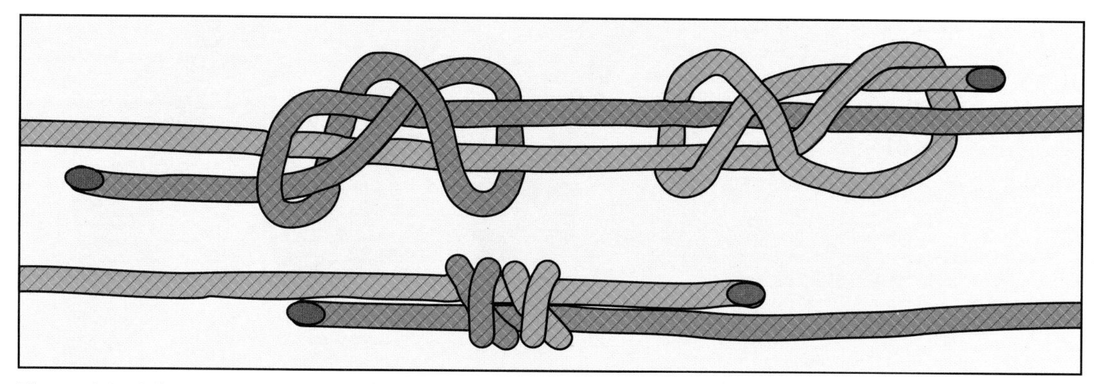
Figure 4.67 A Double Fisherman.

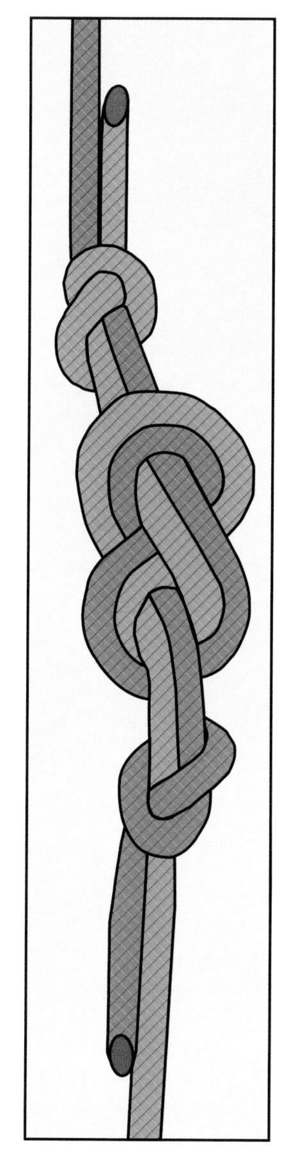
Figure 4.68 A Figure-Eight Follow Through.

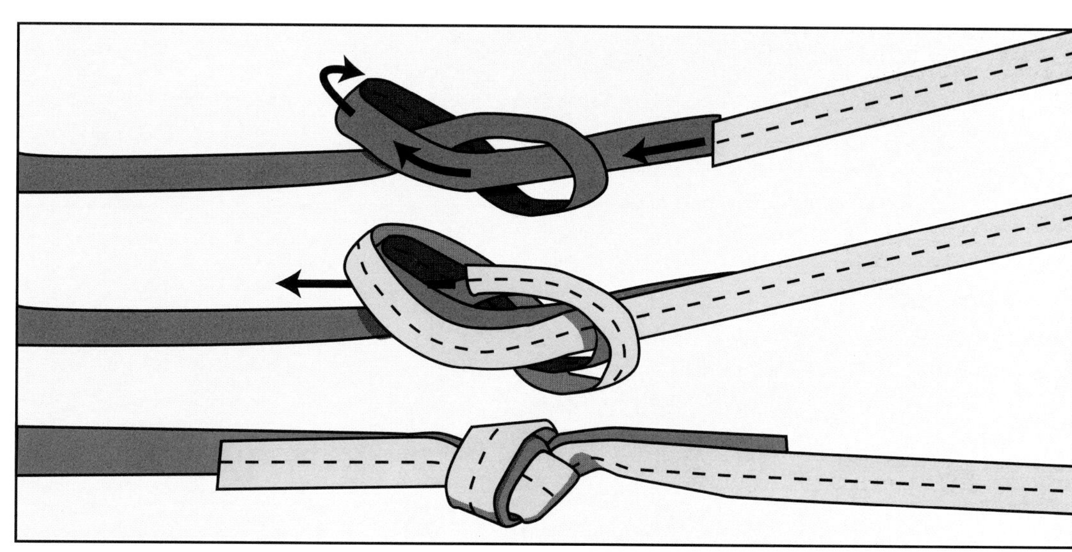
Figure 4.69 A Water Knot.

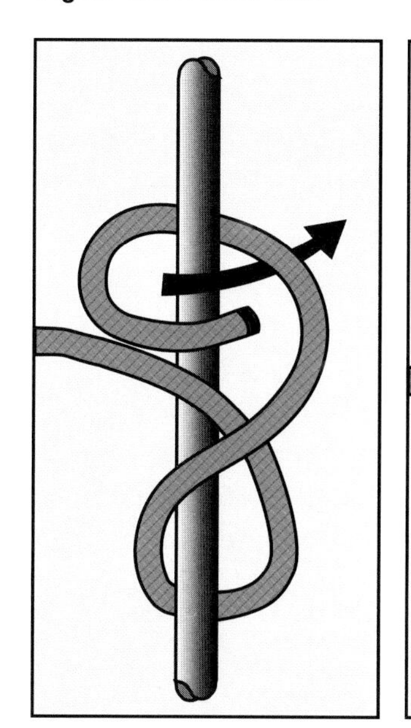
Figure 4.70 A Clove Hitch.

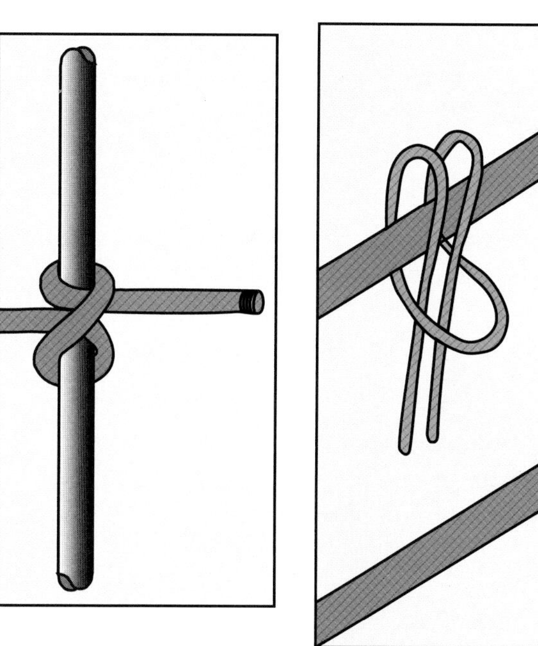
Figure 4.71 A Girth Hitch.

passing a bight in the rope under the object and bringing the bight up and over the object. The free ends of the rope are then passed through the bight, and the knot is dressed (Figure 4.71).

PRUSSIK HITCH

The Prussik is very similar to the Girth Hitch, but it is formed with one or more extra wraps. It is the knot most often used to tie one rope onto another (Figure 4.72). The Prussik will hold securely when loaded, but it can slide easily along the rope to which it is attached when unloaded. This capability makes it possible to use a Prussik as an improvised ascender.

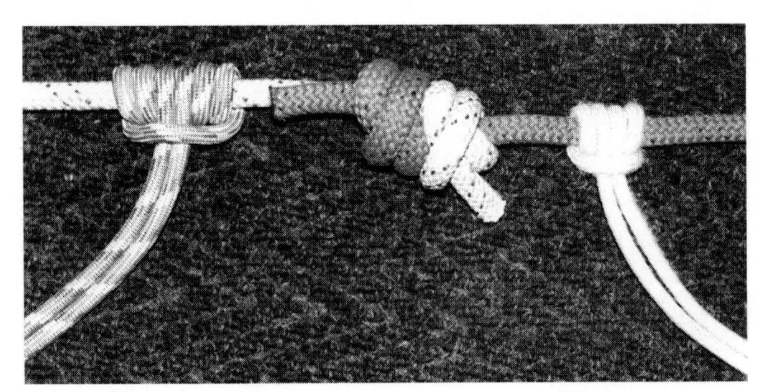

Figure 4.72 Two- and three-wrap Prussiks on either side of a Double Fisherman. *Courtesy of Laura Mauri.*

Loop-Forming Knots

These knots are used when a loop is needed in the middle or the end of a rope or piece of webbing. The most common loop-forming knots used in rescue are the Figure-Eight Follow-Through Loop, Figure-Eight on a Bight, Double-Loop Figure-Eight, Bowline, and the Overhand Loop.

FIGURE-EIGHT FOLLOW-THROUGH LOOP

This knot is one of the easiest ways to form a loop in the end of a rope. It is formed by tying a loose Figure-Eight near the end of the rope. In forming the loop, the end of the rope is folded back, and it follows back through the Figure-Eight in the reverse direction (Figure 4.73). This is the Figure-Eight knot used to tie a rope around or through an object.

FIGURE-EIGHT ON A BIGHT

This knot is a good way to tie a loop in the middle or the end of a rope. It is tied by forming a bight in the end of the rope, or at any point along its length, and then tying a simple Figure-Eight with the doubled part of the rope (bight) (Figure 4.74).

DOUBLE-LOOP FIGURE-EIGHT

This knot can form the basis of the Rescue Knot and can be used in constructing multiple-anchor systems. This knot is relatively complex to tie, and it takes practice to become proficient at tying it (Figure 4.75).

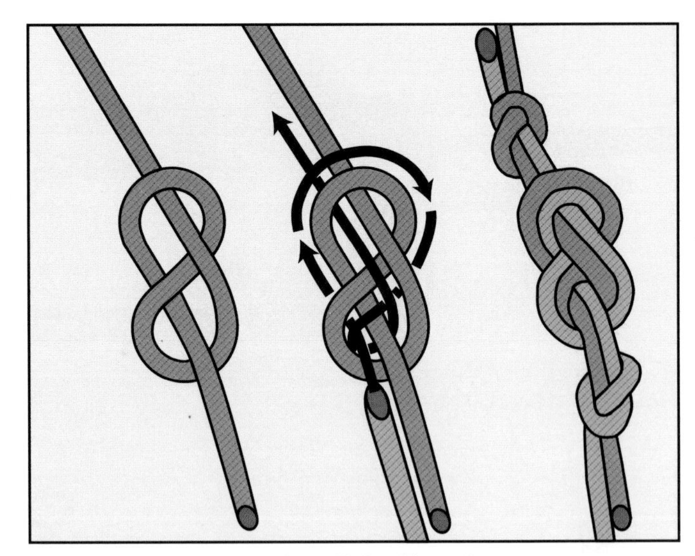

Figure 4.73 Tying a Figure-Eight Follow Through.

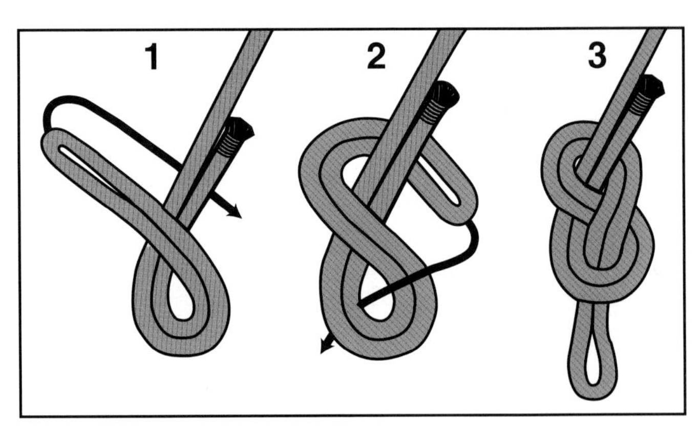

Figure 4.74 Tying a Figure-Eight on a Bight.

BOWLINE

Traditionally, the Bowline was one of the most used knots in the fire service. However, the use of the Bowline has been discontinued by many departments for life safety applications since the advent of synthetic rope. The Bowline is still a useful knot in many applications. For those who are not skilled in tying Figure-Eights, a correctly tied Bowline is safer than another knot tied incorrectly. It can be tied to form a single loop (Figure 4.76). It can also be tied on a bight to form two loops, the basis for the Rescue Knot (Figure 4.77).

OVERHAND LOOP

This knot is tied to form a loop in the end of a piece of webbing (Figure 4.78).

SYSTEMS

Anchor Systems

The variety of ways anchor systems can be configured is limited only by the situation, the

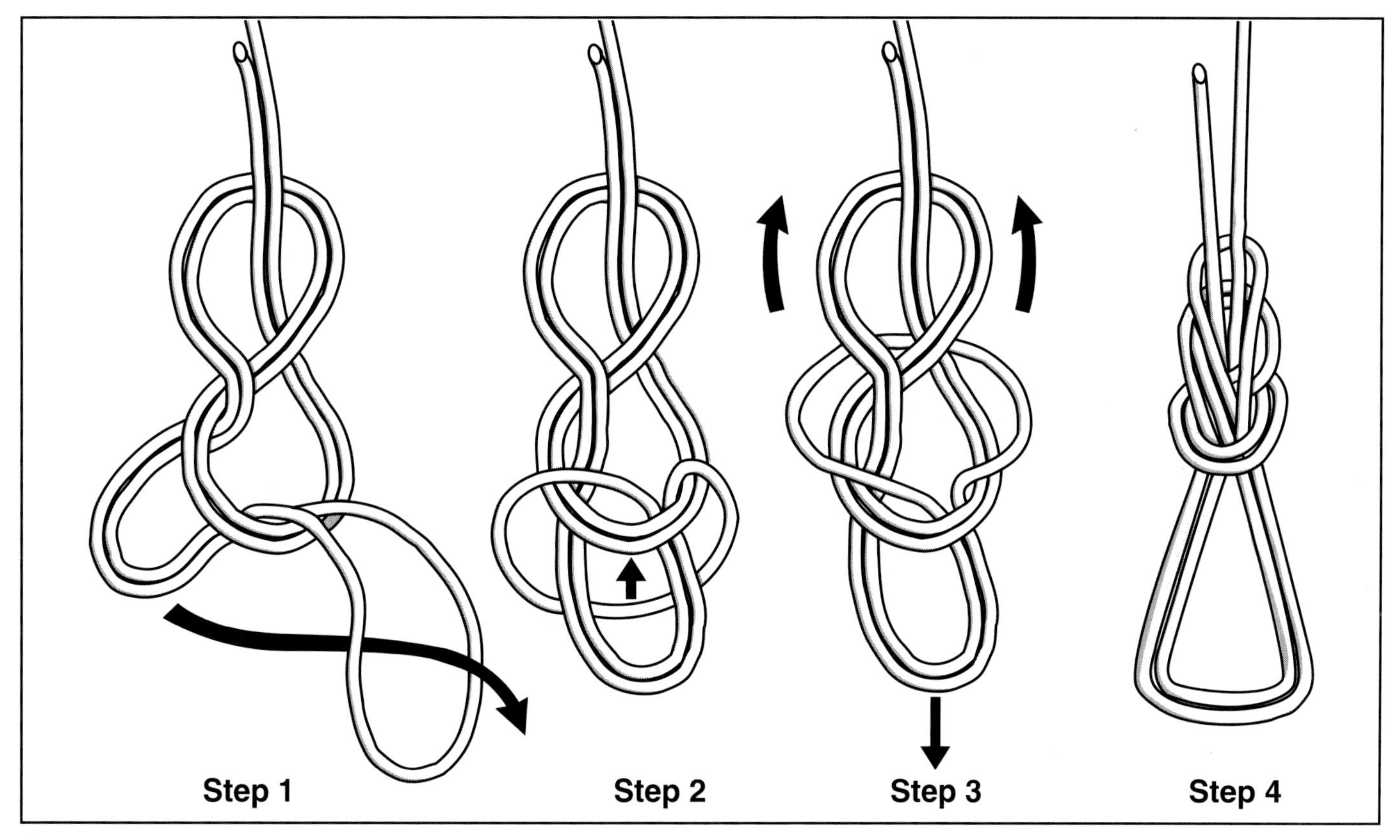

Figure 4.75 Tying a Double-Loop Figure-Eight.

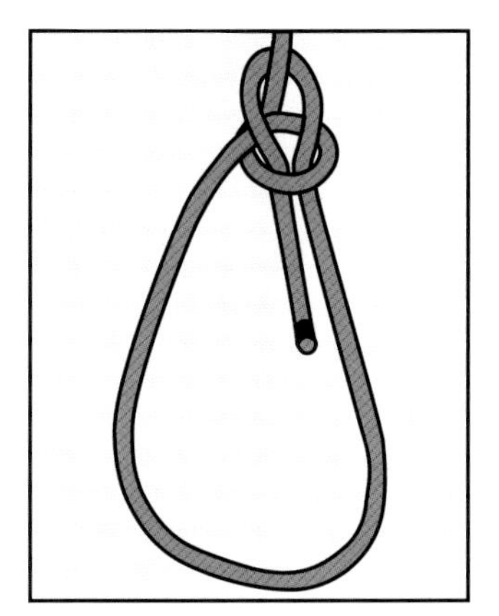

Figure 4.76 A basic Bowline.

Figure 4.77 A Bowline on a Bight.

Figure 4.78 An Overhand Loop.

equipment available, and the imagination of the rigger. However, all anchor systems fulfill the same function: to provide a safe and dependable means of securing the rescue rope to a bombproof anchor point. *Bombproof* is high-angle slang for an absolutely, positively immovable object such as a huge boulder, a large tree, or a fire engine. Anchor systems may also provide redundancy in case of system failure, and/or they may spread the load among two or more less-than-bombproof anchor points. Sharp edges that the rope or webbing must pass over when it is wrapped around an anchor point should be padded to protect the rope or webbing. The most common types of anchor systems used in rescue are single-point and multipoint. In the absence of natural anchor points, a series of pickets can be used for a man-made system (Figure 4.79).

CAUTION: Regardless of what kind of anchor is selected, rescuers must be aware that virtually any anchor can be overloaded and fail if enough pull is applied.

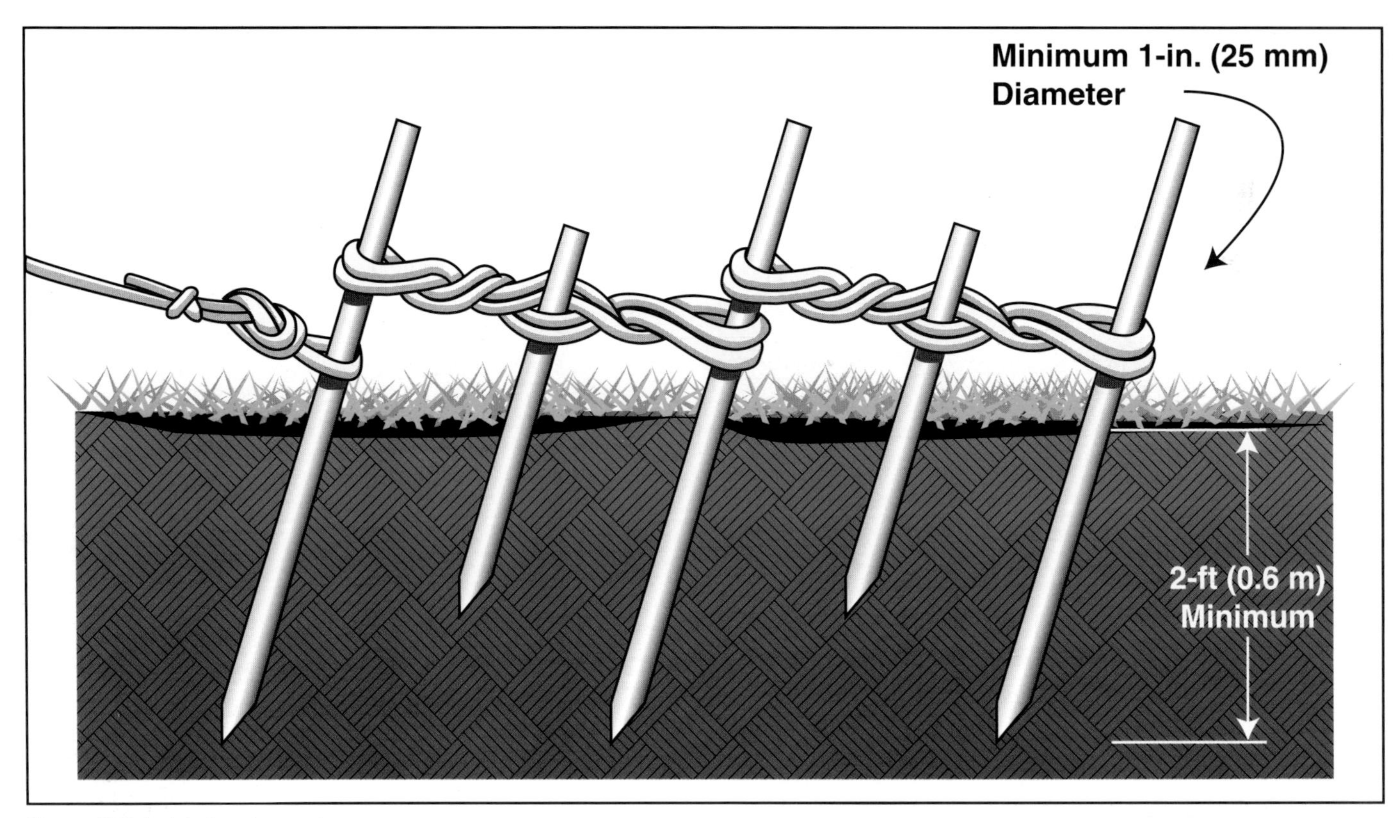

Figure 4.79 A picket anchor system.

SINGLE-POINT ANCHORS

While not considered to be anchor "systems" by purists, single-point anchors are nonetheless an important option in the rescuer's array of skills. Like any other anchor system, single-point anchors provide a safe and dependable means of securing the rescue rope. The most common types of single-point anchors used in rescue are the Tensionless, the Two-Bight, and the Multiwrap anchors.

Tensionless Anchor. This quick and easy anchor provides a dependable anchor with a minimum of equipment. The running end of the rope is wrapped at least four times around an anchor point, such as a tree, in a neat series of wraps (Figure 4.80). As with all anchor systems, the Tensionless Anchor should be applied as low on the anchor point as possible. A Figure-Eight on a Bight is tied in the running end, and a carabiner is snapped into it. The carabiner is then snapped onto the standing part of the rope (Figure 4.81).

Two-Bight Anchor. This simple but effective anchor is very easy and quick to construct (Figure 4.82). It is sometimes called a *Three-Bight Anchor* because the part of the webbing that passes around the anchor point can be seen as a third bight.

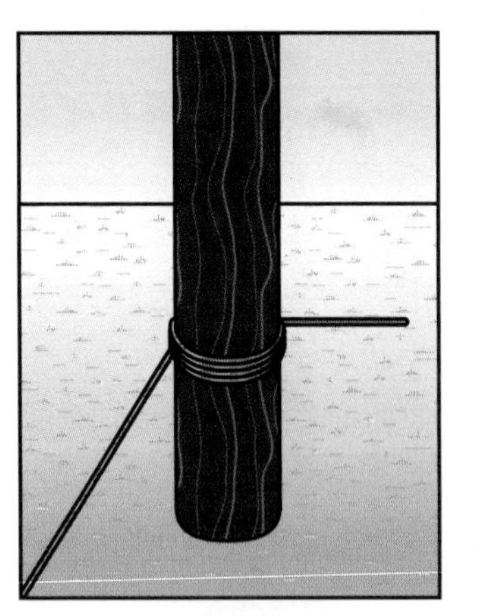

Figure 4.80 The first step in a Tensionless Anchor.

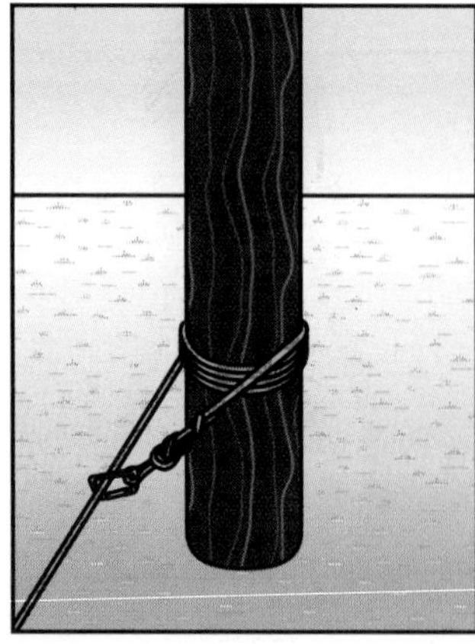

Figure 4.81 The completed anchor system.

Multiwrap Anchor. This anchor is very similar to the Tensionless Anchor described earlier except that the Multiwrap can be constructed of webbing or rope. The webbing or rope is wrapped loosely around an anchor point, and the ends are tied together with a Water Knot or Double Overhand. Two wraps from the middle of the bunch are pulled in the direction of the load, and this snugs the other wraps around the anchor point (Figure 4.83).

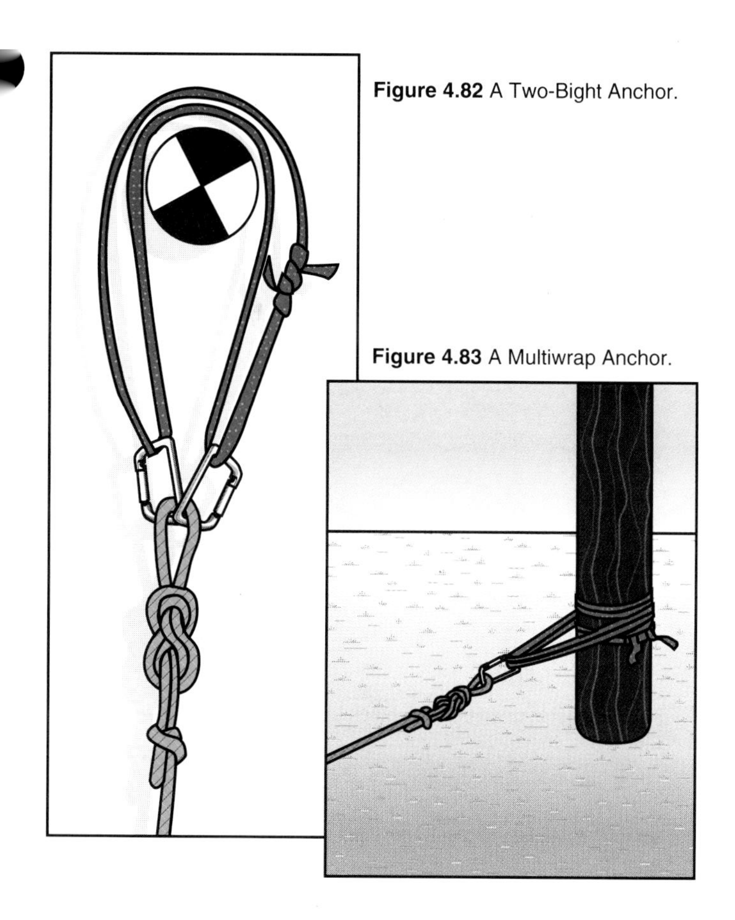

Figure 4.82 A Two-Bight Anchor.

Figure 4.83 A Multiwrap Anchor.

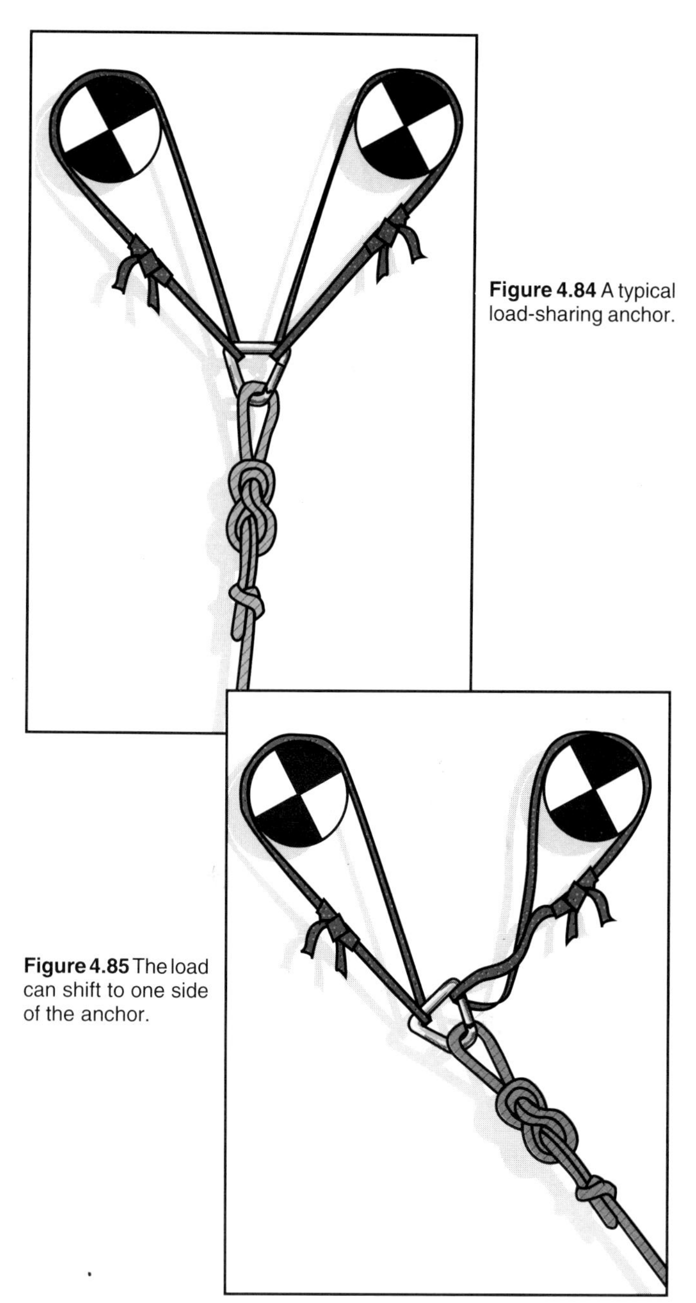

Figure 4.84 A typical load-sharing anchor.

Figure 4.85 The load can shift to one side of the anchor.

LOAD-SHARING ANCHORS

These anchor systems are used when there may be some doubt that one anchor point is sufficient to carry the expected load. Load-sharing anchor systems allow the load to be distributed between two or more anchor points (Figure 4.84). These systems work well as long as the direction of pull remains constant; however, if the direction of pull changes, the entire load can shift to one of the anchors (Figure 4.85). If a change in the direction of pull is anticipated, a better solution in this situation is a self-adjusting anchor system.

SELF-ADJUSTING ANCHORS

These anchor systems are used in the same situations as load-sharing systems — when a single anchor point is not strong enough to support the anticipated load — and when the direction of pull is also likely to change during the rescue operation. A self-adjusting anchor system can be as simple as a quick two-point system or as complex as a three-point system. The situation at hand will dictate which form is most appropriate.

Two-point system. If two adjacent anchor points are available and will support the anticipated load, a two-point self-adjusting anchor system can be constructed very quickly. A length of webbing formed into a loop is crossed over itself to form a figure eight. A carabiner is snapped into each side of the figure eight, and a third is snapped across the intersection in the middle of the figure eight, not around it (Figure 4.86).

CAUTION: In two-point anchor systems, an angle between the anchor points and the point of attachment to the load wider than 90 degrees multiplies the force on each anchor. At 120 degrees, each anchor must support the actual weight of the load (Figure 4.87).

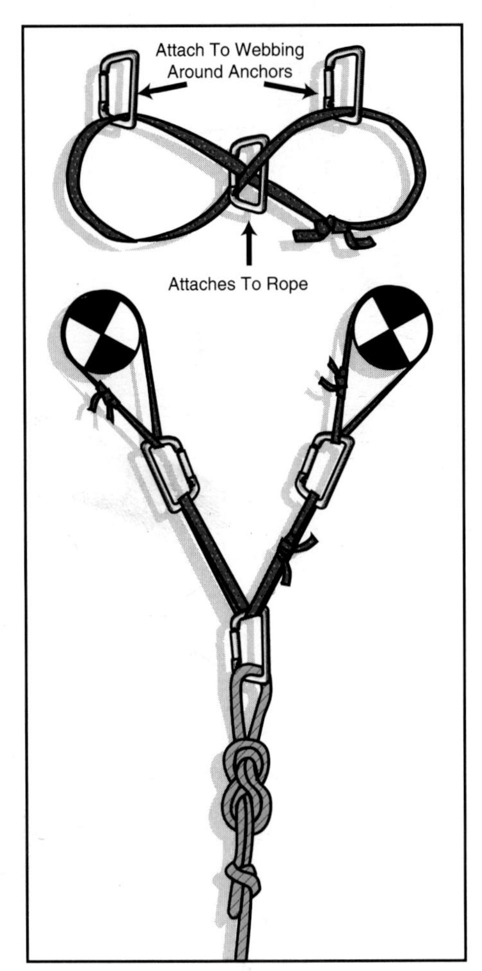

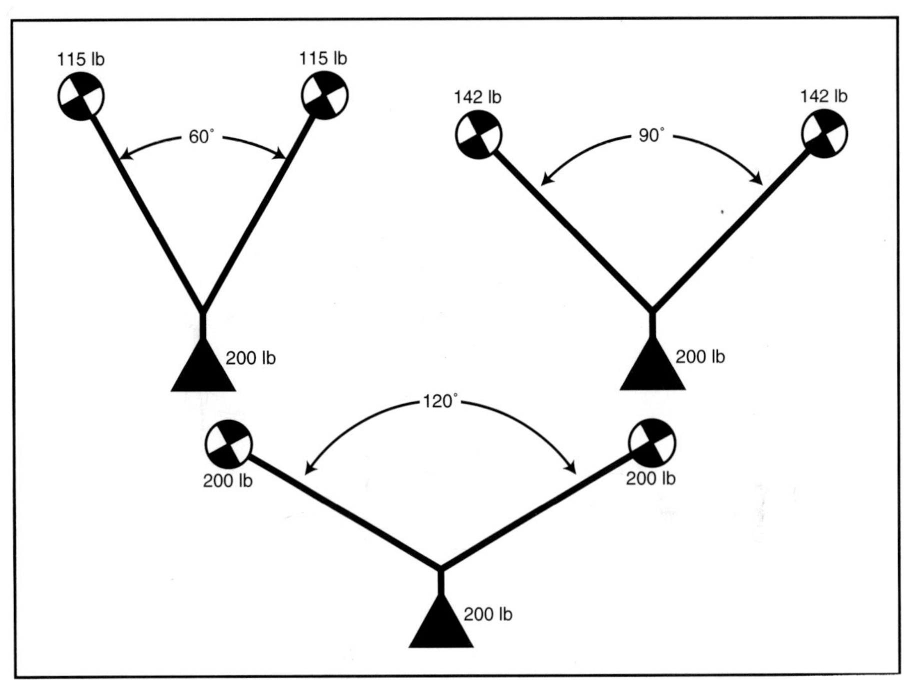

Figure 4.87 The critical angles in a two-point anchor system.

Figure 4.86 A typical two-point self-adjusting anchor system.

Three-point system. If three adjacent anchor points are available and needed, a three-point self-adjusting anchor system can be constructed. First, a length of rope is tied into a loop (Figure 4.88). Then a bight is formed in the loop (Figure 4.89). The four parts of the two bights formed at the bottom are then brought together and tied into a Figure-Eight (Figure 4.90). Five carabiners are then used to construct the anchor system and connect it to the anchor points (Figure 4.91).

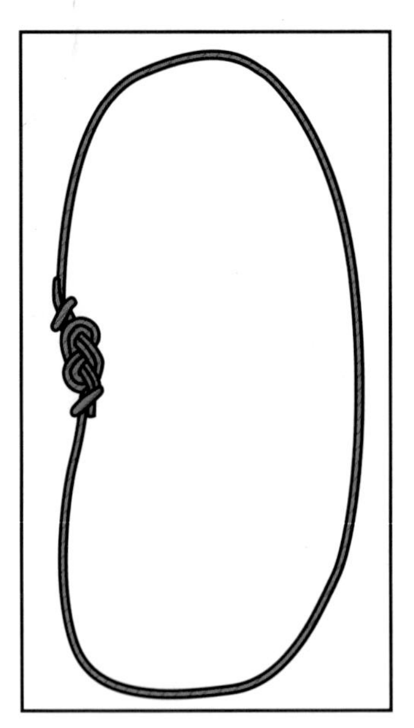

Figure 4.88 A simple loop.

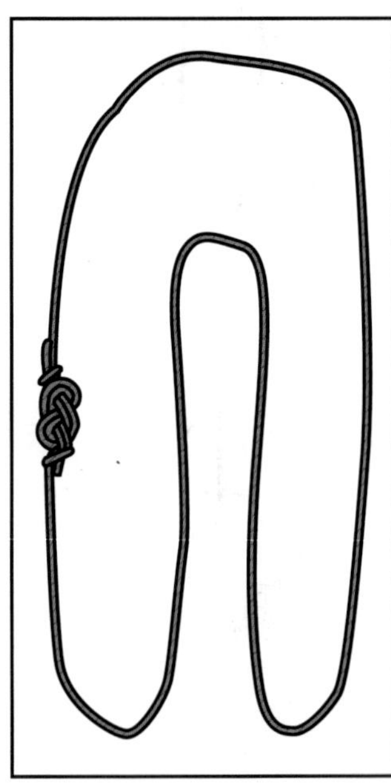

Figure 4.89 A bight in the loop.

Mechanical Advantage Systems

Rescuers often need to construct various types of hauling systems using rope, pulleys, carabiners, and webbing. However, not all hauling systems are mechanical advantage systems. Some hauling systems merely change the direction of pull without adding any mechanical advantage. For example, in a simple system where the haul line is attached to the load and passes through one overhead pulley, it is classified as a 1:1 system. That is, the required pulling force equals the load force so there is no mechanical advantage (Figure 4.92). Fixed, stationary pulleys provide change of direction only; a pulley must move within the system for it to provide any mechanical advantage.

In many situations, the loads that rescuers are required to lift are too heavy to be lifted by a 1:1 system, so systems must be constructed that multi-

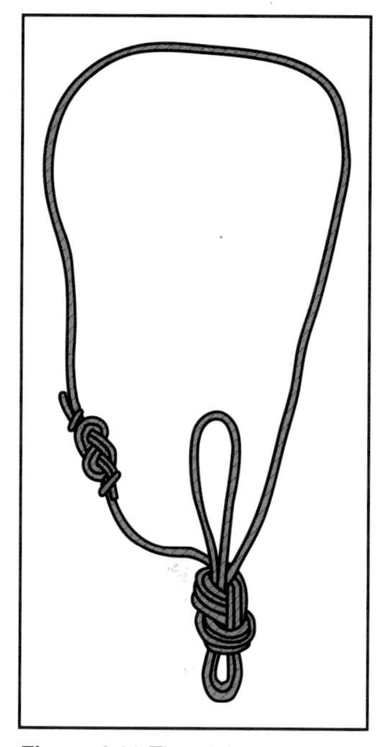

Figure 4.90 The bights are tied in a Figure-Eight.

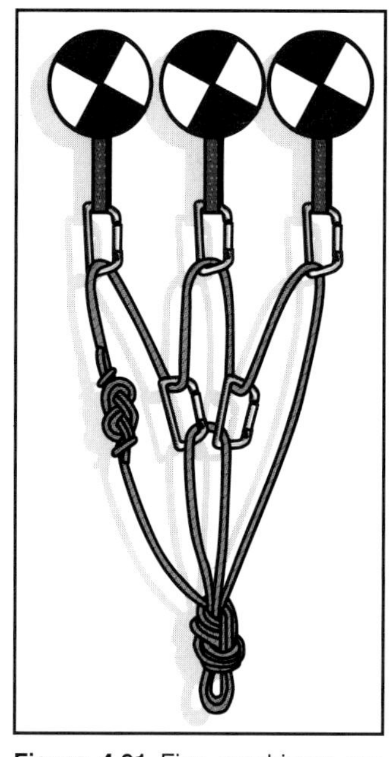

Figure 4.91 Five carabiners are used to construct the system.

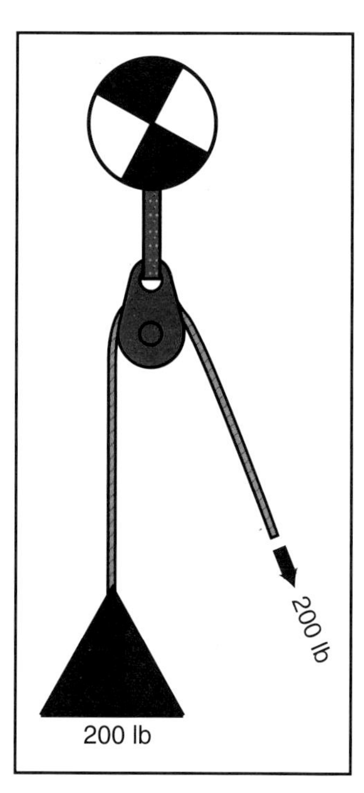

Figure 4.92 A 1:1 system.

ply the pulling force (mechanical advantage) that the rescuers apply. When constructing mechanical advantage systems, care must be taken not to exceed the safe working strength of any part of the system. In general, two types of mechanical advantage systems are used in rescue: simple and compound.

SIMPLE SYSTEMS

To calculate the mechanical advantage provided by a simple system, the following rules apply:

- If the haul rope is attached to the load, the system is odd.
- If the haul rope is attached to the anchor point, the system is even.
- Mechanical advantage is equal to the number of pulleys, plus 1.
- If the last pulley in the system is attached to the anchor, it provides only change of direction and adds no mechanical advantage.

Simple mechanical advantage systems are used mainly for low-angle rescues. These systems typically provide a range of mechanical advantage from 2:1 to 4:1.

2:1 system. This system is the most elementary form of hauling system that provides mechanical advantage (Figure 4.93). It is used when the load is relatively light.

3:1 system. Also known as a "Z-Rig" because of its configuration, this is one of the most often used mechanical advantage systems (Figure 4.94). Because it is relatively easy and quick to construct, this system is used in many different rescue situations.

4:1 system. As heavier loads must be lifted, systems that provide greater mechanical advantage must be constructed. The practical upper limit for a simple system is one that provides a 4:1 advantage (Figure 4.95).

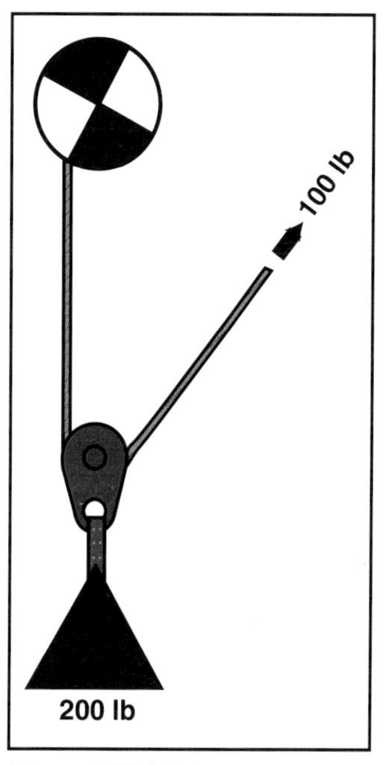

Figure 4.93 A 2:1 system.

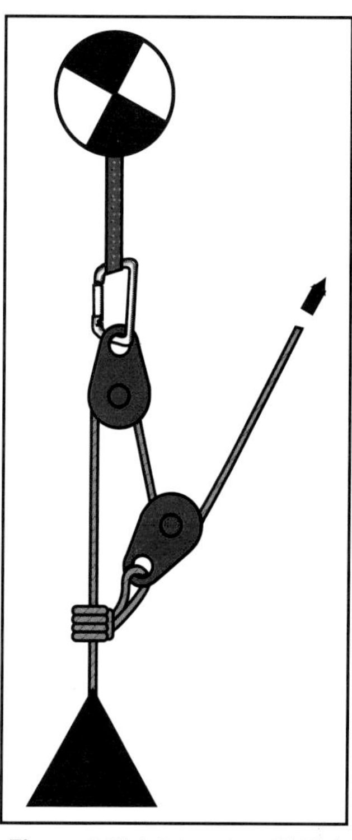

Figure 4.94 A 3:1 system (Z-Rig).

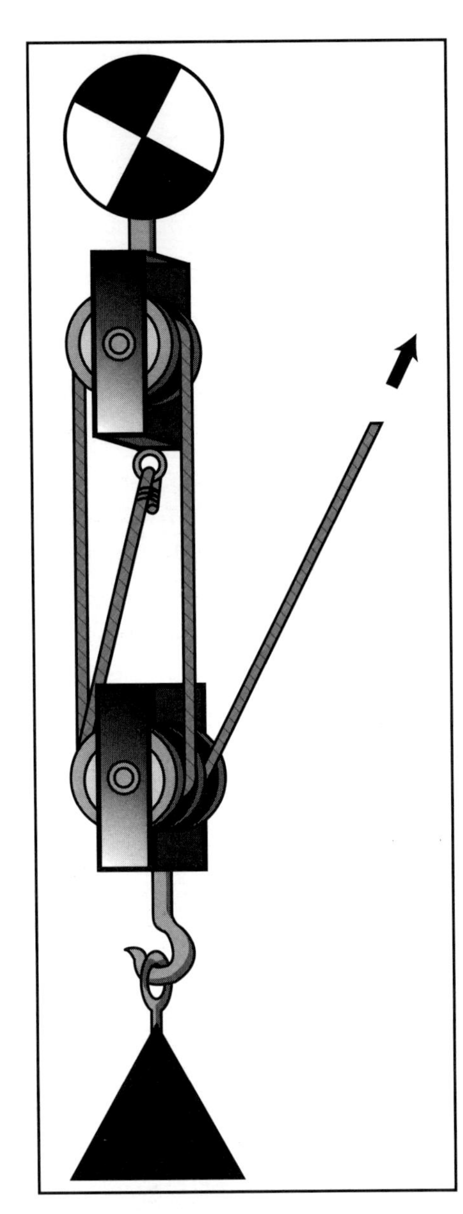

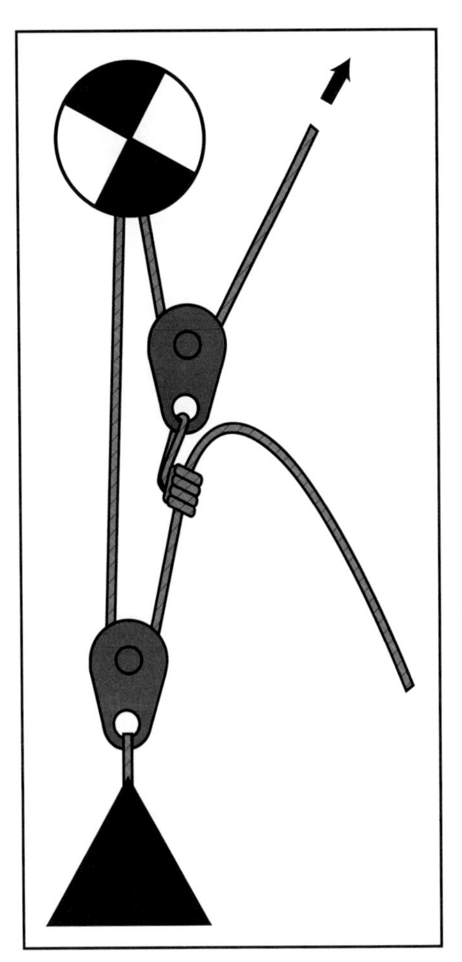

Figure 4.96 A compound 4:1 system.

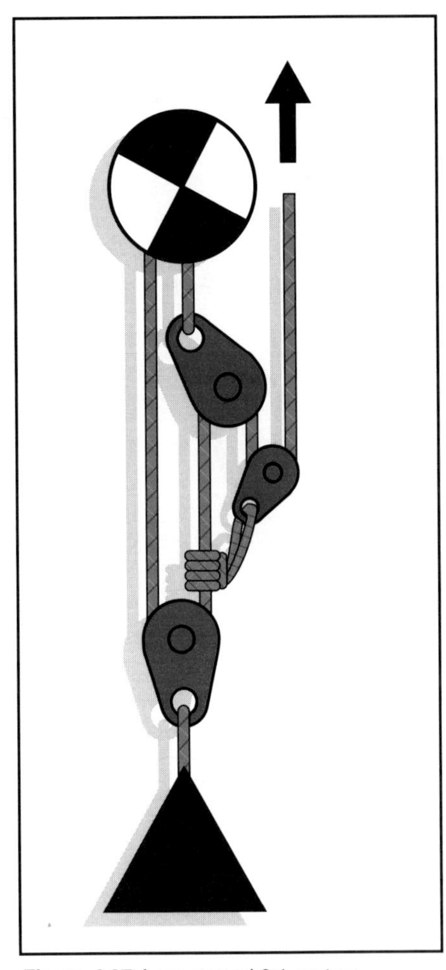

Figure 4.97 A compound 6:1 system.

Figure 4.95 A simple 4:1 system.

COMPOUND SYSTEMS

These systems are used when the lift needed is beyond that provided by a simple system. Compound systems are those in which one simple system is attached to and pulls on another simple system to multiply the mechanical advantage gained. Compound systems typically provide a range of mechanical advantage of from 4:1 to 9:1.

4:1 system. The least complex of all the compound systems, this system is also the quickest and easiest compound system to construct (Figure 4.96).

CAUTION: If it is still necessary to increase the number of haulers to move the load, the load may be too heavy to be moved without compromising the integrity of the system and the safety of the rescuers.

6:1 system. More complex than the 4:1 system, but also more effective, the 6:1 system will lift much heavier loads (Figure 4.97).

9:1 system. The most complex of the hauling systems, and the most time-consuming to construct, the 9:1 system will lift loads heavy enough that the safe working strength of the haul rope may have to be considered (Figure 4.98).

TACTICAL CONSIDERATIONS

The tactical considerations involved in any rope rescue may be broken down into four distinct phases: assessment on arrival, prerescue operations, rescue operations, and termination.

Phase I: Assessment On Arrival

The assessment of the situation that must be done by the first-arriving unit can be divided into a primary and a secondary assessment.

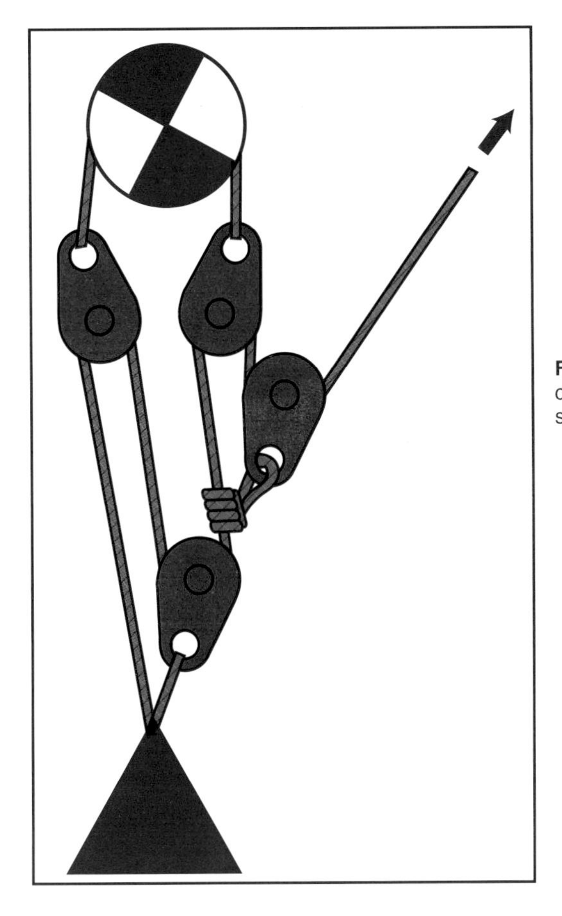

Figure 4.98 A compound 9:1 system.

PRIMARY ASSESSMENT

The primary assessment actually begins with the initial dispatch and continues during response and after arrival. The first-due rescuer must begin to formulate a mental picture of the situation based on the information provided in the dispatch, time of day, weather, and traffic conditions during response to the scene. This mental picture will be further refined and enhanced during the initial size-up.

Information gathering. The process is continued once on scene by attempting to talk to the reporting party (RP) and/or other witnesses. Of critical importance in the primary assessment is gathering any available information about the number, condition, and location of victims. The following questions should be answered:

- Have all occupants been accounted for?
- How many victims are there?
- Is their exact location known?
- Are they injured or merely trapped/stranded?
- Are they conscious, and if so, can they communicate?

Decision making. The answers to these questions help the first-in rescuer make the first critical decision: Can the units on scene or en route handle the situation, or do additional units need to be called? If more resources are needed, they must be requested immediately to get them on scene as soon as possible (Figure 4.99). If he or she has not done so already, the first-in rescuer should assume formal command of the incident because the answers to the initial questions also form the basis for the incident action plan that the IC must begin to develop at this point.

Scene control. If the information gathered during the primary assessment confirms that a legitimate rescue emergency exists, the area surrounding the scene should be cordoned off as de-

Figure 4.99 Additional resources should be called as soon as possible. *Courtesy of Laura Mauri.*

scribed in Chapter 2, Rescue Scene Management (Figure 4.100). Which areas are designated as the hot, warm, and cold zones will depend on the size, nature, and complexity of the rescue situation. The ultimate boundaries of the zones may be affected by the hazards present.

Figure 4.100 The area should be cordoned off as soon as possible. *Courtesy of Laura Mauri.*

SECONDARY ASSESSMENT

The secondary assessment involves some reconnaissance of the scene to gather information about the type of elevation difference involved, its physical characteristics, and the problems to be anticipated. All information gathered during both the primary and secondary assessments help determine the mode of operation.

Type of elevation difference. The type of elevation difference involved can indicate some of the problems that must be dealt with. For example, industrial sites may contain hazardous materials or high-voltage electrical vaults; utility manholes and other below-grade spaces may contain toxic atmospheres; off-road rescues may involve steep slopes with poor footing or sheer cliffs. Off-road rescues may also be mass-casualty incidents and may greatly increase the need for ambulances at the scene. The amount of elevation difference and its physical characteristics may indicate the number of trained rescuers and how much and what types of specialized equipment will be needed to handle the situation safely and efficiently (Figure 4.101).

Hazard assessment. Depending upon the nature of the incident, there may be a number of hazards present on scene. In addition to the hazards associated with the elevation differences, secondary hazards such as downed electrical wires, leaking or spilled hazardous materials, or the potential for fire and/or explosions may have to be mitigated to make the scene safe for rescuers to work in. Also of concern are a lack of adequate anchor points, loose debris or rock that may fall, and sharp edges and abrasive surfaces.

Figure 4.101 A common rope rescue situation.

Mode of operation. Finally, all of the information gathered during the primary and secondary assessments confirms the nature and extent of the rescue problem and helps the IC finalize the incident action plan. The information also helps the IC make one of the most important decisions affecting the action plan: whether it is reasonable to think that the victims are alive and that the operation should be conducted as a rescue or that victims could not be expected to have survived and that the operation should be conducted as a body recovery.

Phase II: Prerescue Operations

During this phase, which may last from a few minutes to several hours, all of the things necessary to make the rescue operation as safe as possible must be done. This includes finalizing the incident action plan, gathering the necessary resources, monitoring and managing the atmosphere in and around the rescue site, making the scene safe for rescuers to work in, and ensuring an adequate communications capability to allow the action plan to be carried out safely.

INCIDENT ACTION PLANS

On small, relatively simple rescue operations the incident action plan need not be in writing, but there *must be a plan*. On larger, more complex operations the plan should be in writing and should reflect an incident management system as described in Chapter 2, Rescue Scene Management. In either case, the plan must be finalized and communicated through channels to everyone involved in the operation.

While the original plan should be flexible enough to accommodate a certain amount of the unexpected, *a backup plan should be available* in case something completely unforeseen occurs to invalidate the original plan. If the information gathered during the primary and secondary assessments was somehow inaccurate or misleading or if some subsequent occurrence, such as a secondary explosion or collapse, changes the situation significantly, a secondary or backup plan should be ready to be implemented. If it suddenly becomes necessary to rescue the rescuers, that plan of operation should be ready.

GATHERING RESOURCES

Resources consist of personnel and equipment, and both are critically important to the success of the operation. If there are too few rescue personnel or if the personnel are insufficiently trained to perform as needed, the best equipment in the world will not get the job done. Likewise, the most highly trained and motivated rescuers will not be able to do what is necessary if they do not have the tools and equipment they need.

The resources gathered at the scene should reflect the incident action plan. But, as stated earlier, if both the personnel and equipment in the initial response are insufficient for the rescue problem at hand, the IC must request additional resources as soon after arrival as possible. The sooner these resources get under way, the sooner they will arrive on scene where they are needed. If the IC is initially unsure about the type and/or amount of equipment that will actually be needed, he or she should call for everything that *might* be needed. Resources that prove to be unnecessary can be returned to quarters while still en route or after they arrive on scene.

Personnel. Depending upon the nature and extent of the rescue problem, the number of rescue personnel needed varies. However, even the rescue of one victim from the top of a water tower can involve at least ten rescuers, including a safety officer, an incident commander, and additional support personnel that may be needed to set up and operate on-scene equipment (Figure 4.102). This list does not include EMS personnel that may be needed to treat and transport the victim to a medical facility. Obviously, as the number of victims and the complexity of the rescue problem increase, the number of rescue personnel needed will increase proportionately.

Equipment. The amount and types of equipment needed also vary with the nature and extent of the rescue problem. The water tower scenario previously described could require a rescue unit, an aerial device, an engine company, plus a command vehicle and an ambulance. More complex incidents could also require specialized units such as lighting units, air units, communications units, or hazardous materials units.

If the incident involves a below-grade rescue, as in a manhole or utility vault, the additional equipment needed to test and manage the atmosphere within the space will have to be obtained. For more information about atmospheric monitoring and ventilating confined spaces, see Chapter 8, Confined Space Rescue.

Figure 4.102 Rope rescue can involve a lot of personnel. *Courtesy of Laura Mauri.*

PREPARING THE SCENE

Even after the necessary resources have been gathered on scene, other steps may still be necessary to prepare the scene. Preparing the scene may be as simple as providing adequate lighting in and around the site, identifying and marking secondary hazards such as downed wires, neutralizing hazards within the controlled zones, and providing fire protection. In structural collapse incidents, preparing the scene may also involve identifying and shoring up areas with a potential for secondary collapse.

Lighting. Because these incidents may occur at any time of the day or night, rescuers must be prepared to provide whatever lighting they need to do their jobs. If rescuers must illuminate areas with flammable atmospheres, the use of explosion-proof lighting equipment will be required. If power is provided by a portable generator, it should be positioned downwind so the exhaust from its engine does not drift into and contaminate the site.

Mitigating hazards. Identifying and mitigating primary and secondary hazards in and around the scene can be an extremely important function. Marking and isolating secondary hazards can prevent rescuers from becoming additional victims by alerting them to the presence of these hazards.

Fire protection. It may be necessary to deploy one or more charged hoselines if there is an actual or potential flammability hazard in or around the scene. Firefighters with appropriate fire extinguishers or charged hoselines will be needed to stand by if spark-producing power tools are used and during cutting operations with oxyacetylene torches or burning bars.

Shoring. If there is any question about the stability of walls, floors, or large pieces of rubble, they will have to be shored up to make it safe for rescuers to work in and around the site (Figure 4.103). See Chapter 6, Structural Collapse Rescue for more information on shoring.

COMMUNICATIONS

Rope rescue incidents can involve a variety of different environments from tanks, vaults, and other confined spaces to high-rise structures of masonry or reinforced concrete with heavy walls, floors, and other barriers to communication. Therefore, the form of communication that will work best can vary considerably. It may involve direct voice communication, a series of tugs on a lifeline, hard-wired phones, or portable radios.

Figure 4.103 Emergency shoring in place. *Courtesy of Mike McGroarty.*

Voice communication. Direct, face-to-face voice communication is to be preferred whenever the physical arrangement of the scene will allow it (Figure 4.104). The chances of miscommunication are greatly reduced if those involved can hear each other's voices, even if they cannot see each other.

Lifeline. While the main purpose of a lifeline attached to the harness worn by members of a rescue team is as a safety line, or providing a means of lowering them to the ground or pulling them up to grade, it can also serve as a primitive means of communication. One or more tugs on a lifeline can communicate a single thought. The acronym OATH is used by some departments to communicate as follows:

1 tug = ***O***K

2 tugs = ***A***dvance

3 tugs = ***T***ake up (eliminate slack)

4 tugs = ***H***elp!

Figure 4.104 Face-to-face communication is best. *Courtesy of Laura Mauri.*

Hard-wired phones. These portable phone systems are very effective over the short distances involved in most rescue situations. These systems have proven their value as a communications medium in military operations and in high-rise fire fighting operations. The main disadvantage is having to lay and maintain the phone line from the command post to the operational location.

Portable radios. These versatile units have been the primary means of emergency scene communications for many years, and advancing technology makes them even more useful in this role. Modern, multichannel units with scanners allow communication on a number of different frequencies (Figure 4.105). Such flexibility allows each incident in progress at the same time to have its own tactical or incident channel. This obviously reduces the amount of traffic that rescuers must monitor and process, and it greatly reduces the chances of miscommunication. Aside from their cost, the primary disadvantage of modern portable radios is that their signal is sometimes incapable of penetrating the mass of some structures (Figure 4.106).

Figure 4.105 A modern multichannel radio. *Courtesy of Laura Mauri.*

Figure 4.106 Radios may be unreliable inside some structures.

RESCUE TEAMS

Those personnel who will actually perform the rescue must be assembled, along with those who will work in direct support of them. Any time rescue personnel or victims are being supported by a rope or rope system, there should also be a safety officer on scene whose sole job is to watch the operation and be alert for any possibly unsafe conditions or practices.

Team leader. The team leader's job is to supervise and direct the actions of the rescue team — not to get directly involved in performing the rescue. Even if the team is composed of only two members, one of the members must act as the team leader. Even if both have portable radios, only the team leader should communicate with the unit supervisor or the IC (Figure 4.107). All actions taken by the team should be coordinated by the team leader and directed toward achieving the objectives established in the incident action plan.

Rescue team. Unlike other forms of rescue that often involve several rescuers at the actual point of delivery, rescue incidents involving elevation differences should limit the number of personnel at risk to as few as possible. In most cases, however, a rope rescue team will be composed of at least two members: a rescuer and a brake tender (belayer). They must be properly dressed and equipped for the conditions in the particular rescue environment and for the work they will be required to perform. Their gear should be checked by the team leader before they are allowed to begin the rescue (Figure 4.108).

Figure 4.107 The team leader communicates for the team.*Courtesy of Laura Mauri.*

Figure 4.108 The team leader checks a member's gear. *Courtesy of Laura Mauri.*

Brake tender (Belayer). Anytime a rescuer or victim is supported by a rope or rope system, a brake tender must be tending the belay line (Figure 4.109). The tender's job is to protect the person on the rope in case of a fall. The recommended belay uses a separate lifeline attached to the rescuer and to a separate anchor point (Figure 4.110).

The belay device may be a triple-wrapped Prussik or a belay plate on a separate safety line attached to the rescuer and/or victim and to a separate bombproof anchor. The belayer must constantly monitor the progress of those on the line and must be ready to engage the belay device instantly should the main line fail.

Figure 4.109 The brake tender protects those on the line. *Courtesy of Laura Mauri.*

Figure 4.110 The belay line attaches to a separate anchor point.

Backup team. There must also be a fully prepared and equipped backup team ready to intervene if the first team gets into trouble. The backup team must be composed of the same number of personnel as the rescue team and should have the same level of equipment, training, and expertise.

Hauling team. In many rescue incidents, relatively heavy loads will have to be raised or lowered. If there is plenty of manpower on the scene, adding more members to the hauling team can reduce the need for the additional equipment required to expand the hauling system (Figure 4.111).

Figure 4.111 Several people may be needed on the haul line. *Courtesy of Laura Mauri.*

Phase III: Rescue Operations

Once all the preparations have been made, the process of actually rescuing victims can begin. This phase of the operation obviously involves rescuers entering a relatively uncontrolled environment, and that necessitates the use of a personnel accounting system. In addition, this phase may involve searching for victims, will certainly involve reaching them, and may involve treating and packaging them prior to moving them to safety.

PERSONNEL ACCOUNTABILITY

The purpose of the accountability system is to ensure that only those who are authorized and properly equipped to enter a hazardous area are allowed to do so and that both their location and their status are known as long as they remain inside the controlled zone. As is true of the incident action plan, the degree of formality of the personnel accounting system should reflect the nature, size, and complexity of the particular rescue problem at hand.

Minimum accountability. Some incidents require only one or two rescuers without respiratory protection working in the hot zone at a time. They are close enough to be in constant visual and verbal contact with the team leader. In these cases, the intent of the accountability system can be met with a minimum amount of information being recorded. Merely writing down the team members' names and their entry and exit times should be sufficient.

Maximum accountability. A more formal system must be employed on larger, more complex incidents when more personnel are required to work inside the controlled zones at once, when rescuers must use respiratory protection, or when they must work out of sight of the team leader. In these cases, the team members' names and/or other identifiers must be recorded as they enter the controlled zones. Their times of entry, SCBA gauge readings, and *projected* exit times must also be written down. Those using supplied-air respiratory equipment may be able to remain in the space for longer periods of time, but the same rules of accountability apply. Each entrant's time of exit must be recorded when he or she leaves the space. As each entrant's projected exit time arrives, the list must be checked to ensure that he or she has exited. If anyone has not, immediate action must be taken to locate them and escort them out of the hazardous area.

REACHING VICTIMS

In order to assist a trapped or stranded victim to safety, one or more rescuers must first reach the victim. This may be the most difficult and time-consuming part of the operation. Regardless of whether victims are above or below grade, if they have access to an interior stairway or exterior fire escape but still need to be rescued, then they must be too incapacitated to walk out on their own. This means they are almost certainly injured or unconscious, or both, and rescuers need to take a trauma kit and a litter with them.

If victims are not this easily accessed, they may or may not be physically incapacitated, but they still cannot help themselves to safety. It could be that they are stranded on some rock outcropping or on an elevated water tank because of an equipment failure (Figure 4.112). Or, it may be that they and their vehicle left the roadway and came to rest at the bottom of a steep slope or even a cliff (Figure 4.113).

Figure 4.112 A victim may be stranded in a precarious place. *Courtesy of Laura Mauri.*

Figure 4.113 Rescuers may have to carry a victim upslope.

STABILIZATION/TREATMENT

Obviously, unless injured victims are in immediate danger from fire or some other threat, they should be stabilized and treated before they can be moved to ground level. Rescuers need to know and follow local medical protocols in the treatment and stabilization of injured victims.

PACKAGING VICTIMS

After local protocols have been followed in treating and stabilizing injured victims, they must still be packaged in a way that will allow them to be safely raised or lowered to ground level. In most cases, injured victims will be packaged in some form of basket litter. Uninjured victims who are merely stranded can be placed in a Class III harness.

Basket litters. Most basket litters are in one of two basic forms: metal frame with molded chicken wire (Stokes) or one-piece molded plastic (Figure 4.114). The procedure for lashing a victim into either type is the same and involves preparing the litter and performing both interior and exterior lashings.

Preparing the litter. Before the victim is placed in a litter, it should be properly prepared. This requires two blankets and a 10-foot (3 m) length of webbing.

Step 1: Place one blanket lengthwise across the head of the litter, allowing one third of its width to extend beyond the top end of the litter (Figure 4.115).

Step 2: Place the second blanket lengthwise in the litter, allowing about one fourth of its length to extend beyond the bottom end of the litter (Figure 4.116).

Step 3: Fold the overhanging portion of the first blanket back onto itself and the second blanket (Figure 4.117).

Step 4: Lay the length of webbing across the litter, just below the midpoint (Figure 4.118).

At this point the litter is ready for the victim to be placed in it. This should be done according to local protocol or as described in an EMS first responder manual.

Figure 4.114 Typical basket litters.

Figure 4.115 The first blanket is placed in the litter.

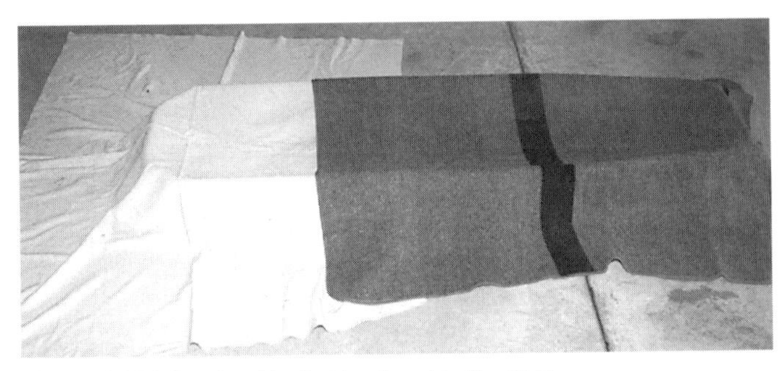

Figure 4.116 Another blanket is placed in the litter.

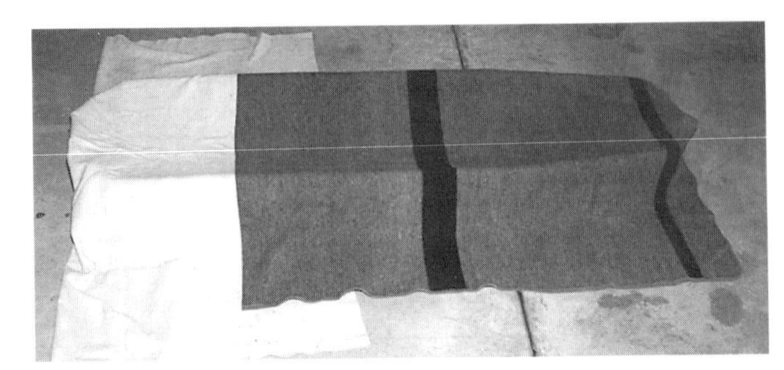

Figure 4.117 The first one is folded over the second.

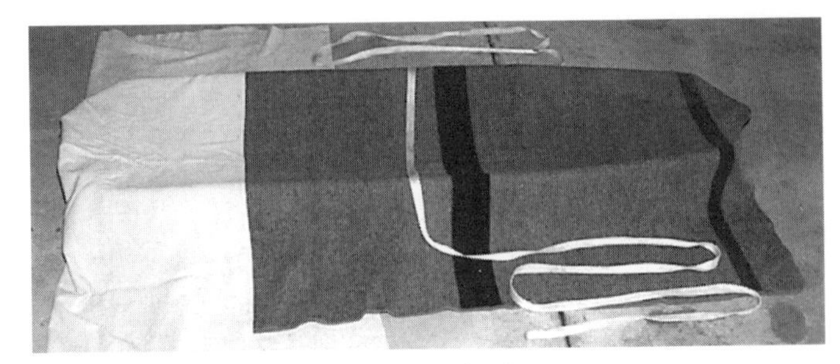

Figure 4.118 Lay the webbing across the litter.

Interior lashing. With the victim supine in the litter, secure the victim's pelvic area using the following steps:

Step 1: Grasp the middle of the webbing and pull the bight thus formed up between the victim's legs, and lay it on the victim's lower abdomen.

Step 2: Bring the ends of the webbing to the center and feed them under and out through the bight in the middle forming a wide "V" (Figure 4.119).

Step 3: Tie a Half Hitch with the ends of the webbing to each side of the bight to reduce cinching up against the victim (Figure 4.120).

Step 4: Bring each leg of the webbing up toward the victim's head, and tie one to each side of the main frame of the litter at the victim's shoulder, taking a full turn around the anchor point and securing it with two Half Hitches and an Overhand Safety (Figure 4.121).

Exterior lashing. Once the victim's pelvic area is secured as just described, the blankets should be folded snugly around the victim. The victim should then be lashed securely into the litter using the following procedure:

Step 1: Feed half of a 20-foot (6 m) length of webbing through the bottom anchor point openings on each side of the litter, leaving half of the webbing's length on each side (Figure 4.122).

Step 2: At the midpoint in the webbing, place a Half Hitch around both of the victim's feet at the instep (Figure 4.123).

Step 3: Bring each leg of the webbing across the litter to the next higher anchor point on the other side of the litter and continue to cross the legs of the webbing over the victim until the anchor points at shoulder level are reached (Figure 4.124).

NOTE: It is recommended that the webbing not be allowed to cross the top side of the litter's main frame rail because this can expose the webbing to abrasion. In a

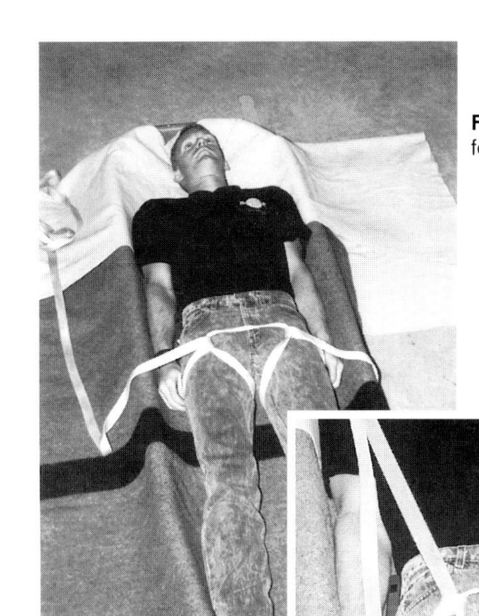

Figure 4.119 The webbing forms a wide V-shape.

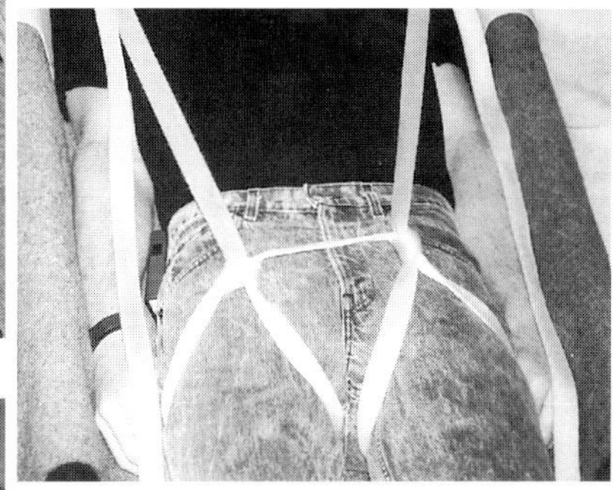

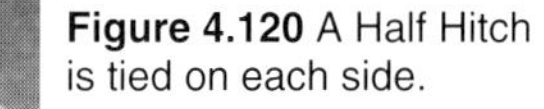

Figure 4.120 A Half Hitch is tied on each side.

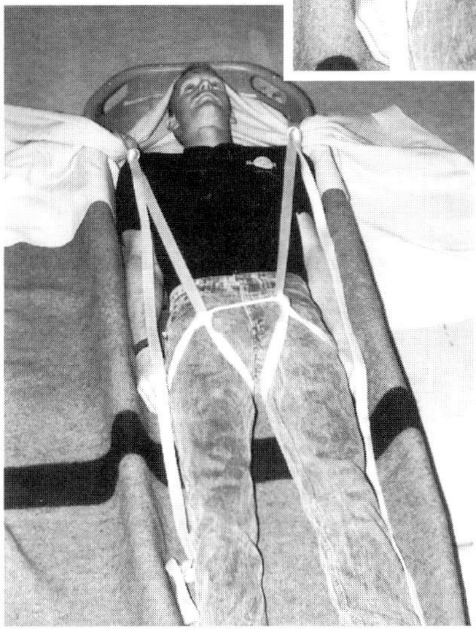

Figure 4.121 Each leg of the webbing is tied off.

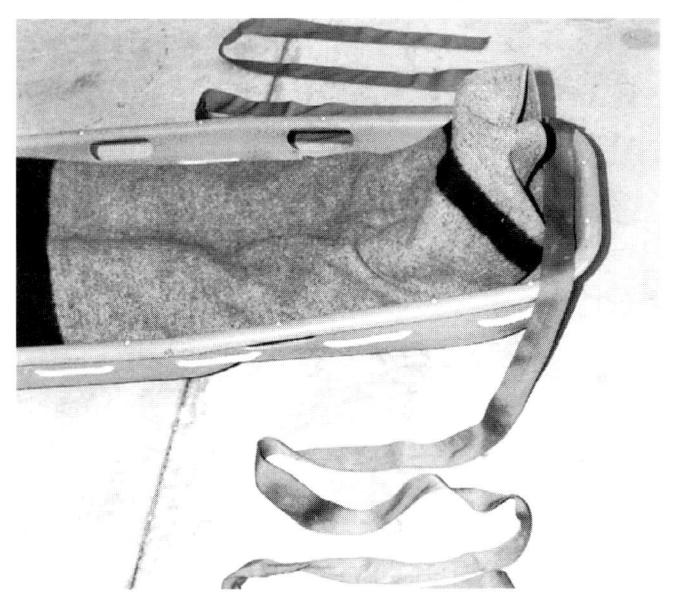

Figure 4.122 Position the webbing at the foot of the litter.

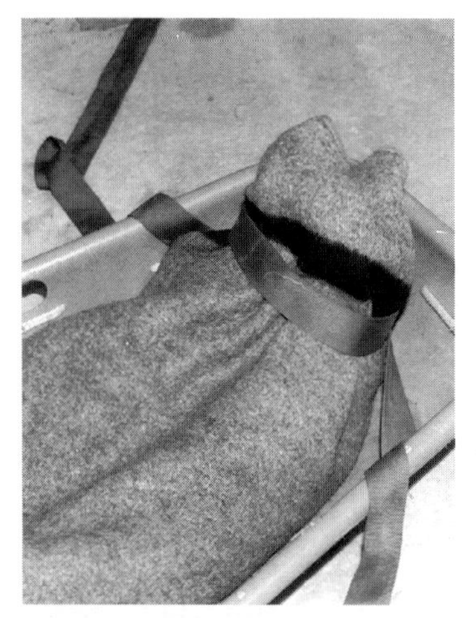

Figure 4.123 The victim's feet are secured.

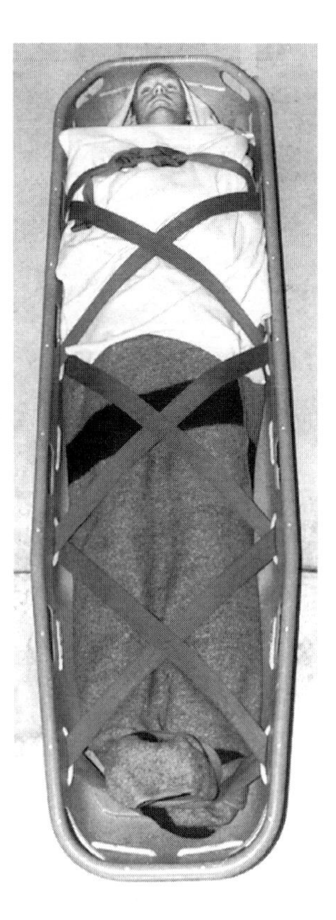

Figure 4.124 Each leg of the webbing crisscrosses the litter and is tied off at the shoulder.

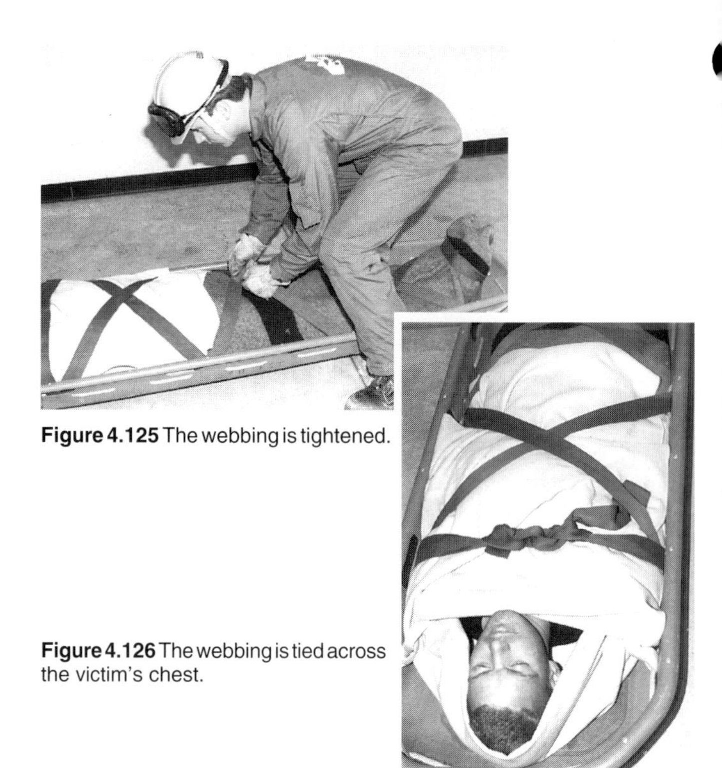

Figure 4.125 The webbing is tightened.

Figure 4.126 The webbing is tied across the victim's chest.

Stokes litter, the webbing can be anchored around the ribs of the litter. Most molded plastic litters have prerigged ropes inside for anchor points, and these should be used to lace the victim into the litter.

Step 4: Beginning at the victim's feet, tighten the webbing by pulling the slack from each section, working up to the victim's shoulder level (Figure 4.125).

Step 5: Tie the ends of the webbing together across the victim's chest with a round turn and two Half Hitches (Figure 4.126).

Litter sling. If the litter does not have a prerigged lifting sling, one will have to be constructed on the spot. Two 15- to 20-foot (4.6 m to 6 m) lengths of rope or webbing and six carabiners will be needed. An improvised sling may be constructed using the following procedure:

Step 1: Clip a carabiner into the anchor point openings on each side of the head and foot ends of the litter.

Step 2: Tie a Figure-Eight on a Bight in the middle of each piece of rope or webbing, and clip them into the two remaining carabiners attached to the haul line.

Step 3: Tie each leg of one piece of rope or webbing into the carabiners at the foot of the litter, using a Figure-Eight on a Bight.

Step 4: Tie each leg of the other piece of rope or webbing to the carabiners at the head of the litter, using the same knot. The legs of the sling attached to the head of the litter should be tied slightly shorter than the ones at the foot so that the head of the litter will be slightly higher when lifted (Figure 4.127).

Harness. Ambulatory and uninjured victims can be assisted to ground level by being placed in a Class II or Class III harness, depending on the level of security required for safety and the level of assistance needed.

RAISING/LOWERING VICTIMS

Once victims have been located, reached, medically assessed and treated if necessary, and packaged for movement, they can then be raised or lowered to ground level.

Low-angle rescue. Low-angle rescues are those where the angle of the slope is such that rescuers do not need to be supported by a lifeline to avoid falling down the slope. These incidents may be either above or below grade, but most are associated with

Figure 4.127 The head of the litter should be slightly higher than the foot.*Courtesy of Laura Mauri.*

vehicle accidents where the vehicle leaves the roadway and comes to rest at the bottom of a steep slope. The vehicle may also be partially or totally submerged in water. Rescue personnel must be able to quickly but safely descend to the vehicle to assess the status of the occupants and to determine the best way to move them to safety. Depending on the level of assistance needed by the victims, one or more of the simple hauling systems previously described may be required.

If a victim must be moved in a basket litter, at least four rescuers may be needed to carry the litter as it is being hauled up or down the slope by a hauling team (Figure 4.128). If the victim is a small child, two rescuers may be sufficient to carry the litter (Figure 4.129).

High-angle rescue. High-angle rescues are those in which the rescuers must be supported by a lifeline to keep from falling. Obviously, these are

Figure 4.128 Rescuers carry a victim upslope. *Courtesy of Laura Mauri.*

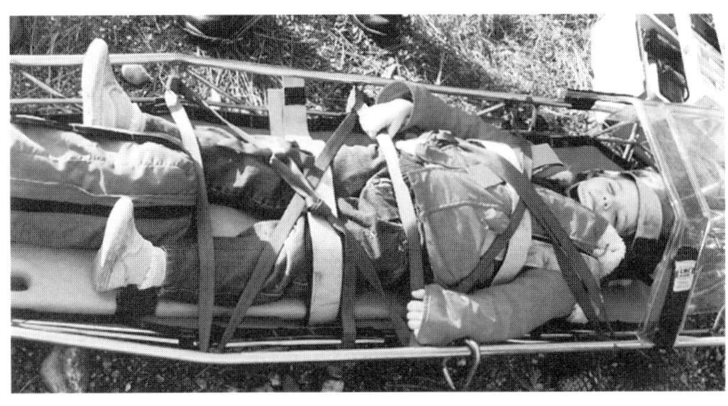

Figure 4.129 Fewer rescuers are needed to carry a small victim. *Courtesy of Laura Mauri.*

more hazardous and complex situations than most low-angle rescues, so additional time may be required to prepare and execute the rescue safely. High-angle rescues may also be either above or below grade and may involve raising or lowering victims.

Rescuing a single uninjured victim who is merely stranded, such as a window washer whose motorized platform has stopped working, may involve a single rescuer rappelling down to him or her with a Class II or Class III harness for the victim. They can both be raised to the rooftop by a hauling team, or they can make a controlled descent to street level.

Rescuing an injured victim from an elevated location may involve two rescuers taking a litter and a trauma kit to stabilize, treat, and package the victim in the litter. The victim and a rescuer can then be hauled up or lowered down as the situation dictates (Figure 4.130).

Below-grade rescue. Rescues from below grade are most often in some sort of confined space. This adds the problem of testing, managing, and monitoring the atmosphere within the space. For more information on atmospheric monitoring equipment and techniques, see Chapter 8, Confined Space Rescue.

Below-grade rescues also often involve access through a manhole. This usually involves the use of a rescue tripod to lower rescuers and equipment into the space and to raise them and the victim out (Figure 4.131). However, the manhole opening may not be large enough to allow a rigid metal or plastic litter to pass through. If the victim cannot be moved in a Class III harness, he or she will have to be packaged in a more flexible litter (Figure 4.132).

Figure 4.130 A rescuer and a victim on line. *Courtesy of SKEDCO, Inc.*

Figure 4.131 A rescuer is lowered into a confined space. *Courtesy of Keith Flood.*

Figure 4.132 A flexible litter may be needed in confined spaces. *Courtesy of SKEDCO, Inc.*

Phase IV: Termination

The termination phase of a rope rescue involves the obvious element of retrieving pieces of equipment used in the operation. But it also involves less obvious elements such as investigating the cause(s) of the incident, releasing the space to those responsible for it, and conducting critical incident stress debriefings with members of the rescue teams.

EQUIPMENT RETRIEVAL

Depending on the size, complexity, and length of time involved in the operation, the job of retrieving all of the various pieces of equipment used may be very easy or very difficult and time-consuming. Under some circumstances, it can also be quite dangerous.

Identifying/collecting. The process of identifying and collecting equipment assigned to the various pieces of apparatus on scene can be made much easier if each piece of equipment is clearly marked; however, nothing should be engraved or stamped into the surface of rope rescue hardware (Figure 4.133). It may be necessary for the driver/operators of the rescue unit and other pieces of apparatus on scene to conduct an inventory of their equipment prior to leaving the scene. If the opera-

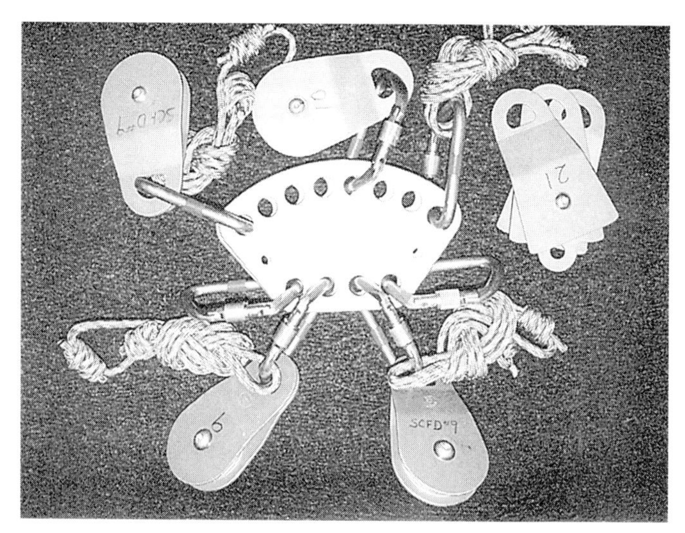

Figure 4.133 Clearly marked equipment is easier to retrieve. *Courtesy of Laura Mauri.*

Figure 4.134 A rescue vehicle in the Demobilization Unit.

tion was large enough for a full incident command/ management system to implemented, the Demobilization Unit will coordinate the recovery of loaned items, such as portable radios, and document lost or damaged pieces of apparatus and equipment (Figure 4.134).

Abandonment. In some cases, the rescue environment is too hazardous to justify sending rescue personnel back to retrieve pieces of equipment — even expensive ones. Rather than putting rescue personnel at risk, it is sometimes advisable to simply abandon the equipment in place. It may be retrievable after the scene has been stabilized or restored to normal, or the cost of replacing the abandoned equipment may be recovered from those responsible for the scene.

INVESTIGATION

All rope rescues should be investigated at some level. At the very minimum, a departmental investigation should be conducted for purposes of reviewing and critiquing the operation. However, if an employee was injured in the incident, it may be investigated by the Occupational Safety and Health Administration (OSHA) and perhaps by other entities such as the employer's insurance carrier. Obviously, if a crime was involved, law enforcement agencies will also investigate the incident.

RELEASE OF CONTROL

Once rescuers respond to the scene of a rope rescue, they assume control of the scene and the immediate surrounding area. Within certain limits, they can deny access to the scene to anyone, including the property owner. Legitimate members of the news media have certain constitutionally protected rights of access, but the interpretation of these rights varies from state to state and from country to country. Rescuers need to be aware of local protocols.

However, the process of releasing control of the scene back to the owner or other responsible party is sometimes not as straightforward as it might seem. The property owner should be escorted on a tour of the scene, or as close to it as possible consistent with safety, and should be given an explanation of any remaining hazards. If the scene is still too hazardous to leave unattended, the owner may be required to post a security guard or erect a security fence around the hazard, or both. The department may require that a written release, describing the hazards and stipulating the conditions the owner must meet, be signed by the owner before the scene is released.

CRITICAL INCIDENT STRESS DEBRIEFING (CISD)

Because the injuries suffered by the victims can sometimes be extremely gruesome and horrific, the members of the rescue teams and any others who had to deal directly with badly injured victims should be *required* to participate in a critical incident stress debriefing (CISD) process. Because individuals react to and deal with extreme stress in different ways — some more successful than others — and because the effects of unresolved stresses tend to accumulate, participation in this type of process should not be optional.

The process should actually start *before* rescuers enter the scene if it is known that conditions exist there that are likely to produce psychological or emotional stress for the rescuers involved. This is done through a prebriefing process wherein the rescuers who are about to enter the scene are told what to expect so that they can prepare themselves (Figure 4.135).

If rescuers will be required to work more than one shift in these conditions, they should go through a minor debriefing, sometimes called "defusing," at the end of each shift. They should also participate in the full debriefing process within 72 hours of completing their work on the incident.

Figure 4.135 A rescue team is briefed before starting its assignment. *Courtesy of Laura Mauri.*

Chapter 4 Review

Directions

The following activities are designed to help you comprehend and apply the information in Chapter 4 of **Fire Service Rescue**, Sixth Edition. To receive the maximum learning experience from these activities, it is recommended that you use the following procedure:

1. Read the chapter, underlining or highlighting important terms, topics, and subject matter. Study the photographs and illustrations, and read the captions with each.
2. Review the list of vocabulary words to ensure that you know the chapter-related meaning of each. If you are unsure of the meaning of a vocabulary word, look up the word in the glossary or a dictionary, and then study its context in the chapter.
3. On a separate sheet of paper, complete all assigned or selected application and review activities before checking your answers.
4. After you have finished, check your answers against those on the pages referenced in parentheses.
5. Correct any incorrect answers, and review material that was answered incorrectly.

Vocabulary

Be sure that you know the chapter-related meanings of the following words.

- accessory cord *(92)*
- ascending *(102)*
- belay *(104)*
- carabiner *(100)*
- cornice *(105)*
- descending *(103)*
- dressed *(106)*
- hardware *(100)*
- imminent *(89)*
- impact loaded *(91)*
- inverting *(97, 98)*
- kernmantle *(91)*
- nominal *(93)*
- packaging *(122)*
- parapet *(105)*
- peripheral *(90)*
- pickets *(109)*
- pivot *(105)*
- pliability *(96)*
- proximity *(89)*
- rappel *(103)*
- redundancy *(109)*
- rescue tripod *(125)*
- serrated *(102)*
- shuttle-loom *(96)*
- software *(90)*
- spiral weave *(96)*
- stabilizing *(122)*
- synthetic fiber *(91)*
- trauma kit *(125)*
- versatile *(105)*

Application Of Knowledge

1. Choose from the following rope rescue procedures those appropriate to your department and equipment, or ask your training officer to choose appropriate procedures. Practice tying each of the knots and attaching rope rescue hardware. Mentally rehearse more complex procedures, or practice these procedures under your training officer's supervision.
   - Place a Class I harness on a victim. *(97)*
   - Place a Class II harness on a victim. *(97)*
   - Place a Class III harness on a victim. *(97)*
   - Construct a Rescue Knot improvised harness and place on a victim. *(97, 98)*
   - Construct a Swiss Seat improvised rescue harness and place on a victim. *(98, 99)*
   - Construct a Swiss Seat-with-shoulder-harness improvised rescue harness and place on a victim. *(99, 100)*
   - Attach anchor, ascending, and descending devices to rope. *(102-104)*
   - Rig single and double pulleys and an edge roller. *(104, 105)*
   - Tie a Becket Bend. *(106)*
   - Tie a Double Fisherman. *(106)*
   - Tie a Figure-Eight Follow Through. *(106)*
   - Tie a Water Knot. *(106)*
   - Tie a Clove Hitch. *(106)*
   - Tie a Girth Hitch. *(106, 107)*
   - Tie a Prussik Hitch. *(107, 108)*
   - Tie a Figure-Eight Follow Through loop. *(108)*
   - Tie a Figure-Eight on a Bight. *(108)*
   - Tie a Double-Loop Figure-Eight. *(108)*
   - Tie a Bowline. *(108)*
   - Tie an Overhand Loop. *(108)*
   - Rig simple and compound mechanical advantage systems. *(112-114)*
   - Given a scenario, perform Phase I arrival assessments. *(114-116)*
   - Given a scenario, perform Phase II prerescue operations. *(116-120)*
   - Stabilize a victim. *(122)*

- Package a victim in a basket litter (to include preparing the litter, lashing in the victim internally and externally, and constructing a litter sling). *(122-124)*
- Raise/lower a packaged victim (low-angle rescue). *(124, 125)*
- Raise/lower a packaged victim (high-angle rescue). *(125)*
- Raise a packaged victim (below-grade rescue). *(125)*
- Perform Phase IV termination procedures. *(126-128)*
- Inspect, clean, and store rope, recording procedure in rope log. *(93-96)*

2. Tour your department's rescue unit to determine the types of rope and webbing available, storage locations and methods, rope log format, and the system your department uses to identify the status of its rescue rope and webbing. *(Local protocol)*

Review Activities

1. Identify the following abbreviations and acronyms:
   - OSHA *(89)*
   - CFR *(89)*
   - RP *(115)*
   - CISD *(127)*
   - NFPA 1983 *(89)*
   - PPE *(89)*
2. Explain the differences between the following pairs:
   - Life safety rope vs. utility rope *(91)*
   - Dynamic rope vs. static rope *(92)*
   - Minimum breaking strength vs. maximum working load *(91, 92)*
   - Flat webbing vs. tubular webbing *(96)*
   - Spiral weave construction vs. shuttle-loom tubular nylon webbing construction *(96)*
   - Mechanical advantage vs. hauling system *(112, 113)*
   - Simple mechanical advantage system vs. compound mechanical advantage system *(113, 114)*
   - Primary assessment vs. secondary assessment *(115, 116)*
   - Minimum personnel accountability system vs. maximum personnel accountability system *(121)*
3. List the three reasons for not using a piece of life safety rope and removing it from service. *(91)*
4. List typical lengths, sizes, major characteristics, minimum breaking strength, maximum working load, and typical uses for kernmantle rope used in rescue operations. *(91, 92)*
5. List typical widths, breaking strengths, and lengths of webbing used in rescue operations. *(96)*
6. List the recommended color-coding system for rescue webbing. *(96)*
7. Compare and contrast Class I, Class II, and Class III harness. *(97)*
8. Identify the following load-bearing rope hardware:
   - Carabiner *(100, 101)*
   - Tri-link *(101)*
   - Rescue ring *(101)*
   - Swivel *(101)*
   - Anchor plate *(102)*
9. State the safety rule of thumb for a carabiner that has been dropped from at least waist height onto a hard surface. *(100)*
10. Compare and contrast the functions and uses of cams (Gibbs ascenders) and handled ascenders. *(102, 103)*
11. Explain the differences between free-running and spring-loaded cams. *(103)*
12. Identify the following descending devices:
    - Figure 8 plates (rescue 8s) *(103, 104)*
    - Rappel racks (brake bar racks) *(104)*
    - Belay plates *(104, 105)*
13. Explain the general uses of pulleys in rope rescue operations. *(104, 105)*
14. Compare the rope rescue uses for single-sheave pulleys, double-sheave pulleys, and edge rollers. *(105)*
15. Define the term *bombproof. (109)*
16. Explain the rescue uses, advantages, and disadvantages of the following group of knots:
    - Pulling Prussik *(92, 93)*
    - Tandem Prussik *(93)*
    - Ratchet Prussik *(93)*
17. Explain the rescue uses, advantages, and disadvantages of the following group of knots:
    - Rescue Knot *(97, 98)*
    - Swiss Seat *(98, 99)*
    - Swiss Seat with shoulder harness *(99, 100)*
18. Explain the rescue uses, advantages, and disadvantages of the following group of knots:
    - Becket Bend (Sheet Bend) *(106)*
    - Double Fisherman (Double Overhand) *(106)*
    - Figure-Eight Follow Through *(106)*
    - Water Knot *(106)*
    - Clove Hitch *(106)*
    - Girth Hitch *(106, 107)*
    - Prussik Hitch *(107, 108)*

19. Explain the rescue uses, advantages, and disadvantages of the following group of knots:
    - Figure-Eight Follow Through Loop *(108)*
    - Figure-Eight on a Bight *(108)*
    - Double-Loop Figure-Eight *(108)*
    - Bowline *(108)*
    - Overhand Loop *(108)*
20. Identify the following single-point anchors:
    - Tensionless Anchor *(110)*
    - Two-bight Anchor *(110)*
    - Multiwrap Anchor *(110)*
21. Compare and contrast the two-point and three-point self-adjusting load-sharing anchor systems. *(111)*
22. List the rules for calculating mechanical advantage in a simple mechanical advantage system. *(113)*
23. Describe 2:1, 3:1, and 4:1 simple mechanical advantage systems. *(113)*
24. Describe 4:1, 6:1, and 9:1 compound mechanical advantage systems. *(114)*
25. List in order the tactical considerations involved in any rope rescue. *(114)*
26. List questions that should be asked during the primary assessment. *(115)*
27. List the major parts of a secondary assessment. *(116)*
28. Discuss the importance of incident action plans. *(116, 117)*
29. Explain how the acronym OATH is used to communicate via lifeline. *(118)*
30. List those areas that the IC should consider when gathering needed resources (equipment and personnel) for a rescue operation. *(117)*
31. List those areas that the IC should consider when preparing the scene for a rescue operation. *(117, 118)*
32. Discuss the pros and cons of hard-wired phones versus portable radios versus lifeline communication systems. *(119)*
33. Identify the roles and responsibilities of the following rescue personnel:
    - Team leader *(119)*
    - Rescue team *(119, 120)*
    - Brake tender *(120)*
    - Backup team *(120)*
    - Hauling team *(120)*
34. Explain the guidelines for identifying/retrieving equipment used in the rescue operation. *(126, 127)*
35. Explain when it is better to abandon rather than retrieve a piece of equipment. *(127)*
36. Describe the variables for releasing control of the scene back to the owner or other responsible party. *(127)*
37. Discuss critical incident stress, prebriefings, and "defusings." Why should participation in a full critical incident stress debriefing within 72 hours of the incident be mandatory for all involved rescuers? *(127, 128)*

Questions And Answers

Fireground Search And Rescue

Chapter 5
Fireground Search And Rescue

INTRODUCTION

Fire departments were originally organized to protect life and property from fire. However, the mission of most fire departments has been expanded to include rescuing people from a wide range of hazardous environments. These include confined spaces, collapsed structures, bodies of water, very high or very low places, and many other dangerous situations. But, the vast majority of search and rescue operations conducted by firefighters are on the fireground. Even though thousands of people die in fires each year in the United States and Canada, many more are successfully rescued by firefighters. Fireground search and rescue often includes all of the hazards presented by the other types of rescues combined in one operation. To be successful in fireground search and rescue, rescue units need to have the following:

- Sufficient manpower
- Proper equipment
- Information about the situation
- A logical rescue plan coordinated with correct suppression techniques
- The training, courage, and determination to carry out the rescue plan

This chapter covers fireground search and rescue techniques in buildings, as well as search and rescue safety and the use of personal alert safety system (PASS) devices. Also covered are incident management, personnel accountability, rescuing trapped firefighters, various means of short-distance transfer of injured persons, and rescues from upper floors.

BUILDING SEARCH

Regardless of how small a structure fire may look upon arrival, the fire department must always do a thorough search of the building. Even in relatively minor fires, there may be occupants in the building who are incapable of exiting on their own for any number of reasons. Not locating a victim until after a "minor" fire is extinguished or, worse yet, missing a victim entirely is unacceptable.

Building Size-Up

While size-up is initially the responsibility of the first-arriving officer, all members of the rescue team should look at the entire building and its surroundings as they approach it. Careful observation will give them some indication as to the size of the fire, whether or not the building is likely to be occupied, the probable structural integrity of the building, and some idea of the amount of time it will take to effectively search the structure. Their initial exterior size-up will help them maintain their orientation within the building. They should identify their alternate escape routes (i.e., windows, doors, fire escapes) *before* they enter the building. Once inside, their specific location can sometimes be confirmed by looking out windows.

To obtain information about those who might still be inside and where they might be found, as well as to obtain information about the location and extent of the fire, firefighters should first question occupants who have escaped the fire (Figure 5.1). If possible, all information should be verified, but in any case, firefighters should not assume that all occupants are out until the building has been searched by fire department personnel. Because

Figure 5.1 A firefighter questions occupant about the fire.

Figure 5.2 A firefighter checks in with the Accountability Officer.

neighbors may be familiar with occupants' habits and room locations, they may be able to suggest where occupants are likely to be found, or a victim may have been seen near a window prior to the fire department's arrival. Information on the number and location of victims should be relayed to the incident commander (IC) and all incoming units.

Search and rescue personnel should always use whatever personnel accountability system the department has adopted, and in no case should they enter a building without first informing the responsible officer outside the structure (Figure 5.2). This practice will assist in tracking all on-scene personnel and will provide a means of alerting those in charge if contact is lost with anyone who has entered the building.

Conducting A Search

There are two objectives of a building search: finding victims (searching for life) and obtaining information about the extent of the fire (searching for fire extension).

In most structure fires, the search for life will require that two types of searches be done: a primary search and a secondary search. A *primary search* is a rapid but thorough search that is performed either before or during fire suppression operations. It is often done under extremely adverse conditions, but it must be done expeditiously. During the primary search, members must be sure to search the known or likely locations of victims as rapidly as conditions allow, moving quickly to search all affected areas of the structure as soon as possible. The search team(s) can verify that the fire conditions are as they appeared from the outside or report any surprises they may encounter. A *secondary search* is conducted *after* the fire is under control and the hazards are somewhat abated, and it should be conducted by personnel other than those who conducted the primary search. It is a very thorough, painstaking search that attempts to ensure that any remaining occupants have been found.

PRIMARY SEARCH

During the primary search, rescuers should always use the buddy system — working in teams of two or more. By working together, two rescuers can conduct a search quickly while maintaining their own safety.

Primary search personnel should always carry forcible entry tools with them whenever they enter a building and throughout the search (Figure 5.3). Valuable time will be lost if rescuers have to return to their apparatus to obtain this equipment. Also, tools used to force entry may be needed to force a way out of the building if rescuers become trapped.

Depending on conditions within the fire building, rescuers may be able to search while walking in an upright position, or they may have to crawl on their hands and knees (Figure 5.4). If there is only light smoke and little or no heat, walking is the most rapid means of searching a building. Searching on hands and knees can increase visibility (beneath the smoke) and will reduce the chances of tripping or falling into stairways or holes in floors. Movement in this position is much slower than when walking, but it is usually noticeably cooler near the floor.

When searching within a structure, rescuers should move systematically from room to room, searching each room completely, while constantly listening for sounds from victims. On the fire floor, firefighters should start their search as close to the fire as possible and then search back toward the entrance door. This procedure allows the search team to reach those in the most danger first — those who would be overtaken by any fire extension that might occur while the rest of the search was in progress. Since those who are a greater distance from the fire are in less immediate danger, they can wait to be reached as the team moves back toward safety.

Figure 5.4 Search teams may have to proceed on all fours.

It is very important for rescuers to search all areas such as bathrooms, bathtubs, shower stalls, closets, under beds, behind furniture, attics, basements, and any other areas where children may hide and where either infirm or disoriented victims may be found (Figure 5.5). Rescuers should search the perimeter of each room, and they should extend their arms or legs or use the handle of a tool to reach completely under beds and other furniture (Figure 5.6). When the perimeter has been searched, they should then search the middle of the room.

Figure 5.3 Search/rescue personnel should always carry forcible entry tools with them.

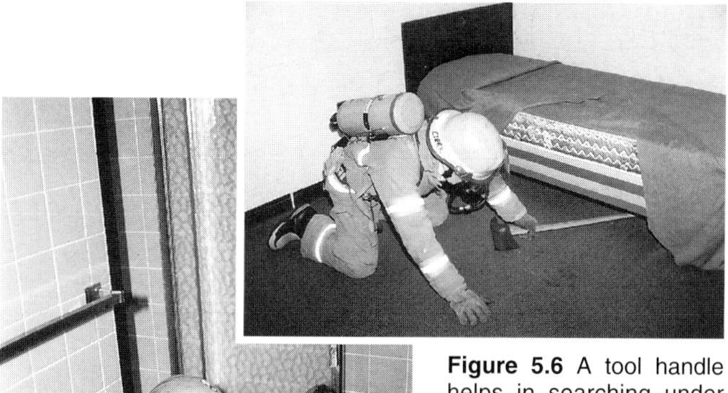

Figure 5.6 A tool handle helps in searching under furniture.

Figure 5.5 Every area must be searched.

During the primary search, visibility may be extremely limited, so rescuers may have to identify objects by touch — touch may provide the only clue to what type of room the team is in. Visibility being obscured by smoke should be reported through channels to the IC because it may indicate a need for additional ventilation.

Rescue teams should maintain radio contact with their supervisor and periodically report their progress and their needs in accordance with departmental general operating guidelines (Figure 5.7). Informing the IC of any areas that have not been completely searched is especially important so that additional search teams can be assigned to these areas if necessary.

During the primary search, negative information is just as important as positive information to ensure a complete search. If the search has to be aborted for any reason, the officer in charge should be notified immediately and the search resumed as soon as possible.

Figure 5.7 Search team leader reports progress.

VENT, ENTER, SEARCH (VES)

In certain situations, some departments use a technique they call Vent, Enter, Search (VES) to supplement the required primary interior search. VES is used where the fire has cut off the normal means of egress from some part of the building and where credible reports indicate that one or more victims are trapped in a room that is cut off by the fire but that can be accessed from the outside. Since victims trapped in areas filled with smoke and superheated gases cannot survive for very long, the decision to use VES may be justified.

VES involves forcing entry from the exterior into rooms where trapped victims are most likely to be found, such as bedrooms. This most often involves breaking a bedroom window (venting) and entering through the window to search the room (Figure 5.8). To avoid drawing fire to the room because of the open window, the firefighters should locate and close the bedroom door as soon as possible. As with any other ventilation technique, VES should be coordinated with the attack and primary search efforts.

CAUTION: When VES is used, positive-pressure ventilation (PPV) must be delayed until the fire is knocked down or all victims have been removed from the building.

Firefighters performing VES must be aware of the possibility of initiating a backdraft in the room when they break the glass, so they must be alert for the usual signs of backdraft conditions. Even when backdraft conditions are not present, a room filled with superheated smoke can still flashover, so firefighters must evaluate the conditions before deciding to attempt entry. If conditions within the room make entry for a search too dangerous, firefighters should still probe inside the window with a tool handle as victims are often found near windows (Figure 5.9). If a victim is found, it may justify attempting to quickly enter through the window opening far enough to pull the victim out.

SECONDARY SEARCH

After the initial fire suppression and ventilation operations have been completed, personnel other than those who conducted the primary search are assigned to conduct a secondary search of the

Figure 5.8 Firefighters prepare to initiate VES through a bedroom window.

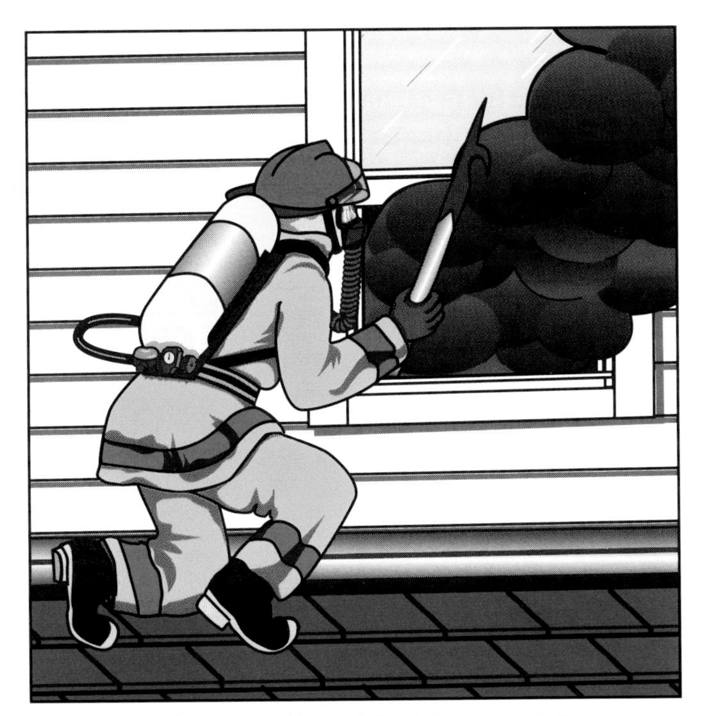

Figure 5.9 Firefighters should search for victims near windows.

fire building. During the secondary search, speed is not as critical as thoroughness (Figure 5.10). The secondary search is conducted just as systematically as the primary search to ensure that no rooms or spaces are missed. As in the primary search, any negative information, such as the fire beginning to rekindle in some area, is reported immediately.

MULTISTORY BUILDINGS

When searching in multistory buildings, the most critical areas are the fire floor, the floor directly above the fire, and the topmost floor (Figure 5.11). These floors should be searched immediately because this is where any remaining occupants will be in the greatest jeopardy because of rising smoke, heat, and fire. The majority of victims are likely to be found in these areas. Once these floors have been searched, the intervening floors should be checked also.

Figure 5.10 Firefighters must conduct a thorough secondary search.

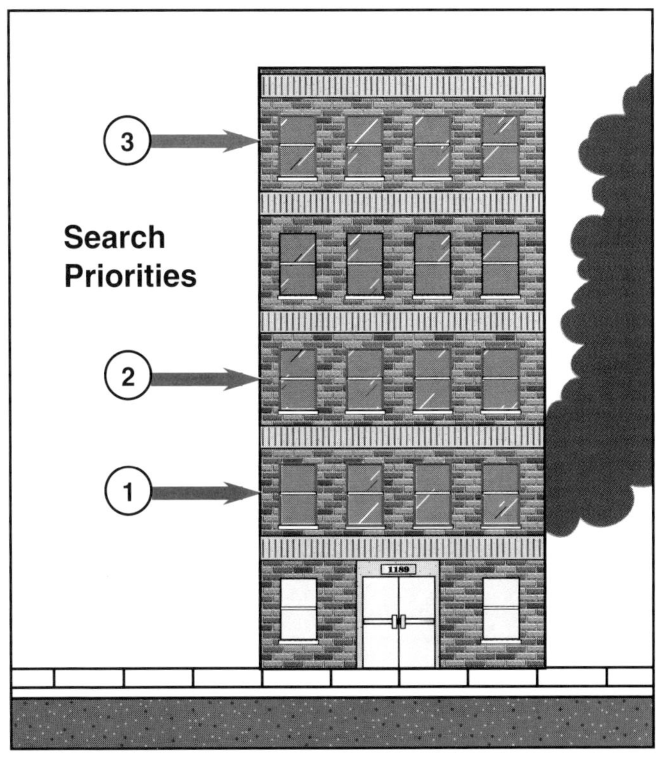

Figure 5.11 These areas have the highest search priority in multistory buildings.

During the primary search, doors to rooms not involved in fire should be closed to prevent the spread of fire into these rooms. Exits, hallways, and stairs from fire buildings should be kept as clear as possible of unused hoselines and other equipment to facilitate the egress of occupants and to reduce the trip hazard (Figure 5.12).

While still the source of much debate within the fire service, some departments insist that search and rescue personnel have a charged hoseline with them on all floors. Because advancing a charged hoseline during a search is a time-consuming process that may unnecessarily delay and impede the primary search, other departments make this an option based on conditions. Firefighters must be guided by their department's policy.

Figure 5.12 Exit stairways should be kept clear of trip hazards.

SEARCH METHODS

When rooms, offices, or apartments extend from a center hallway, teams should be assigned to search both sides of the hallway, but each team should search only the rooms on one side of the hallway. If there is only one search team, they should search down one side of the hallway and back up the other side. If two teams are available, each can take one side of the hallway.

Entering the first room, the searchers turn right or left and follow the walls around the room until they return to the starting point (Figure 5.13). As rescuers leave the room, they turn in the same direction they used to enter the room and continue to the next room to be searched. For example, if they turned left when they entered the room, they turn left when they leave the room. When removing a victim, however, rescuers must turn opposite the direction used to enter the room. It is important that rescuers exit each room through the same doorway they entered to ensure a complete search. This technique may be used to search most buildings, from a one-story single-family residence to a large high-rise building.

In most cases, the best method of searching small rooms is for one member to stay at the door while another member searches the room. The searcher remains oriented by maintaining a more-or-less constant dialogue with the member at the door. The searcher keeps the member at the door informed of the progress of the search. When the room search is completed, the team rejoins at the doorway, closes and marks the door (see following section), and proceeds to the next room. When searching the next room, the partners exchange their roles of searching the room and waiting at the door.

This method reduces the likelihood of team members becoming lost within the room, which reduces some of the stress of the situation. When searching relatively small rooms, this technique is often quicker than when both members search together because the searcher can move along more quickly without the fear of becoming disoriented.

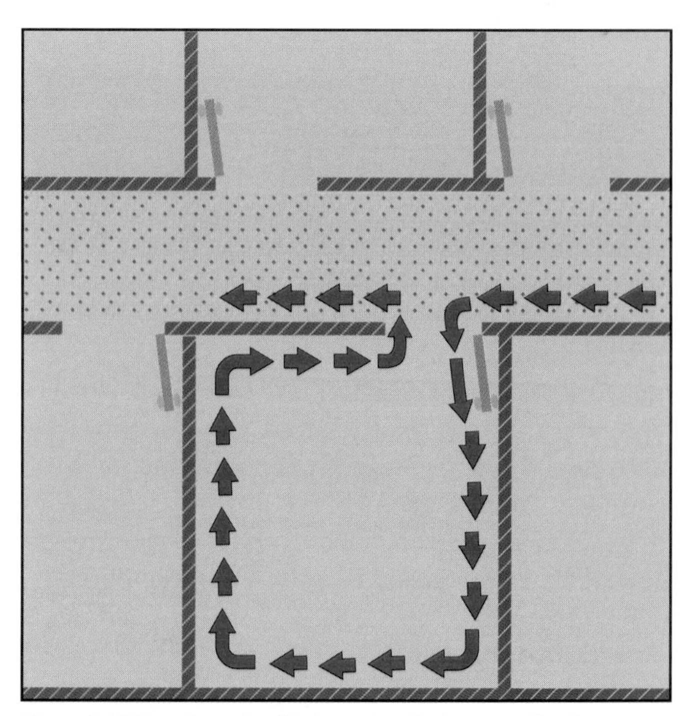

Figure 5.13 Searchers should always turn in the same direction.

MARKING SYSTEMS

Rooms that have been searched should be marked to avoid duplication of effort. It is a good idea for search teams to use a two-part marking system. The rescue team affixes half of the mark when entering the room and completes the mark when exiting the room (Figure 5.14). This alerts other rescuers that a room is being or has been searched by another team. If the search team becomes lost, this mark will serve as a starting point for others to begin looking for them.

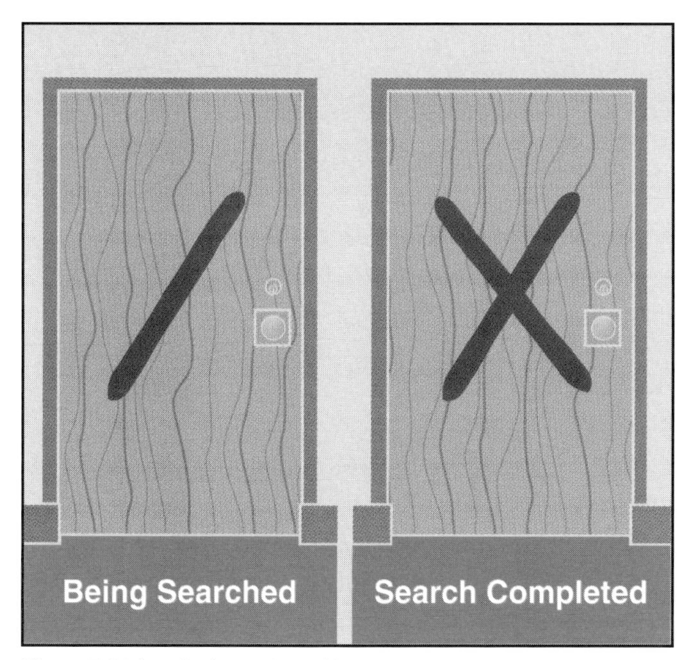

Figure 5.14 A typical search marking system.

Several methods of marking searched rooms are used by the fire service: chalk or crayon marks, masking tape, specially designed door markers, and latch straps over doorknobs (Figure 5.15). Latch straps also serve the secondary function of preventing a rescuer from being locked in a room. Methods that might contribute to fire spread by blocking doors open with furniture or those that require subsequent searchers to enter the room to find the marker, are not recommended. General operating guidelines usually dictate the method of marking; however, any method used must be known to and clearly understood by all personnel who may participate in the search.

Figure 5.15 This latch strap indicates that the room has been searched.

SAFETY

While searching for victims in a fire, rescuers must always consider their own safety. Fireground commanders also must consider the hazards to which rescuers may be exposed while performing search and rescue. A conscious decision should be made about whether the search teams need to take protective hoselines with them. Personnel must be properly trained and equipped with the necessary tools to accomplish a rescue in the least possible time. Typical search and rescue tools include such items as rope (to use as a lifeline or in rescuing victims), marking devices (to indicate which rooms have already been searched), and forcible entry tools (to aid in entry and egress and to enlarge the sweep area when searching) (Figure 5.16). Safety is the primary concern of rescuers because hurried, unsafe rescue attempts may have serious consequences for rescuers as well as victims.

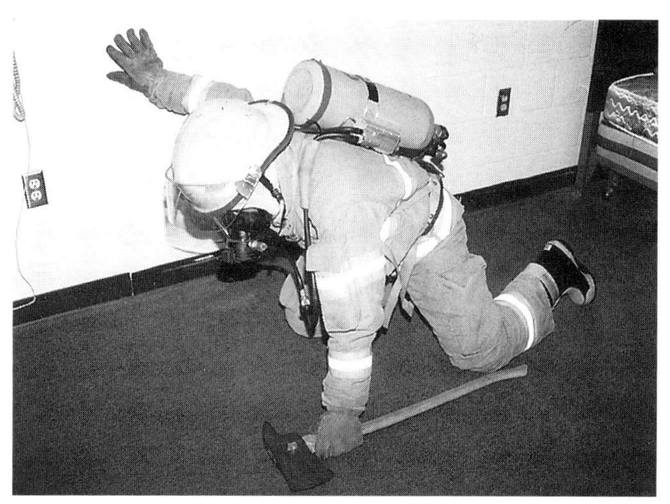

Figure 5.16 Searchers keep tools with them throughout the search.

Safety During Building Searches

Every time a firefighter or rescuer responds to a fire, a human life may be in jeopardy. In order to assess the degree to which someone may be threatened, a search is initiated as soon as possible. While rescuers must work quickly, they must also operate safely and with sound judgment if they are to fulfill their assignment and avoid becoming victims themselves.

As personnel search a multistory building, especially when visibility is limited because of smoke and/or darkness, they must always be alert for weakened or hazardous structural conditions, especially the floors. They should continually feel the floor in front of them with their hands or a tool to ensure that the floor is still intact (Figure 5.17). Otherwise, they may blindly crawl into an open elevator shaft, a stairway, an arsonist's trap, or a hole that may have burned through the floor. Personnel on or directly below the fire floor should also be alert for signs that the floor/ceiling assembly above them has weakened.

If searchers become disoriented and lose their direction during a search, they should stay calm. Following a wall will usually lead them to a door, possibly the one by which they entered or one leading into another room (Figure 5.18). Using a latch strap on the entry door will make it easier to identify the entry door in zero-visibility conditions. Once the entry door is found, turning in the direction opposite the one used to enter the room will lead the rescuers toward the exit. A door to another room may lead into an area of safe refuge or may show another way out.

Figure 5.17 A searcher checks the floor ahead.

Figure 5.18 Following a wall helps searchers remain oriented.

If searchers can locate a hoseline, they can crawl along it and feel the first set of couplings they come to. The female coupling will be toward the nozzle and the male toward the water source. The male coupling has lugs on its shank; the female does not (Figure 5.19). Following the hoseline will lead them either to an exit or to the nozzle team. If they find a window, they can signal for assistance by straddling the windowsill and turning on their PASS device (see Personal Alert Safety Systems section), by using their flashlight, by yelling and waving their arms, or by throwing objects out the window. However, under no circumstances should firefighters throw out their helmets or any other parts of their protective ensemble.

Some departments equip their firefighters with a personal escape bag containing a couple of carabiners, some webbing to use as a harness, and a length of kernmantle rope (Figure 5.20). This equipment gives them a better chance to save themselves if trapped by fire on upper floors of a building.

When searching within a fire building, personnel should be very cautious when opening doors. They should feel the top of the door and the doorknob to determine the heat level (Figure 5.21). If the door is excessively hot, it should not be opened until a charged hoseline is in position. Firefighters should not remain in front of the door while opening it. They should stay to one side, keep low, and slowly open the door. If there is fire behind the door, staying low will allow the heat and combustion products to pass over their heads.

NOTE: Some departments insist that their firefighters keep their gloves (and all other parts of the protective ensemble) on when in a burning building; others allow them to remove a glove to feel a door for heat. Firefighters should be guided by local protocols.

If an inward-opening door is difficult to open, firefighters should not kick the door to force it open because a victim may have collapsed just inside the door. Kicking the door may injure the victim further, and it is neither a safe nor a very professional way to force a door. The door should be slowly pushed open and the area behind it checked for possible victims.

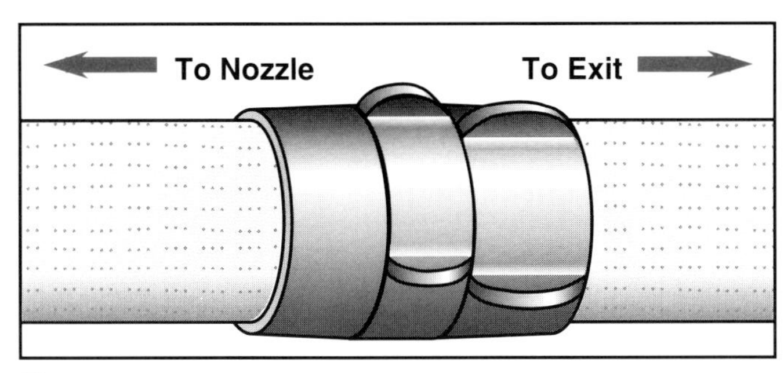

Figure 5.19 Hose couplings will indicate the direction toward the exit.

Figure 5.21 A firefighter checks a door for heat.

Figure 5.20 Some firefighters carry self-rescue bags.

Safety Guidelines

The following is a list of safety guidelines that should be used by search and rescue personnel in any type of search operation within a building:

- Do not enter a building in which the fire has progressed to the point where viable victims are not likely to be found.
- Attempt entry only *after* ventilation is accomplished when backdraft conditions exist.
- Work from a single operational plan. Crews should not be allowed to freelance.
- Maintain contact with Command, which has control over search/rescue teams.
- Constantly monitor fire conditions that might affect search teams and individual firefighters.
- Have a rapid intervention team constantly available to help firefighters or teams in need of assistance.
- Use the established personnel accountability system without exception.
- Be aware of the secondary means of egress established for personnel involved in the search.
- Wear full personal protective equipment, including SCBA and a PASS device.
- Work in teams of two or more and stay in constant contact with each other. Rescuers are responsible for themselves and each other.
- Search systematically to increase efficiency and to reduce the possibility of becoming disoriented.
- Stay low and move cautiously while searching.
- Stay alert — use all senses.
- Continually monitor the structure's integrity.
- Feel doors for excessive heat before opening them.
- Mark entry doors into rooms and remember the direction turned when entering the room. To exit the room and the building, turn in the opposite direction.
- Maintain contact with a wall when visibility is obscured. Working together, search team members can extend their reach by using ropes or straps.
- Have a charged hoseline at hand whenever possible when working on the fire floor (or the floor immediately below or above the fire) as it may be used as a guide for egress as well as for fire fighting.
- Coordinate with ventilation teams before opening windows to relieve heat and smoke during search.
- Close the door, report the condition, and be guided by the group/sector supervisor's orders if fire is encountered during a search.
- Immediately inform the group/sector supervisor of any room(s) that could not be searched, for whatever reason.

- Report promptly to the supervisor once the search is complete. Besides giving an "all clear," also report the progress of the fire and the condition of the building.

Personal Alert Safety Systems

Personal alert safety systems (PASS) can be of great assistance in locating downed rescuers, and they should be turned on anytime entry into a hazardous area is necessary (Figure 5.22). These devices produce a loud audible signal whenever body movement stops for more than 30 seconds. They can also be manually activated if the wearer is in need of assistance. NFPA 1982, *Standard on Personal Alert Safety Systems (PASS) for Fire Fighters*, covers the design and testing of these devices. To ensure their proper operation, the battery must be changed regularly, and they should be tested at the start of each work shift.

HIGH-RISE SEARCH AND RESCUE

Fire and rescue problems in any building above the reach of fire department aerial apparatus are a challenge to all fire departments. All of the usual difficulties with search, rescue, and extinguishment are compounded in high-rise incidents. The amount of time and energy needed to get personnel and equipment up to the level of the fire can severely tax any department's resources. Preincident planning can make high-rise fire and rescue incidents much easier to cope with.

Figure 5.22 PASS devices can save firefighters' lives.

The key to effective high-rise operations is having enough manpower and equipment soon enough to get ahead of the fire (Figure 5.23). Once the needed resources are on scene, the most critical functions are organization and coordination. To keep track of the resources that a high-rise fire will require and to use them to maximum effectiveness, an incident management system must be implemented with the arrival of the first unit and used throughout the incident. Personnel assigned to search and rescue must know and understand the management system being used in order to fulfill their responsibilities effectively. The necessary familiarity with the incident management system is achieved through adequate training, through frequent practice, and by routinely using the system on every incident.

Figure 5.23 High-rise operations require many resources.

Incident Command/Management

The first things that must be done at any high-rise structure fire are to establish a command post and to implement an incident command/management system. The command post should be located outside the fire building and at least 200 feet (60 m) away to be clear of falling glass and debris.

One of the earliest assignments in the high-rise management system is lobby control. The officer assigned to this critical function takes a position in the main elevator lobby near the elevators and the communication equipment (Figure 5.24). "Lobby" (ICS designator for the lobby control officer) controls and coordinates the use of the building's elevators and the movement of manpower and equipment up to the fire floor(s) from this location. Rescue teams passing through the lobby may be used to carry spare equipment to "Staging," usually located two floors below the fire floor, while on their way to their assigned area (Figure 5.25).

Rescue teams must be briefed on the operational plan for the incident so they can conduct their activities according to the plan (Figure 5.26). Following the operational plan prevents companies from freelancing. Freelancing tends to waste manpower through inefficient, uncoordinated activity and may compromise firefighter safety by placing personnel in untenable areas.

Figure 5.24 The lobby must be controlled as soon as possible.

Figure 5.25 Rescue personnel carry spare equipment to Staging.

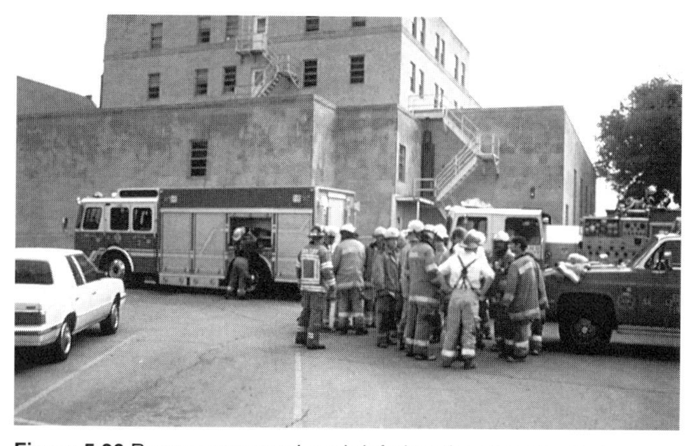

Figure 5.26 Rescue personnel are briefed on the operational plan.

Search Procedures

At a high-rise incident, it is recommended that at least two teams be assigned to the primary search on each fire floor, the floor immediately above the fire floor, and the topmost floor. Smoke may not always reach the topmost floor but may stratify at some level below (Figure 5.27). Any floor where smoke has accumulated should be searched as soon as possible. The teams on the topmost and other upper floors should work down, and those on lower floors should work up until they meet at some point above the fire floor. They should then report the results of their search: Transmit either an "all clear" for the floors that have been searched or advise the supervisor of the status of any rescues performed or in progress.

Search teams on the fire floor of a high-rise building must remain alert for fire traveling above their heads in the plenum or cockloft. This may allow fire to get behind them and cut off their escape route (Figure 5.28). Also, many high-rise offices are subdivided by movable wall panels into

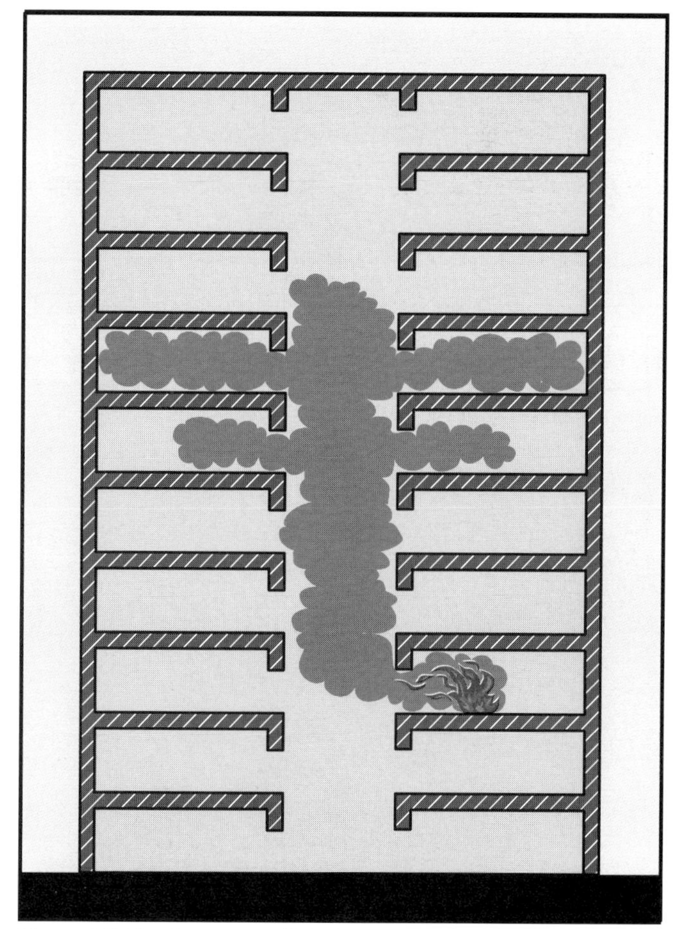

Figure 5.27 Smoke may stratify below the top of the building.

complex mazes of little cubicles. These walls, which are often about 6 feet (2 m) high, obstruct the search team and obscure their vision but allow fire, heat, and smoke to pass freely between the wall panels and the ceiling. When this potential exists, the search team should pay out a lifeline attached to a solid object in the stairwell so that they can quickly retrace their steps to safety if fire conditions suddenly deteriorate (Figure 5.29). Teams on the fire floor should be sure to check any restrooms or offices along the outer perimeter of the floor as occupants of these areas can be completely unaware of the fire until their escape route has been cut off.

First-arriving attack/search teams are usually assigned to the fire floor; those arriving next, to the floor above the fire and to the topmost floor (Figure 5.30). Others are assigned to the intervening floors as they arrive. These incidents require a large commitment of personnel, so close coordination of search/rescue teams is especially important. The search/rescue effort requires sufficient manpower to conduct the necessary searches in time to save lives. Large floor areas, maze-like conditions, numerous locked doors to be forced, and heavy smoke all slow these operations and require additional manpower. Four to six searchers per floor may be needed under these conditions. Under better conditions, one team might be expected to cover three to five floors. In addition, an adequate number of attack teams must be deployed to allow control of the stairways and to have a positive effect on the fire. A firefighter should be assigned to each door between stairwells and fire-involved areas to prevent occupants from opening them and allowing fire spread to upper floors.

Figure 5.28 Search teams may be endangered if fire gets behind them.

Figure 5.29 A search team pays out a lifeline.

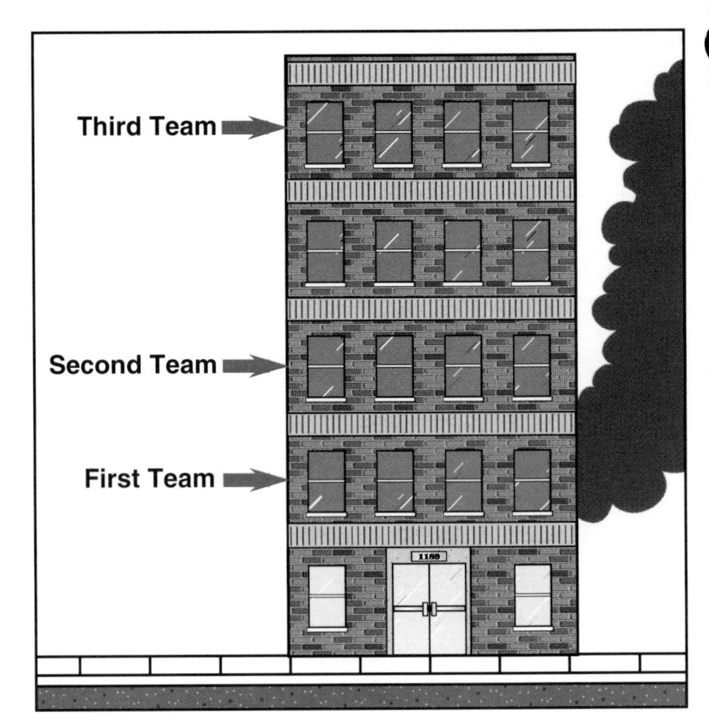

Figure 5.30 Typical search team assignment sequence.

In most high-rise buildings there are at least two stairways serving all floors. One of the primary duties of the search/rescue supervisor is to designate one as the evacuation stairway. This procedure precludes that stairway being used for fire suppression as the hoselines would block the doors open and allow the stairwell to fill with heat and smoke.

Search teams should be provided with master keys to the areas they are assigned to search or they should take hydraulic forcible entry tools if master keys are not available (Figure 5.31). Also, if available, they should review a copy of the floor plan of their assigned search area.

Figure 5.31 Search teams need master keys or forcible entry tools.

VICTIM REMOVAL

An ambulatory or semiambulatory victim may only require help to walk to safety — walking being probably the least laborious of all transportation methods. One or two rescuers may be needed, depending on how much help is available and the size and condition of the victim (Figure 5.32).

NOTE: The procedures described in this section include the steps for various carries for short-distance transfers. The specific steps involved in taking spine injury precautions and other preparation for transfer precautions are not covered. See an EMS first responder manual for these steps.

The victim is not moved before treatment is provided unless there is an immediate danger to the victim or to rescuers. Emergency moves are necessary under the following conditions:

- There is fire or danger of fire in the immediate area.
- Explosives or other hazardous materials are involved.
- It is impossible to protect the accident scene.
- It is impossible to gain access to other victims who need immediate life-saving care.
- The victim is in cardiac arrest and must be moved to a different area (to a firm surface, for instance) so that rescuers can administer cardiopulmonary resuscitation (CPR).

Figure 5.32 Occupants may only need to be escorted to safety.

The chief danger in moving a victim quickly is the possibility of aggravating a spinal injury. In an extreme emergency, however, the possible spinal injuries become secondary to the goal of preserving life.

If it is necessary to perform an emergency move, the victim should be pulled in the direction of the long axis of the body — not sideways. Jackknifing the victim should also be avoided. If the victim is on the floor, pull on his or her clothing in the neck or shoulder area (Figures 5.33). It may be easier to pull the victim onto a blanket and then drag the blanket. There are a number of carries and drags that may be used to move a victim from an area quickly; these are described later in the chapter.

It is always better to have two or more rescuers when attempting to lift or carry an adult. One rescuer can safely carry a small child, but two, three, or even four rescuers may be needed to safely

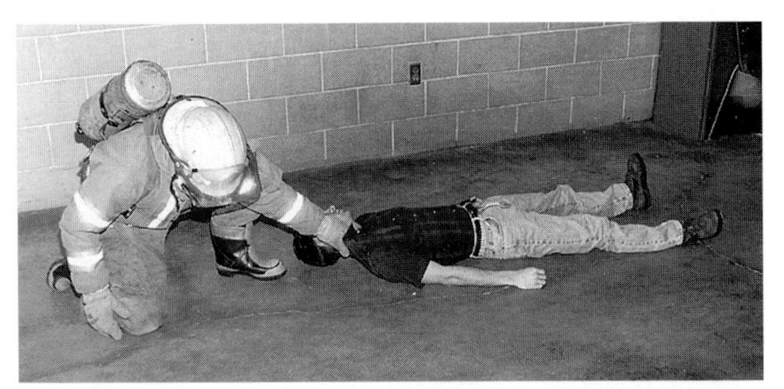

Figure 5.33 One way to move an unconscious victim in an emergency.

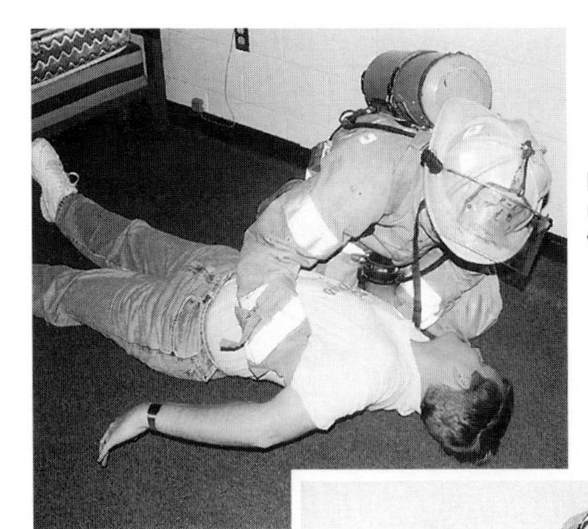

Figure 5.34 An unconscious adult is very difficult to lift.

lift and carry a large adult. An unconscious victim is always more difficult to lift; he or she is unable to assist in any way, and a relaxed body becomes "dead weight" (Figure 5.34).

Figure 5.35 Rescuers should keep their backs straight when lifting.

It is not easy for inexperienced people to lift and carry a victim correctly. Their efforts may be uncoordinated, and they usually need close supervision to avoid further injury to the victim. Rescuers helping to carry a victim should guard against losing their balance. They should lift as a team and with proper technique to avoid jostling the victim unnecessarily. Lifting incorrectly is also one of the most common causes of injury to rescuers. Rescuers should always remember to keep their backs straight and lift with their legs, not their backs (Figure 5.35).

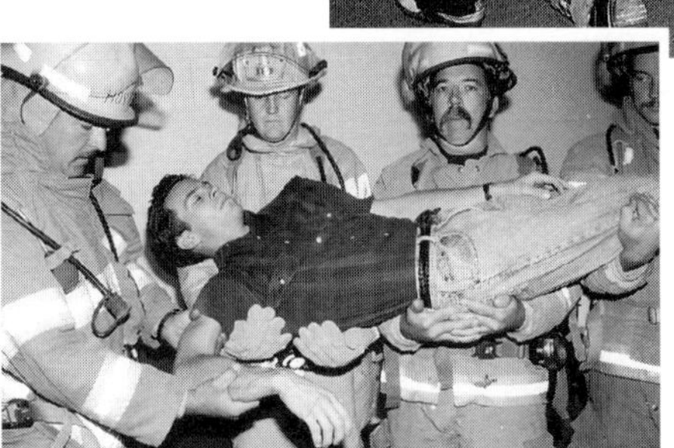

Figure 5.36 Close coordination between rescuers is necessary to avoid aggravating the injury.

If immobilization of a fracture is not feasible until the victim has been moved a short distance, one rescuer should support the weight of the injured part while others move the victim (Figure 5.36).

Cradle-In-Arms Lift/Carry

This lift/carry is effective for carrying children or very small adults if they are conscious. It is usually not practical for carrying an unconscious adult because of the weight and relaxed condition of the body.

Step 1: Place one arm under the victim's arms and across the back, and place the other arm under the victim's knees (Figure 5.37).

Step 2: Keeping the back straight, lift the victim to about waist height and carry the victim to safety (Figure 5.38).

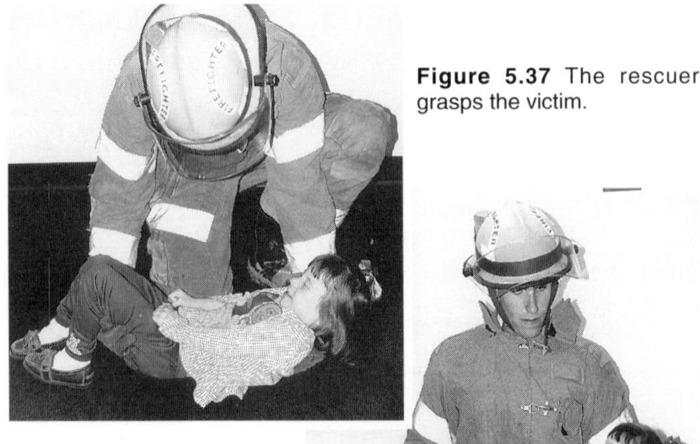

Figure 5.37 The rescuer grasps the victim.

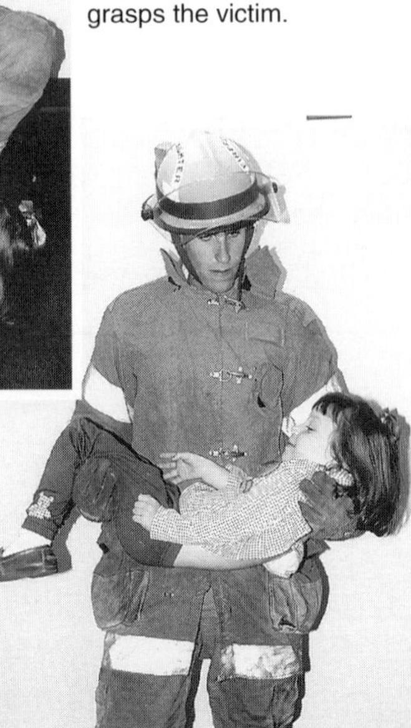

Figure 5.38 The rescuer lifts the victim to waist height.

Seat Lift/Carry

This lift/carry can be used with a conscious or an unconscious victim and is performed by two rescuers.

Step 1: Both rescuers raise the victim to a sitting position and link arms across the victim's back (Figure 5.39).

Step 2: Both rescuers then reach under the victim's knees to form a seat (Figure 5.40).

Step 3: Both rescuers then stand, lift the victim (using their legs), and move the victim to safety (Figure 5.41).

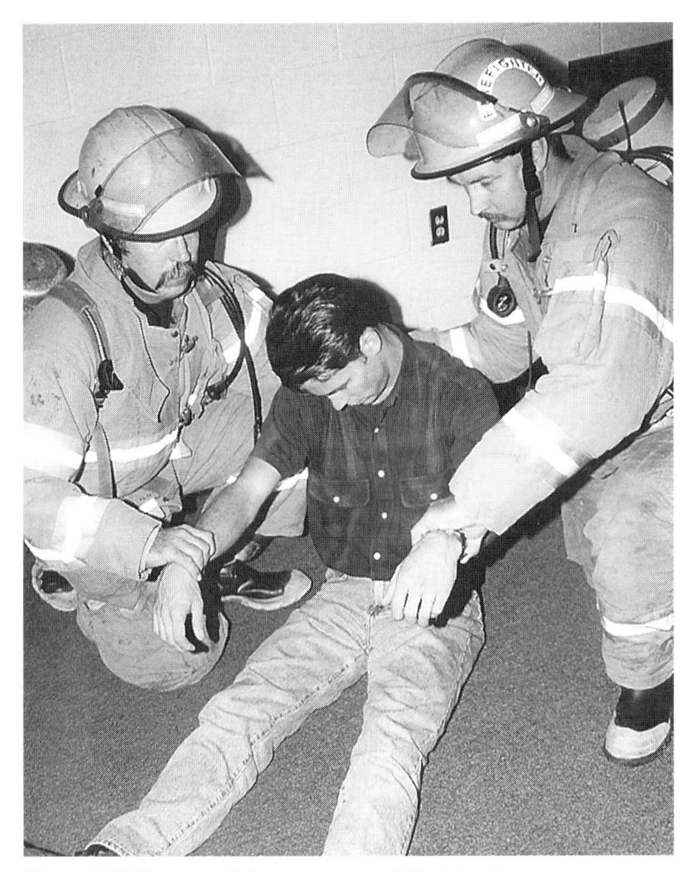

Figure 5.39 Rescuers link arms across victim's back.

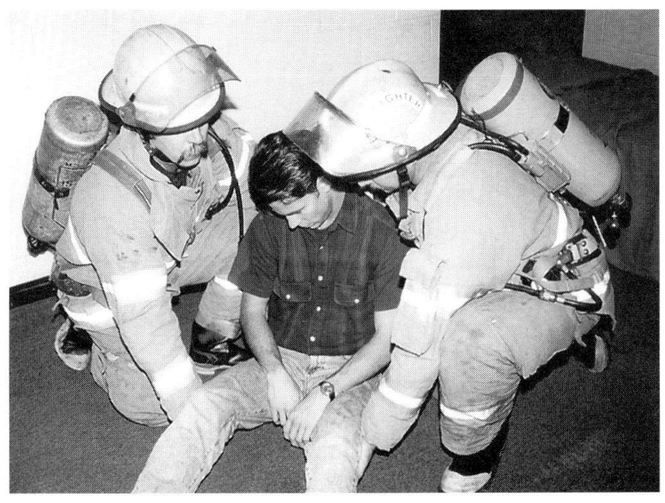

Figure 5.40 Rescuers pick up victim's knees.

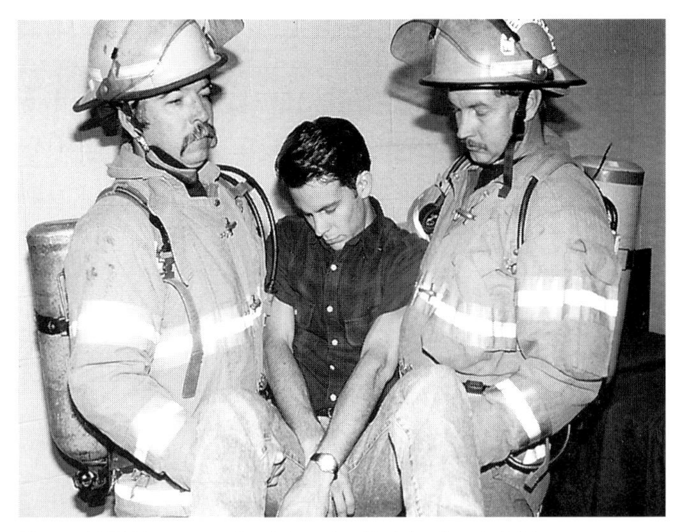

Figure 5.41 Rescuers stand, keeping their backs straight.

Two- Or Three-Person Lift/Carry

Many victims are more comfortable when left in a supine position, and this lift/carry is an effective way to lift a victim who is lying down. The two- or three-person lift/carry is often used for moving a victim from a bed to a gurney, especially when the victim is in cramped quarters. If the victim is small, two rescuers may be sufficient for the carry; if large, three rescuers may be needed.

Step 1: Position the gurney so that the victim can be carried to it and placed on it with the least amount of movement. This may require leaving the gurney in the fully up position (Figure 5.42).

Step 2: Rescuers position themselves on the side of the victim that is easiest to reach and/or that will facilitate placing the victim on the gurney. One rescuer takes a position at the head and upper torso of the victim, one at the waist and legs, and if necessary, one at the lower legs (Figure 5.43).

Step 3: Keeping their backs straight, rescuers crouch or kneel as close to the victim as possible. The rescuer at the head places one hand under the victim's head and the other hand and arm under the victim's upper back. At their respective positions, the other rescuers place their arms under the victim (Figure 5.44).

Step 4: On the signal from the rescuer at the head, all rescuers carefully roll the victim toward their chests (Figure 5.45).

Step 5: Again, on the signal from the rescuer at the head, all rescuers stand while holding the victim against their chests and carry the victim to the desired location (Figure 5.46).

Step 6: Reverse the above procedures, again on the signal of the rescuer at the victim's head, to place the victim on the gurney (Figure 5.47).

With a smaller victim, two rescuers can perform this lift. One rescuer supports the victim's head and upper back, and the other rescuer supports the victim's torso and legs.

Figure 5.45 Rescuers roll the victim to their chests.

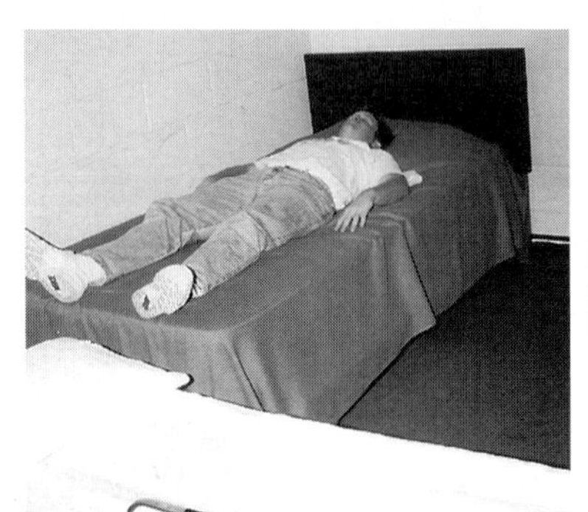

Figure 5.42 The gurney is placed close to the victim's location.

Figure 5.43 Rescuers position themselves next to the victim.

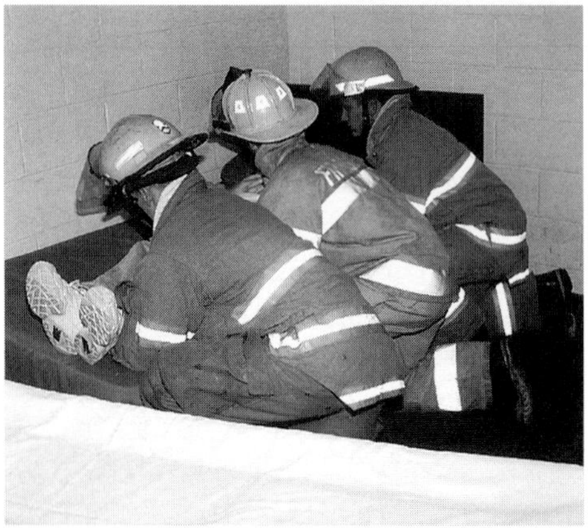

Figure 5.44 They slide their arms under the victim.

Figure 5.46 They stand in unison.

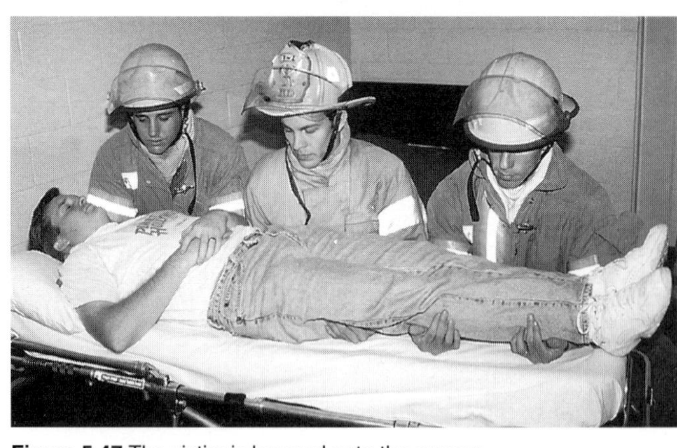

Figure 5.47 The victim is lowered onto the gurney.

Extremities Lift/Carry

The extremities lift/carry is used on either a conscious or an unconscious victim. This technique requires two rescuers and should be performed as follows:

Step 1: Turn the victim (if necessary) so that he or she is supine.

Step 2: One rescuer kneels at the head of the victim, and the second rescuer stands between the victim's knees (Figure 5.48).

Step 3: The rescuer at the head supports the victim's head and neck with one hand and places the other hand under the victim's shoulders, while the second rescuer grasps the victim's wrists (Figure 5.49).

Step 4: The rescuer holding the victim's wrists pulls the victim to a sitting position; the other rescuer assists by gently pushing on the victim's back. (Figure 5.50).

Step 5: The rescuer at the victim's head reaches under the victim's arms and grasps the victim's wrists as the other rescuer releases them. The rescuer grasps the victim's left wrist with the right hand and right wrist with the left hand (Figure 5.51).

Step 6: The rescuer between the victim's knees turns around, kneels down, and slips his or her hands under the victim's knees (Figure 5.52).

Step 7: On a command by the rescuer at the victim's head, both rescuers stand and move the victim (Figure 5.53).

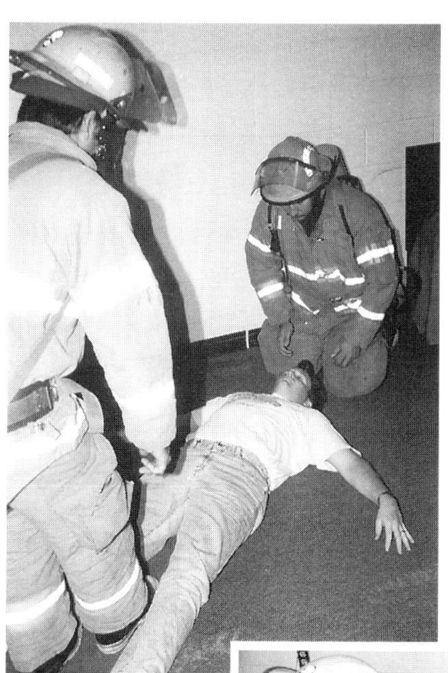

Figure 5.48 The rescuers take their positions.

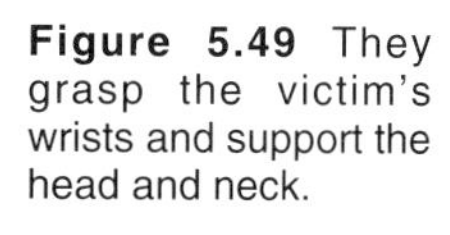

Figure 5.49 They grasp the victim's wrists and support the head and neck.

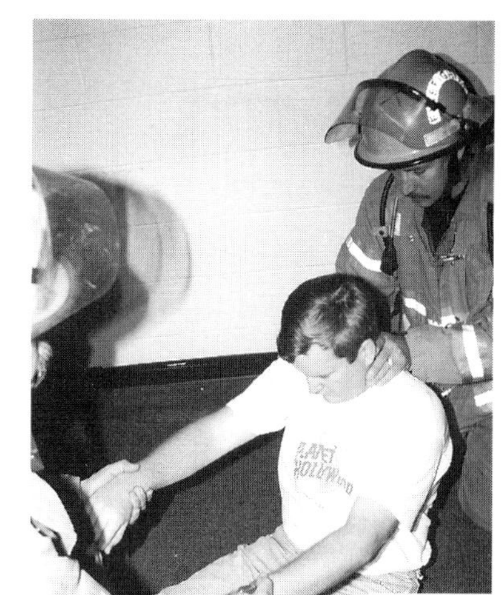

Figure 5.50 Working in unison, they push/pull the victim into a sitting position.

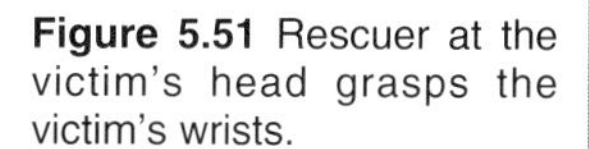

Figure 5.51 Rescuer at the victim's head grasps the victim's wrists.

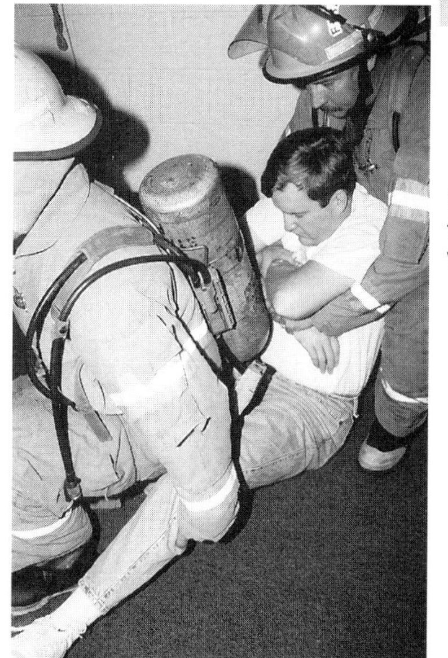

Figure 5.52 Other rescuer turns around and grasps the victim's knees.

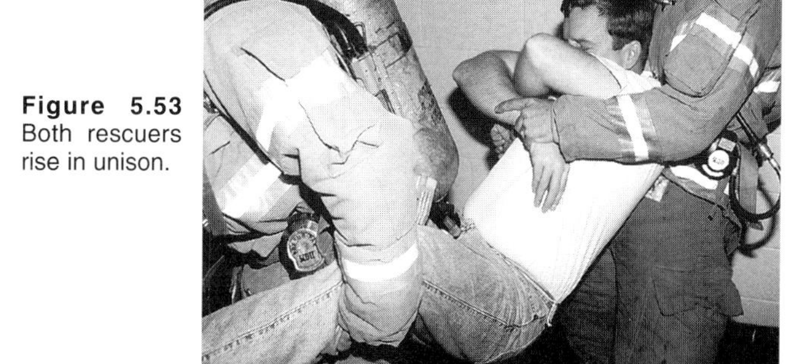

Figure 5.53 Both rescuers rise in unison.

Chair Lift/Carry

The chair lift/carry is used for either a conscious or an unconscious person. Be sure that the chair used is sturdy; do not attempt this carry using a folding chair. The two methods for executing the chair carry are as follows:

METHOD 1

Step 1: Turn the victim (if necessary) so that he or she is supine (Figure 5.54).

Step 2: One rescuer lifts the victim's knees until the knees, buttocks, and lower back are high enough, and the second rescuer slips a chair under the victim (Figure 5.55).

Step 3: Both rescuers raise the victim and chair to a 45-degree angle (Figure 5.56).

Step 4: Lifting the seated victim, one rescuer carries the legs of the chair, and the other carries the back of the chair (Figure 5.57).

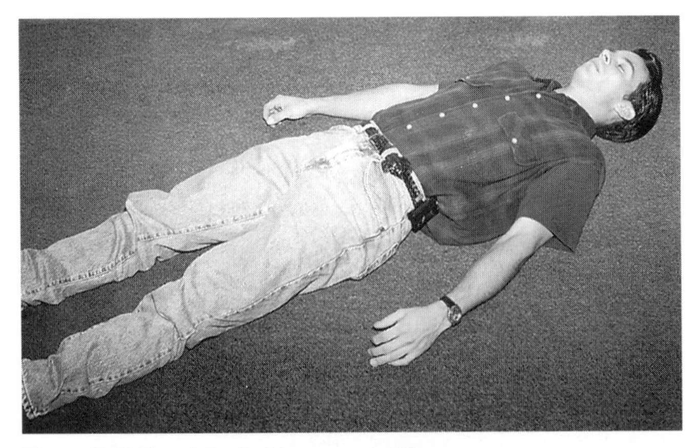

Figure 5.54 The victim in the proper position.

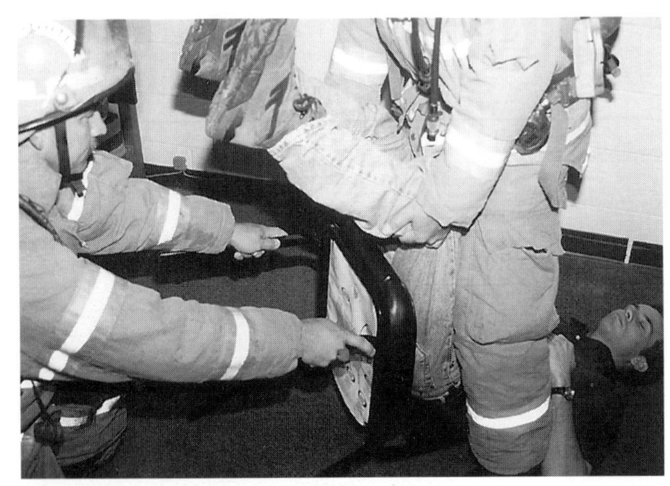

Figure 5.55 A chair is slipped under the victim.

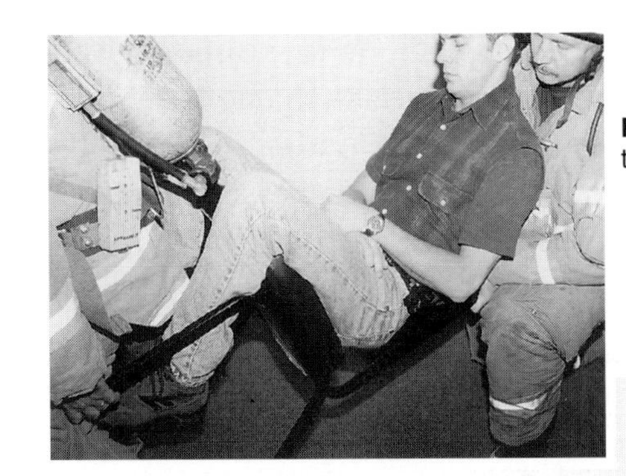

Figure 5.56 They raise the chair and the victim.

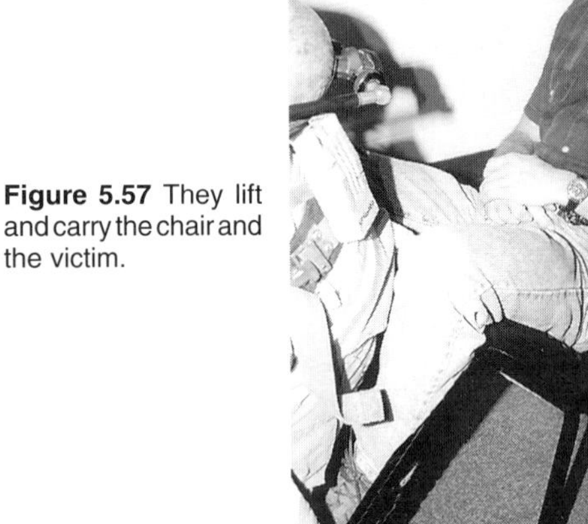

Figure 5.57 They lift and carry the chair and the victim.

METHOD 2

Step 1: Place the victim in a sitting position (Figure 5.58).

Step 2: One rescuer reaches under the victim's arms and grasps the victim's wrists. The other rescuer straddles the victim's lower legs and grasps the victim's legs under the knees (Figure 5.59).

Step 3: Both rescuers gently lift and place the victim onto the chair (Figure 5.60).

Step 4: Both rescuers raise the victim and chair to a 45-degree angle (Figure 5.61).

Step 5: Lifting the seated victim, one rescuer carries the legs of the chair, and the other carries the back of the chair (Figure 5.62).

Incline Drag

This drag is used to move a victim down a stairway or incline and is very useful for moving an unconscious victim.

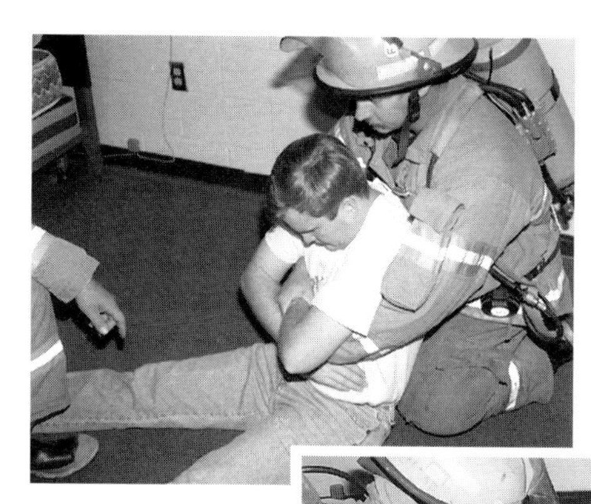

Figure 5.58 Rescuers place the victim in a sitting position.

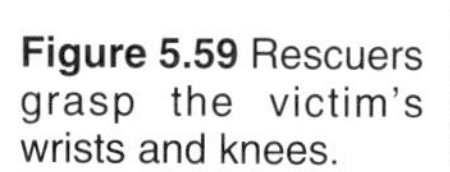

Figure 5.59 Rescuers grasp the victim's wrists and knees.

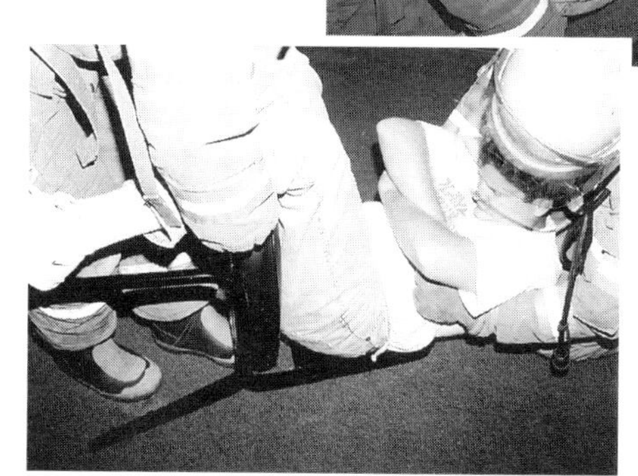

Figure 5.60 They lift the victim onto a chair.

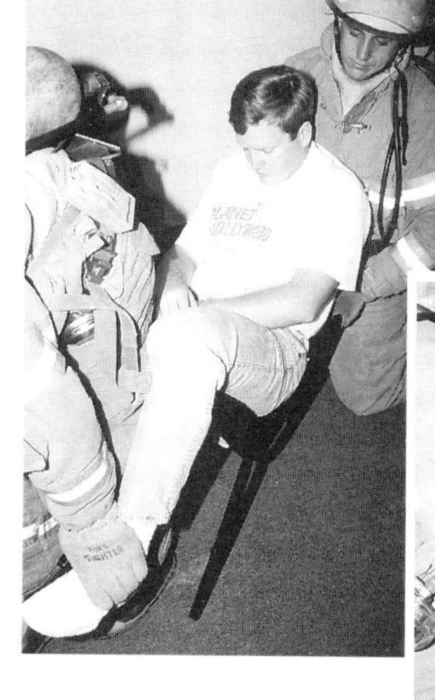

Figure 5.61 They tilt the chair.

Figure 5.62 They pick up the chair and the victim.

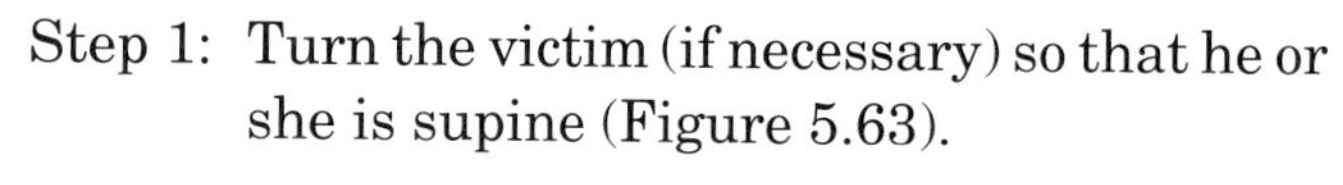

Step 1: Turn the victim (if necessary) so that he or she is supine (Figure 5.63).

Step 2: Kneel at victim's head (Figure 5.64).

Step 3: Supporting the victim's head and neck, lift the victim's upper body into a sitting position (Figure 5.65).

Step 4: Reach under the victim's arms and grasp the victim's wrists (Figure 5.66).

Step 5: Stand up (Figure 5.67). The victim can now be eased down a stairway or ramp to safety.

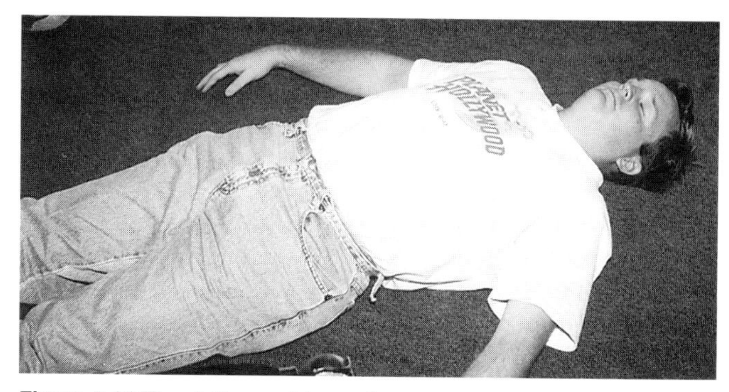

Figure 5.63 The victim must be supine.

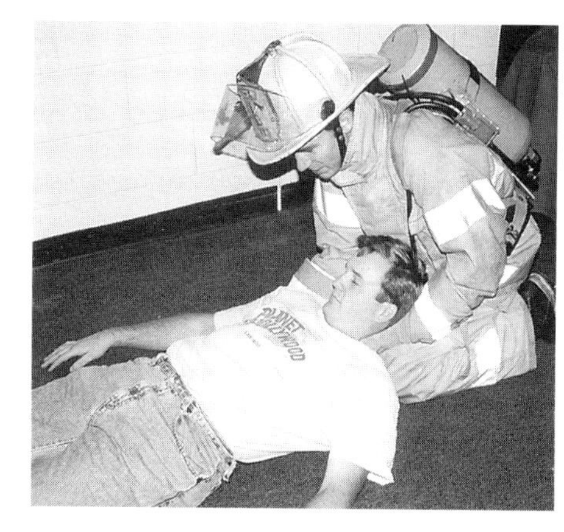

Figure 5.64 The rescuer kneels at the victim's head.

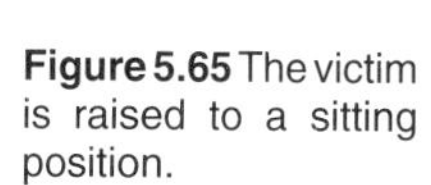

Figure 5.65 The victim is raised to a sitting position.

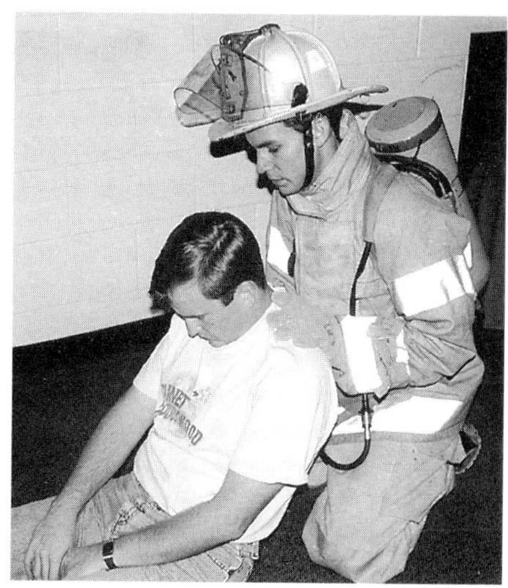

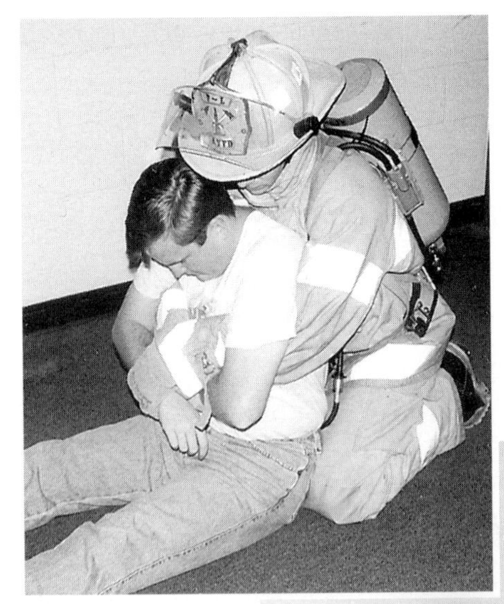

Figure 5.66 Rescuer grasps the victim's wrists.

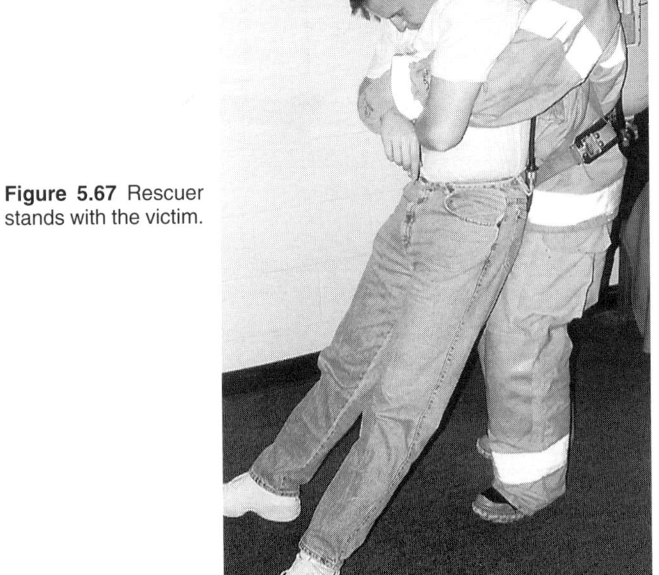

Figure 5.67 Rescuer stands with the victim.

Blanket Drag

This drag is implemented by one rescuer using a blanket, rug, or sheet. The steps are as follows:

Step 1: Spread a blanket next to the victim, making sure that it extends above the victim's head (Figure 5.68).

Step 2: Kneel on both knees at the victim's side opposite the blanket, and extend victim's arm above his or her head (Figure 5.69).

Step 3: Roll victim up against your knees (Figure 5.70).

Step 4: Pull the blanket against the victim, gathering it slightly against the victim's back (Figure 5.71).

Step 5: Gently allow victim to roll onto the blanket, and straighten the blanket out on both sides. Wrap the blanket around the victim and tuck the lower ends around the victim's feet (Figure 5.72).

Step 6: Pulling the end of the blanket at the victim's head, drag the victim to safety (Figure 5.73).

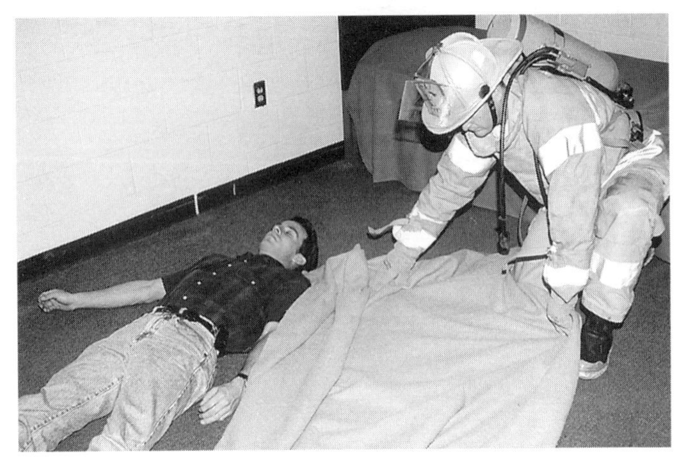

Figure 5.68 Rescuer spreads a blanket next to the victim.

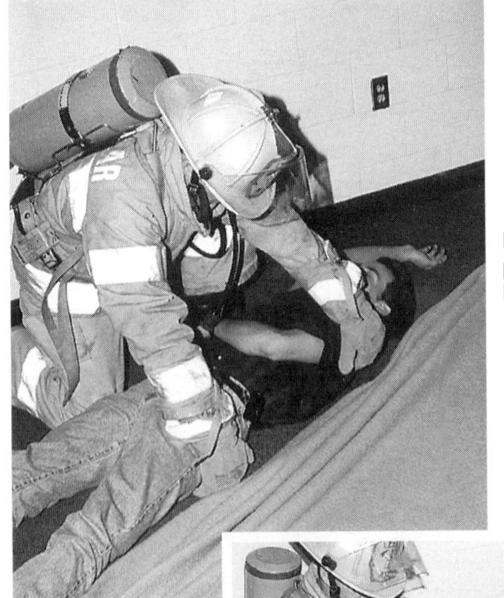

Figure 5.69 Rescuer prepares victim to be logrolled.

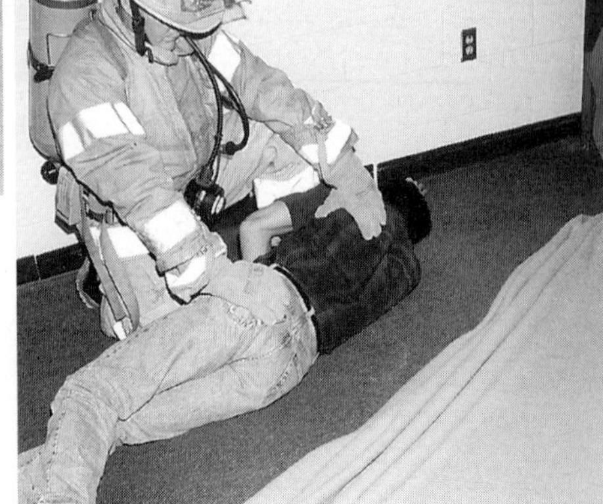

Figure 5.70 Rescuer logrolls victim.

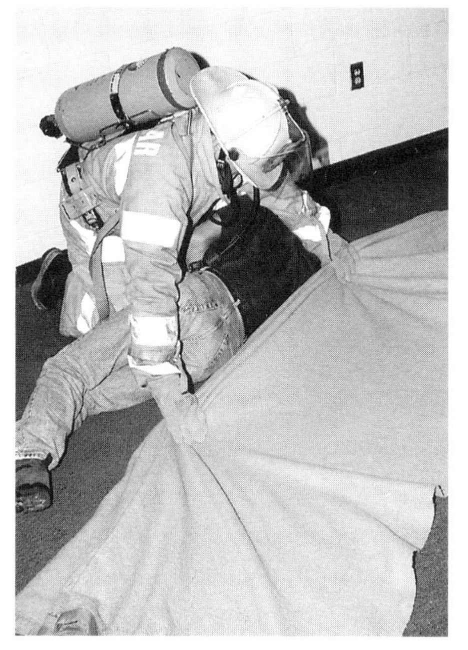

Figure 5.71 Blanket is pulled to the victim's back.

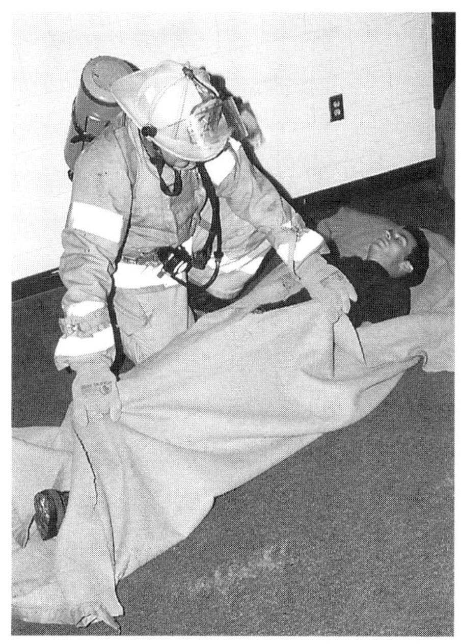
Figure 5.72 Victim is wrapped in the blanket.

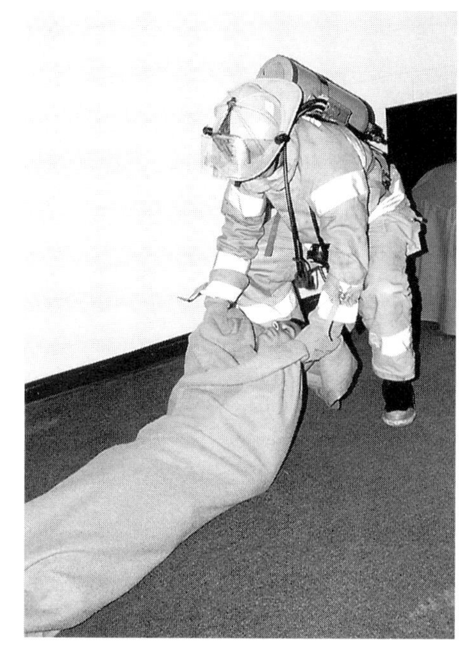
Figure 5.73 Rescuer drags victim to safety.

Basket Litter

The procedures described in this section apply to both metal and plastic or fiberglass basket litters. These procedures are slightly different than those in Chapter 4, Rope Rescue, because they are applied to different situations. The procedures described here are sufficient for short-distance transfers but not for full-scale rescue operations. As in other rescue procedures described in this manual, those that follow do not include the emergency medical steps necessary to prepare a victim for transfer. For pretransfer emergency medical procedures, see an EMS first responder manual.

To secure a victim for short-distance transfer in a basket litter, two blankets and 50 feet (15.5 m) of rope are needed. The procedure is as follows:

Step 1: Prepare the litter for the victim as described in Chapter 4.

Step 2: Place the victim in the litter.

Step 3: Fold the lengthwise blanket around the victim's feet and legs.

Step 4: Fold the other blanket around the victim's upper body, folding the top corners in and wrapping the ends across the chest. If the litter has straps, fasten them at this point.

Step 5: Run the lashing rope through the openings at the foot of the litter, and adjust the rope until an equal amount is on each side of the litter (Figure 5.74).

Step 6: Secure the victim's feet. Put a Half Hitch over both of the victim's feet at the instep (Figure 5.75). Tighten the Half Hitch until it is snug.

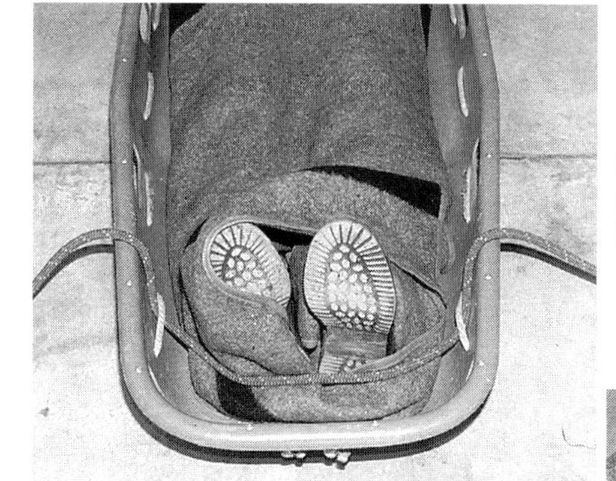
Figure 5.74 Lashing rope is run through the litter.

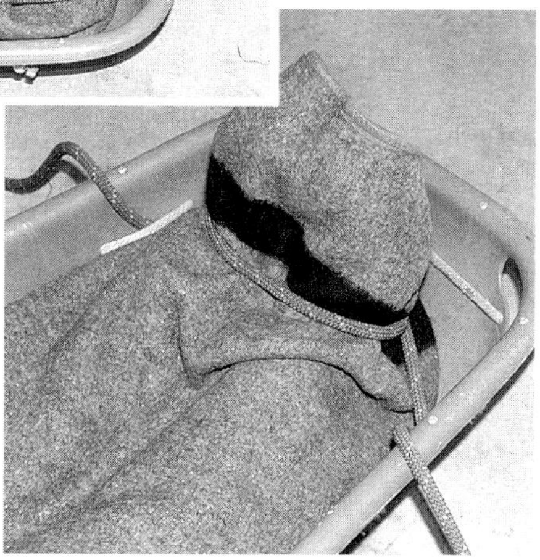

Figure 5.75 Victim's feet are secured.

NOTE: A Clove Hitch can be used in place of the Half Hitch.

Step 7: Pass the right-hand part of the rope over and around the right rail, across the litter, and through the closest opening on the left side. Repeat this procedure on the other side of the litter with the left-hand part of the rope. (**NOTE:** If a wire basket litter is used, the rope should be threaded through the "D" openings.)

Continue lacing back and forth across the litter up to the victim's shoulders (Figure 5.76). With two rescuers, it is easier to lace if each rescuer passes the rope through the opening, gathers the excess rope, and hands the rope to the other rescuer to lace on the opposite side.

Step 8: At the victim's shoulders, secure the rope on each side with a Clove Hitch and a safety (Figure 5.77). Protect the victim's face when handling ropes at that end of the basket.

Step 9: Starting at the victim's feet, tighten the lashing by pulling up on the ropes where they cross, working up from the feet to the head. Adjust all knots until they are snug.

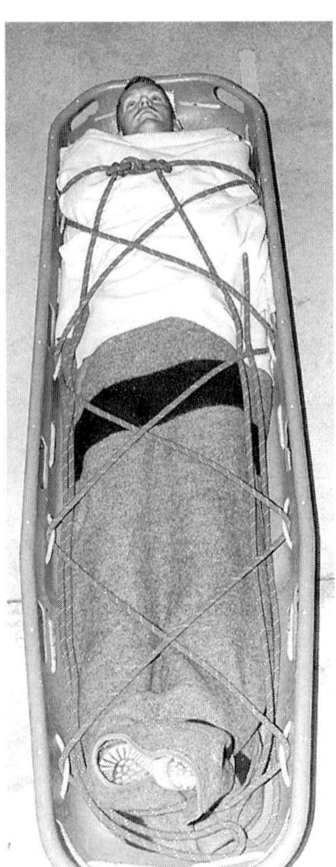

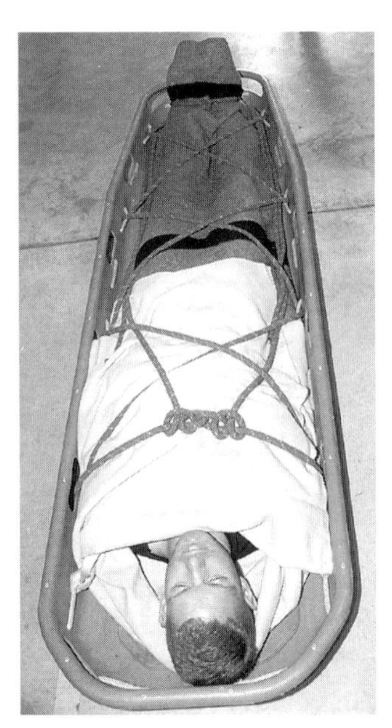

Figure 5.77 Lashing rope is secured on each side at the victim's shoulders, and excess rope is tucked away.

Figure 5.76 The victim is lashed into the litter.

Step 10: Secure the excess rope by tucking it under the lashing.

RESCUES FROM UPPER FLOORS

The type of assistance given to those on upper floors of buildings depends upon the number of victims, the physical condition of those victims, the condition of the building and its stairways, and the rescue resources (personnel and equipment) available. This section addresses a variety of procedures involved in moving a victim from one floor or level of a building to another.

If available, the building's stairways or fire escapes are the preferred means of egress. If not, aerial devices are next best. If aerial devices are not available either, firefighters will have to use all their training and ingenuity to employ whatever resources are available for rescues from these locations. With the exception of using an air cushion (perhaps the most dangerous of the rescue evolutions), it is usually preferable to use the methods discussed in this section instead of the more difficult, dangerous, and time-consuming rope rescue methods described in Chapter 4.

Stairways

When available, the stairway is the easiest and safest means of removing occupants from upper floors. This fact is another reason why firefighters should protect stairways during a fire and keep them clear for use as an exit. There are several common methods used to assist victims down stairways, including merely walking next to victims and supporting them (Figure 5.78). Assistance may also involve either carrying or dragging them down the stairs by one of the methods described earlier in this chapter. Reasons to use the stairs are as follows:

- Occupants may be capable of exiting unassisted if they are directed to the appropriate stairway.
- In a given amount of time, more people can be removed or guided down a stairway than down a ladder.
- Some people are reluctant to get on a ladder, even at relatively low heights.

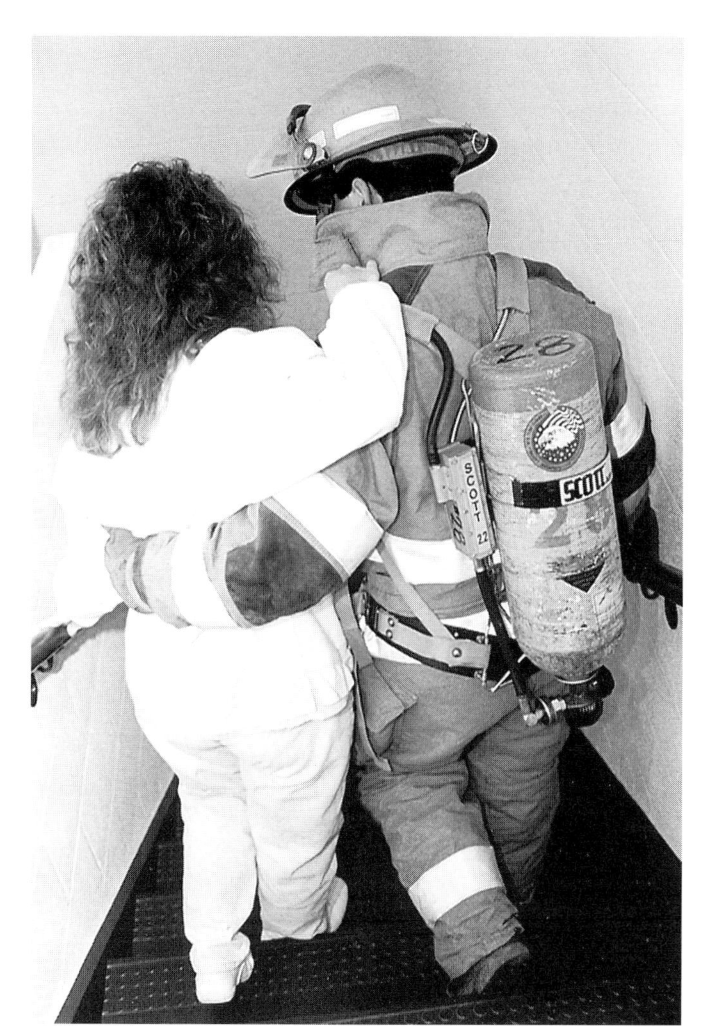

Figure 5.78 A victim is assisted down the stairs.

- Elderly occupants and others who could walk down a stairway might have to be carried down a ladder.
- People on ladders are exposed to falling objects.
- There is less likelihood of someone falling on stairs than on ladders.
- Moving an unconscious victim down a stairway is easier and safer for rescuers than moving the victim down a ladder.

Fire Escape Stairs

Exterior fire escape stairs found on many older buildings are more difficult to use for rescue purposes than conventional interior or exterior stairs. Many of the problems associated with using ladders also apply to fire escapes. The steepness of fire escape stairs makes the use of handrails almost imperative, making it more difficult for rescuers to carry or drag victims (Figure 5.79). Inclement weather can make exterior fire escapes very slippery, and people afraid of heights may be reluctant to use them. Also, weathering and lack of maintenance can make them structurally unsound. However, circumstances on the fireground may make the use of fire escapes the only available option.

Some fire escapes extend completely to the sidewalk level, which is ideal. In most, however, the lowest flight is hinged at the top and is held in a horizontal position by a counterweight (Figure 5.80). It is designed to swing down into position when someone steps on it from above, but the landing area is sometimes blocked by either parked vehicles or materials stored there. Firefighters can lower the flight of stairs from the sidewalk by pulling it down with a pike pole.

Figure 5.79 Some fire escapes may be difficult to descend.

Figure 5.80 A counterweighted fire escape ladder.

Figure 5.81 A victim is provided with a safety line.

Figure 5.82 A safety line can be controlled by a firefighter heeling the ladder.

Ladders

If fire conditions permit, it is preferable to remove occupants by taking them down an interior stairway or even up to the roof and across to an adjacent building where the stairway is safe. In some cases, however, using ladders will be the most appropriate means of removing occupants of the fire building.

SAFETY LINES

To increase safety, aid rescuers, and give additional support to victims, safety lines should be attached to victims during ladder rescues (Figure 5.81). If a rescue harness is not available, an approved rescue knot as described in Chapter 4 can be used. During descent, the safety line is controlled by one of the rescuers inside the building. If the rescuers inside the building are not available to control the safety line, it can be done by the firefighter heeling the ladder (Figure 5.82). The rescuer takes one end of the safety line up the underside of the ladder, leaving the other end with the firefighter on the ground (Figure 5.83). At the top of the ladder, the rescuer passes the rope under the last rung below the windowsill, and into the window opening. The rope is then attached to the rescue harness or a rescue knot is tied on the victim. The firefighter heeling the ladder takes up the slack in the safety line and maintains a slight tension on the safety line during the descent. If the victim loses consciousness or falls while on the ladder, the rescuer on the ladder and the firefighter on the ground will be able to control the victim's descent safely to the ground.

Figure 5.83 The safety line is passed up and under the ladder.

AERIAL DEVICES

For rescues above the third floor, aerial devices are generally used. Initially, the tip of the device is brought in above the window and then lowered to the correct height. If the ladder approaches the window from below, a panicky victim may try to jump onto the ladder before it is in position.

Conscious victims should be closely escorted down the ladder by rescuers who should maintain a more-or-less continuous conversation with them to coach and reassure them as they descend the ladder. Rescuers descend first, keeping both arms around the victim in case the victim slips or loses consciousness (Figure 5.84). The dialogue helps to reduce victims' anxiety and helps rescuers detect early signs that victims may be losing consciousness. Unconscious victims will have to be packaged in a litter or carried down the ladder if conditions preclude using a litter (Figure 5.85).

CAUTION: When aerial devices are fully extended at low angles, they can be overloaded if too many rescuers and/or victims are placed on them at one time.

Figure 5.84 A rescuer escorts a victim down an aerial ladder.

Figure 5.85 An unconscious victim is lowered down an aerial ladder.

GROUND LADDERS

When it is known in advance that a ground ladder will be used for rescue through a window, the ladder tip is raised only to the sill (Figure 5.86). This allows the victim easier access to the ladder. The ladder is heeled, and all other loads and activity removed from it during rescue operations. Since even healthy, conscious occupants will probably be unaccustomed to climbing down a ladder, care must be exercised to keep them from slipping and possibly hurting themselves. To bring victims down a ground ladder, at least four firefighters are needed: two inside the building, one or two on the ladder, and one heeling the ladder.

Several methods for lowering conscious or unconscious victims are as follows:

- Conscious victims are lowered feet first from the building onto the ladder (Figure 5.87).
- An unconscious victim is held on the ladder in the same way as a conscious victim except that the victim's body rests on the rescuer's supporting knee (Figure 5.88). The victim's feet are placed outside the rails to prevent entanglement.

Figure 5.86 Ladder tip is placed just below window sill.

Figure 5.87 Victim is lowered onto ladder feet first.

- A similar way to lower an unconscious victim involves using the same hold by the rescuer described in the previous paragraph, but the victim is turned around to face the rescuer (Figure 5.89). This position reduces the chances of the victim's limbs catching between the rungs.
- An unconscious victim is supported at the crotch by one of the rescuer's arms and at the chest by the other arm (Figure 5.90). The rescuer may be aided by another firefighter.
- A conscious or unconscious victim is cradled in front of the rescuer, with the victim's legs over rescuer's shoulders, and the victim's arms draped over the rescuer's arms (Figure 5.91). If the ladder is set at a slightly steeper-than-normal climbing angle, the unconscious victim's head can be tilted forward to avoid hitting each rung during descent. This method is also very effective with extremely heavy victims, whether conscious or unconscious (Figure 5.92).
- Another method of removing extraordinarily heavy victims involves several firefighters. Two ground ladders are placed side by side. One firefighter supports the victim's waist and legs. A second firefighter on the other ladder supports the victim's head and upper torso (Figure 5.93).
- Small children who must be brought down a ladder can be cradled across the rescuer's arms (Figure 5.94).

Litters

The handles of tools, such as pike poles and shovels, can be used to temporarily modify a litter to allow it to slide down a ladder. With military-style litters, the handles are run through the bottom "D" rings. With a basket litter or a backboard, they are lashed across the bottom. A rope is secured to the head of the stretcher with a Figure-Eight Follow Through.

Rescuers slide the litter onto the ladder as tension is maintained on the rope from an anchor point inside the building. When the litter is on the ladder, it is guided down by another rescuer on the ladder (Figure 5.95).

Figure 5.88 Victim is supported by rescuer's knee.

Figure 5.89 The victim faces the rescuer.

Figure 5.90 Another way to rescue an unconscious victim.

Figure 5.91 Victim is cradled between the rescuer and the ladder.

Figure 5.92 An effective method for handling very heavy victims.

Figure 5.93 Two ladders and two rescuers are needed for this method.

Figure 5.94 A small child is cradled across the rescuer's arms.

Figure 5.96 Guidelines attached for horizontal descent.

Figure 5.95 A rescuer guides the litter down the ladder.

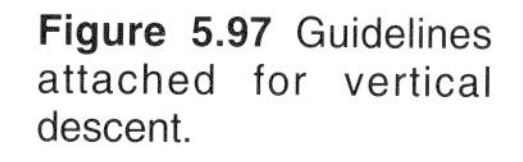

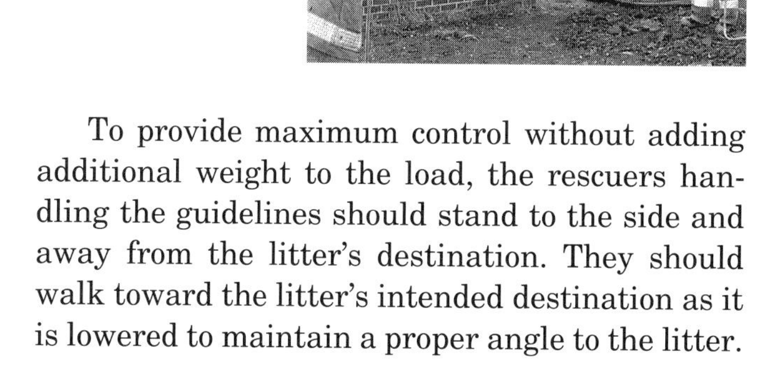

Figure 5.97 Guidelines attached for vertical descent.

When a litter containing a victim must be lowered from an upper floor, guidelines may be necessary to control the swing of the litter to keep it away from walls or other obstacles.

How the litter is suspended will determine where the guidelines should be attached. If it is being lowered in a horizontal position, the guidelines should be attached at the head and at the foot of the litter (Figure 5.96). If the litter is being lowered in a vertical position, the guidelines should be attached to each side (Figure 5.97).

To provide maximum control without adding additional weight to the load, the rescuers handling the guidelines should stand to the side and away from the litter's destination. They should walk toward the litter's intended destination as it is lowered to maintain a proper angle to the litter.

Bridging Gaps

In a life-threatening emergency, ladders can be used to bridge spaces between buildings or damaged areas of floors or roofs. However, all ladder manufacturers recommend against using ladders in any way except that for which they were designed, and doing so may void all manufacturer's warranties and certifications.

WARNING

Using a ladder in any way other than that recommended by the manufacturer places all liability for accidents or injuries completely on the user, and IFSTA does NOT recommend such use.

However, if saving someone's life demands that a ladder be used to bridge a gap, the following procedure will reduce the likelihood of accidents or injuries occurring in the process. Boards or planks should be placed on the rungs to spread the load and to make passage over the ladder safer (Figure 5.98). If an extension ladder is used, the sections should be lashed to prevent accidental extension or retraction of the ladder during use. To bridge a gap with a ground ladder, the following procedure can be used:

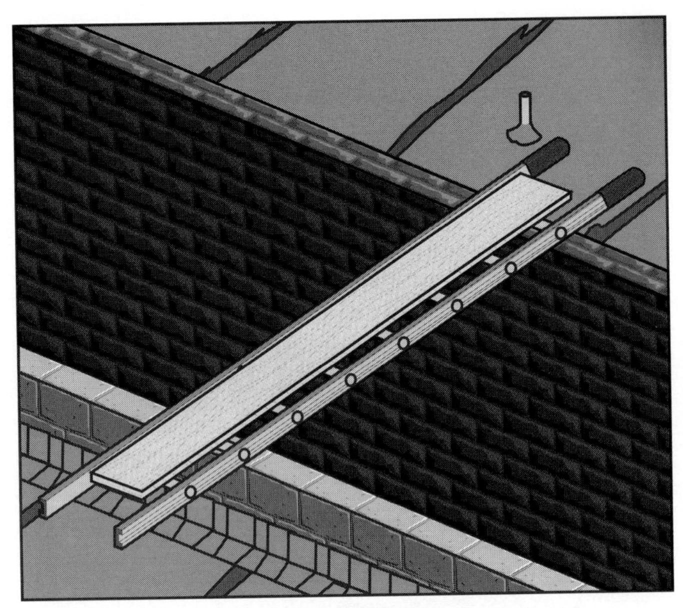

Figure 5.98 Planks are used to spread the load on the ladder.

Step 1: Place the ladder flat on the surface with ladder tips toward the gap (Figure 5.99).

Step 2: With a rescuer on each side of the ladder and at the edge of the gap, begin to slide the tip of the ladder across the gap while a third rescuer holds the other end of the ladder down (Figure 5.100).

Step 3: Continue sliding the ladder until the tip reaches the other side of the gap. If the gap is too wide, lines can be tied to the tip of the ladder to prevent it from nosing down into the gap (Figure 5.101).

Figure 5.99 Ladder is positioned at the edge of the gap.

Jumping Situations

Firefighters are sometimes faced with situations in which people have been forced to windows, balconies, or ledges by heat, smoke, and fire in their offices or apartments. They are often high enough above the street level to be beyond the reach of available aerial devices.

Firefighters should use a bullhorn or public-address system to make sure these stranded individuals can hear instructions because any movement could cause the potential jumper to slip and fall (Figure 5.102). It is important for firefighters to establish and maintain a dialogue with potential jumpers to reassure them that someone knows of their situation and that help is on the way. Firefighters should continually repeat instructions and update these people about what is being done to rescue them.

SUICIDE ATTEMPTS

One of the most difficult situations to which firefighters may be called is that in which someone is threatening to jump from an upper floor as a means of suicide. Because of their specialized training and because suicide is classified as a crime, the

Figure 5.100 Firefighters slide ladder tip across the gap.

Figure 5.101 Lines keep ladder tip up.

overall situation will usually be handled by law enforcement officers, but firefighters may become involved because of their special equipment and rescue skills.

A potential jumper's feelings should never be taken lightly, and every threat should be taken seriously. The mental state of the potential jumper may vary from agitation and anger to near hysteria. However, once people have made the decision to end their life, they are usually quite calm and composed; this is a signal that they may jump at any moment. Firefighters can sometimes gain the

Figure 5.102 A firefighter communicates with a victim.

confidence of a potential jumper because firefighters are perceived as less threatening than police; however, they should be cautious about crowding potential jumpers or making any overt gestures or sudden moves. Until those trained to intervene directly arrive on scene, firefighters can set up an air safety cushion while another firefighter keeps the potential jumper engaged in conversation (Figure 5.103). A firefighter can express a genuine concern that if the person jumps, firefighters or others on the ground may be injured. This concern sometimes appeals to the person's desire to avoid harming anyone else in the process of doing away with himself or herself.

AIR SAFETY CUSHION

Because of a variety of problems, life nets are rarely used anymore. They are not reliable at the greater heights of many buildings today, and people jumping cannot be instructed how to jump the best way. They may jump too soon or land on their heads, and rescuers may be injured from improper holding methods. A life net is carried on some ladder apparatus, but the person most likely to successfully jump into it is the trained firefighter.

Figure 5.103 A rescuer talks to a potential jumper.

A more recent device, an air-inflated cushion, is available to increase the chances of a successful jump from up to ten stories. However, the possibility of several people jumping at the same time or of a jumper partially or completely missing the bag make this a dangerous operation at best.

The air-cushion bag has two air cells. The lower cell is airtight and absorbs twice the kinetic energy absorbed by the upper cell. A special air release system in the upper cell lets some of the compressed air vent upon impact to relieve the excess pressure. Fans continuously provide replacement air for the cell. Jumps can be made every five to seven seconds, and the impact energy will still be absorbed sufficiently. The air-cushion bag can even be set up on parked cars or other obstructions.

The air cushion is designed so that any object landing on the cushion is engulfed and held, rather than bounced. This means that two or more falling objects will not collide or be thrown from the bag. The bags are inflated at low air pressures, so a small cut in the bag is not critical.

Chapter 5 Review

Directions

The following activities are designed to help you comprehend and apply the information in Chapter 5 of **Fire Service Rescue**, Sixth Edition. To receive the maximum learning experience from these activities, it is recommended that you use the following procedure:

1. Read the chapter, underlining or highlighting important terms, topics, and subject matter. Study the photographs and illustrations, and read the captions with each.
2. Review the list of vocabulary words to ensure that you know the chapter-related meaning of each. If you are unsure of the meaning of a vocabulary word, look up the word in the glossary or a dictionary, and then study its context in the chapter.
3. On a separate sheet of paper, complete all assigned or selected application and review activities before checking your answers.
4. After you have finished, check your answers against those on the pages referenced in parentheses.
5. Correct any incorrect answers, and review material that was answered incorrectly.

Vocabulary

Be sure that you know the chapter-related meanings of the following words:

- backdraft *(138)*
- carabiners *(142)*
- cockloft *(145)*
- expeditious *(136)*
- flashover *(138)*
- freelance *(143)*
- gurney *(149)*
- imperative *(157)*
- inclement *(157)*
- kernmantle rope *(142)*
- lifeline *(146)*
- obscured *(138)*
- plenum *(145)*
- preclude *(146)*
- protocol *(142)*
- semiambulatory *(147)*
- stratify *(146)*
- supine *(149)*
- untenable *(145)*
- webbing *(142)*

Application Of Knowledge

1. Describe your department's fireground personnel accountability system. *(Local protocol)*
2. Explain the method used in your department to mark searched rooms. *(Local protocol)*
3. Choose from the following search and rescue procedures those appropriate to your department and equipment, or ask your training officer to choose appropriate procedures. Mentally rehearse the procedures, or practice the chosen procedures under your training officer's supervision.
   - Conduct a primary search of a one-story building. *(136-138)*
   - Conduct a primary search of a multistory building with a central hallway. *(140)*
   - Perform VES. *(138)*
   - Perform a secondary search of either a one-story or a multistory building. *(138-141)*
   - Perform a primary search of a high-rise building. *(144-147)*
4. Choose from the following victim lifts, carries, and transfers those that are appropriate to your department and equipment, or ask your training officer to choose appropriate procedures. Mentally rehearse the procedures, or practice the chosen procedures under your training officer's supervision.
   - Perform a cradle-in-arms carry. *(148)*
   - Perform a seat lift/carry. *(148, 149)*
   - Perform a two- or three-person lift/carry. *(149)*
   - Perform a extremities lift/carry. *(150, 151)*
   - Perform a chair lift/carry. *(152)*
   - Perform an incline drag. *(152, 153)*
   - Perform a blanket drag. *(154)*
   - Secure a victim in a basket litter. *(155, 156)*
5. Choose from the following methods of bringing conscious and unconscious victims down ground ladders those that are appropriate to your department and equipment, or ask your training officer to choose appropriate procedures. Mentally rehearse the procedures, or practice the chosen procedures under your training officer's supervision.
   - Bring a conscious victim down a ladder (feet first). *(159)*
   - Bring an unconscious victim down a ladder (supported by rescuer's knee, facing away from rescuer). *(159, 160)*
   - Bring an unconscious victim down a ground ladder (supported by rescuer's knee, facing rescuer). *(160)*
   - Bring an unconscious victim down a ground ladder (supported at crotch and chest). *(160)*

- Bring an unconscious victim down a ground ladder (cradled in front of rescuer). *(160)*
- Bring a very heavy victim down a ground ladder. *(160)*
- Bring a small child down a ground ladder. *(160)*

6. Chose from the following procedures those that are appropriate for your department and equipment, or ask your training officer to choose appropriate procedures. Mentally rehearse the procedures, or practice the chosen procedures under your training officer's supervision.
   - Use a ground ladder to assist in lowering an upper-story victim on a litter. *(160, 161)*
   - Use a ground ladder to bridge a gap. *(162)*
   - Use an aerial device to rescue a victim beyond ground ladder reach. *(159)*
   - Use a bullhorn or public-address system to instruct and reassure visible upper-story victims. *(162)*
   - Engage a potential suicide in conversation. *(162, 163)*
   - Set up an air safety cushion. *(163, 164)*

Review Activities

1. Identify the following abbreviations and acronyms:
   - CPR *(147)*
   - PASS *(135 and 144)*
   - ICS *(145)*
   - VES *(138)*
   - PPV *(138)*
2. Name the five essentials of successful fireground search and rescue. *(135)*
3. Describe the information that the rescue team's initial exterior size-up should provide. *(135)*
4. Name the two objectives of a building search. *(136)*
5. Distinguish between a primary search and a secondary search. *(136 & 138)*
6. Explain what a buddy system is and why it is required during a primary search. *(136, 137)*
7. Write a brief list of guidelines for conducting a primary search. *(136-138)*
8. Explain when VES is used by some departments. *(138)*
9. Explain why the fire floor, the floor directly above the fire floor, and the topmost floor should be searched first in multistory buildings. *(139)*
10. Discuss the pros and cons of requiring rescuers to have a charged hoseline with them on all floors during the search and rescue procedure. *(140)*
11. Explain the search methods recommended when rooms extend from a center hallway: *(140)*
    - One search and rescue team
    - Two search and rescue teams
12. Describe the search patterns rescuers should use when entering, searching, and exiting normal-sized rooms. *(140)*
13. Explain the recommended method for searching small rooms. *(140)*
14. Describe the advantages of searching small rooms using the method described in Activity 13. *(140)*
15. Explain the two-part marking system. *(141)*
16. List methods of marking searched rooms. *(141)*
    - Recommended methods
    - Methods *not* recommended
17. List the purposes of the following typical search and rescue tools: *(141)*
    - Rope
    - Marking devices
    - Forcible entry tools
18. Explain why members of search and rescue teams should continually feel the floor in front of them with a hand or tool. *(142)*
19. Describe techniques search team members should use to find their way out of a building and to signal for assistance when they become disoriented and lose their direction during a search. *(142)*
20. List safety guidelines that should be used by search and rescue personnel in any type of search operation within a building. *(143, 144)*
21. Explain the purpose and operation of a PASS. *(144)*
22. Describe where the incident management command post should be located at a high-rise structure. *(144)*
23. Explain what is meant by the terms *lobby* and *staging* in the ICS. *(145)*
24. Describe how the two teams assigned to the primary search in a high-rise incident should conduct their search. *(145, 146)*
25. Answer the following questions:
    - For what hazard should search and rescue teams on the fire floor of a high-rise building be especially alert? Why? *(145)*
    - When should high-rise search and rescue teams use a lifeline? Why? *(146)*
    - Why is it important that rescue teams check perimeter rooms on the fire floor of high-rise structures? *(146)*
    - Why do high-rise incidents require more personnel than are required for the ordinary residential structure? *(146)*

- When conditions are poor, how many searchers should be assigned per floor of a high-rise structure? *(146)*
- What is one of the primary duties of the search and rescue supervisor in regard to stairways? *(146)*
- What types of tools may be provided to high-rise search teams? *(146, 147)*

26. Define *emergency move. (147)*
27. List conditions under which emergency moves are necessary. *(147)*
28. List general guidelines for moving victims. *(147, 148)*
29. List (from most to least acceptable) procedures for rescuing victims from upper floors. *(156)*
30. Explain why stairs are the easiest and safest means of rescuing victims from upper stories. *(156, 157)*
31. Explain the disadvantages of using fire escape stairs to interior stairs for rescue from upper stories. *(157)*
32. Explain the purpose of safety lines in ladder rescue from upper stories. *(158)*
33. Describe the procedure used with aerial devices for rescuing conscious and unconscious victims above the third floor. *(159)*

Questions And Notes

6

Structural Collapse Rescue

Chapter 6
Structural Collapse Rescue

INTRODUCTION

Severe weather, earthquakes, explosions, or faulty building design and construction can result in the partial or the total collapse of buildings. Structural collapse often results in fatalities or serious injuries to the building occupants. Hurricane Andrew in 1992; the major earthquakes in California in 1989 and 1994, and in Kobe, Japan in 1995; the 1995 department store collapse in Seoul, Korea; and the 1995 bombing of the federal building in Oklahoma City (Figure 6.1) were all examples of sudden, violent building collapse that resulted in many deaths and injuries — some of them to fire/rescue personnel (Figure 6.2). Structural collapse can occur in any jurisdiction at any time, and the fire department is usually the first to respond to these incidents. Therefore, rescue personnel must always be ready to locate and free victims from collapsed structures in the safest and most efficient way possible.

The lessons learned during actual emergency incidents may or may not apply to other incidents. Emergencies occur at random intervals, so rescuers cannot rely on experience alone to prepare them to deal with structural collapse incidents. Rescuers must have a solid foundation of classroom study and hands-on training that is periodically reinforced by realistic and challenging training exercises (Figure 6.3). Many states, provinces, and regional training systems offer excellent courses in various rescue disciplines. Rescue personnel should take advantage of these training opportunities.

Structural collapse incidents can involve a wide variety of situations and demand a broad range of different skills. This chapter addresses those situations and skills that are unique to structural collapse rescues. It also refers the reader to other chapters for more information about situations and skills that apply to structural collapse as well as other rescue situations. The chapter reviews the applicable national standards, pre-incident planning for structural collapse, types of construction, types of collapse, hazards associated with structural collapse, and the equipment used in structural collapse rescues. Also discussed are tactical considerations such as size-up, scene and incident management, stabilizing collapsed structures, rescue techniques, and incident termination.

Figure 6.1 The Murrah Building in Oklahoma City.

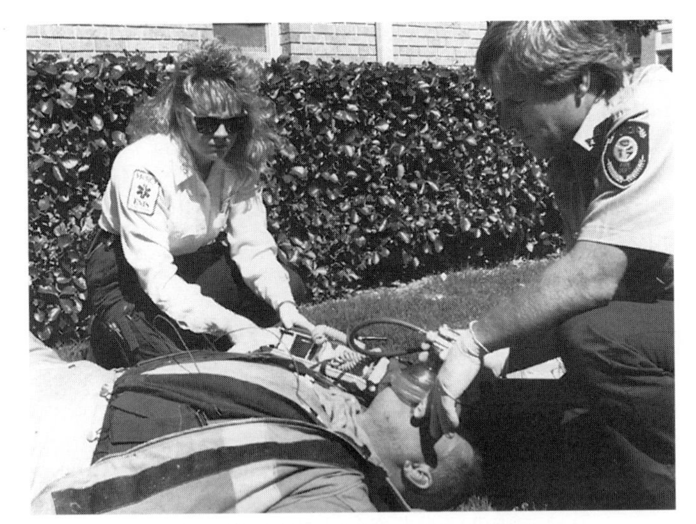

Figure 6.2 An injured firefighter receives medical attention.

Figure 6.3 Firefighters need structural collapse training.

BACKGROUND

Prior to a major structural collapse occurring within their response district, fire/rescue personnel should familiarize themselves with the nature and scope of the potential problems they may encounter. This includes becoming familiar with the applicable standards, with the collapse potential in the district, with types of construction and types of collapse, and with the resources available for mitigating such incidents.

Standards

One of the most important resources in the field of structural collapse rescue is NFPA 1470, *Standard on Search and Rescue Training for Structural Collapse Incidents*. While this standard is a training document, it addresses all of the essential operational elements of structural collapse rescue. NFPA 1470 can be used as a checklist of things to consider in the planning and delivery of rescue services at collapsed structures.

Another important document with which rescuers should be familiar is the urban search and rescue manual published by the Federal Emergency Management Agency (FEMA). Entitled *Urban Search and Rescue Response System — Operational System Description and Mission Operational Procedures*, this manual outlines in great detail the organization and operation of FEMA urban search and rescue resources. It specifies the procedures for requesting FEMA assistance and describes the operational guidelines under which these resources will operate.

Pre-Incident Planning

Both of these structural collapse rescue documents stress the importance of pre-incident planning for structural collapse incidents. The planning consists of four elements: hazard survey, plans development, resource identification, and training.

HAZARD SURVEY

Because the makeup of most response areas is in a constant state of flux as new structures are built and old ones removed, the hazard survey must be an ongoing process. The response district should be toured and a list compiled of the types of structures that exist within the district (Figure 6.4). In general, the survey will identify the types of construction, the number of stories in the major buildings within the district, and those buildings that have basements or subbasements. It will also be used to identify target hazards such as those with hazardous materials in use or in storage or those with extraordinary life hazards such as hospitals, schools, apartments, and rest homes (Figure 6.5). Structures of the types known to be prone to collapse, such as elevated highways, bridges and overpasses, tilt-up concrete buildings, and multilevel parking garages, will be identified (Figure 6.6). Considering the history of the area, the most likely causes of future collapses will also be considered because this factor may translate into useful planning information regarding the potential numbers of victims and where they are likely to be located.

PLANS DEVELOPMENT

Starting with those for buildings and other structures identified as target hazards during the hazard survey, operational plans should be devel-

Figure 6.4 Firefighters should tour their response district frequently.

Figure 6.5 A typical target hazard.

Figure 6.6 Parking garages are collapse-prone structures.

oped for each structure or each class of structures. Those structures most likely to entrap large numbers of people in the event of a collapse should be addressed first. The process of developing the plans will involve one or more visits to the structures to inspect them thoroughly and make notes. During these visits, the structures should be viewed with a "worst-case" collapse scenario in mind. The visits will help to answer the following questions, which address the most critical elements of an operational plan:

- What is the most likely mechanism of collapse?
- How many people are likely to be trapped?
- What is the likelihood of secondary collapse?
- Are secondary hazards likely to be created?
- How much and what type of debris will be created?
- How accessible will the site be after a collapse?
- How much and what types of equipment will be needed?
- How many rescue and support personnel will be required?
- Where will the resources be staged?
- Where are the shutoffs for primary and secondary utilities?

Not every structural collapse incident will be of sufficient size or complexity to require mutual aid. The answers to these questions form the basis of an operational plan because they focus attention on what is most likely to happen and, therefore, what resources will be needed to cope with the anticipated problems. Preparing to deal with a collapse caused by an explosion, for example, may be far different than preparing for one caused by an earthquake. An explosion is likely to be localized and confined to one jurisdiction; so if mutual aid forces are needed, they are likely to be available. After an earthquake, however, nearby mutual aid forces may be committed to their own rescue incidents and be unavailable to respond outside of their own boundaries (Figure 6.7). The likelihood of secondary collapse may be greater following an earthquake because of aftershocks. However, secondary

Figure 6.7 The hazards in the surrounding area must be known.

Figure 6.8 The resource list must be kept up to date.

hazards, such as broken gas lines or spilled hazardous materials, can also create the possibility of secondary explosions.

RESOURCE IDENTIFICATION

When the various target hazards have been thoroughly studied and site-specific plans have been developed, many different types of resources may be identified as necessary to implement the plans when needed. The process of matching the resources needed with those available will start within the jurisdiction. All of the resources available internally are identified and listed. The names and phone numbers (both home and work) of the contact persons for these resources must also be listed. The list should be reviewed periodically and updated at least annually (Figure 6.8).

If the numbers of apparatus, equipment, and personnel that may be needed exceed those available internally, the number of those available from external sources must be determined. This may involve negotiating mutual aid agreements if they do not already exist or involve expanding, updating, and redefining those that do exist. As mentioned earlier, plans should also be made for obtaining needed resources when those nearby are unavailable. Ensuring the availability of needed resources may also necessitate negotiating agreements with private contractors for cranes, dozers, and other heavy equipment (Figure 6.9). These agreements should also be reviewed and reaffirmed periodically.

In addition to the site-specific plans, general plans for dealing with large numbers of mutual aid resources should also be made. Such things as how to feed dozens of rescue and support personnel called in from outside the jurisdiction must be anticipated and accommodated within the plan. Financial agreements between the jurisdiction and one or more fast-food restaurants may have to be entered into in advance if this need is to be met on

Figure 6.9 Heavy equipment from other sources may be needed. *Courtesy of Steve George.*

short notice (Figure 6.10). Arrangements should be made for the required number of portable sanitary facilities to be placed at the collapse site and in staging areas (Figure 6.11). A source of folding cots and disposable sleeping bags for mutual aid crews may have to be located as well (Figure 6.12). Sources of fuel, lubricants, tires, and other spare parts for mutual aid vehicles should be identified, and arrangements should be made to have them available on a 24-hour basis. A source of a large number of portable radios with one or more common frequencies should be identified. The availability of search dogs and electronic search devices should also be determined.

Figure 6.10 Arrangements for feeding large numbers of personnel must be made.

Figure 6.11 Adequate sanitary facilities must be provided.

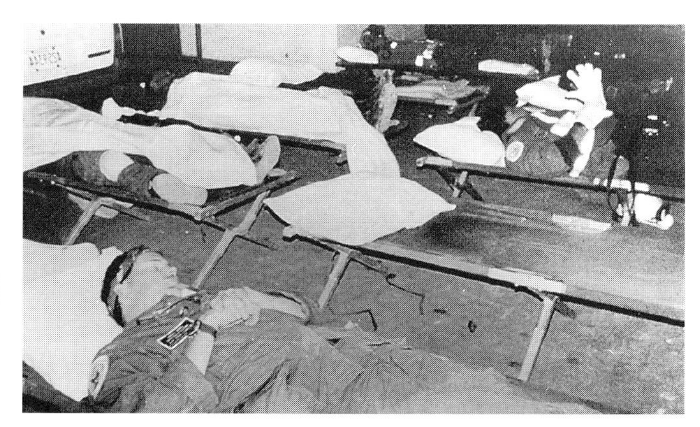

Figure 6.12 Sleeping arrangements must be made.

TRAINING

The pre-incident planning process is not complete without the necessary training to test the operational plan and ensure its safe and efficient implementation. Everyone who may be involved in an actual incident must be briefed on the plan in general and on their roles in particular (Figure 6.13). Regardless of how well the plan is written, it must be tested in one or more realistic exercises to see whether it works as well in practice as it does in theory (Figure 6.14). These exercises give those involved an opportunity to see the plan in operation and to test their own understanding of the plan and what it requires of them. Any deficiencies in the plan that show up in an exercise can be discussed and corrected.

Types Of Construction

Structural collapse incidents occur in one or more of four general types of construction: light frame, heavy wall, heavy floor, and precast concrete.

Figure 6.13 Rescuers are briefed on the operational plan.

Figure 6.14 Emergency plans need to be tested.

LIGHT FRAME

Materials used for this type of construction are generally lightweight and provide a high degree of structural flexibility to applied forces such as earthquakes, hurricanes, tornadoes, etc., which makes them less likely to collapse. These structures are typically constructed with skeletal structural frame systems of wood or light-gauge steel components that support the floor or roof assemblies (Figure 6.15). Examples of this type of construction are wood frame structures used for residential occupancies, multiple low-rise occupancies, and light commercial occupancies up to four stories in height. Light-gauge steel frame buildings include commercial business and light manufacturing occupancies and facilities.

Figure 6.15 A typical light frame structure.

HEAVY WALL

Materials used for this type of construction are generally heavy and utilize an interdependent structural or monolithic system. These types of materials and their assemblies tend to make the structural system inherently rigid. This type of structure is usually built without a skeletal structural frame. It uses a heavy-wall support-and-assembly system to support the floor and roof assemblies. Structures with tilt-up concrete construction are typically one to three stories in height and consist of multiple monolithic concrete-wall-panel assemblies (Figure 6.16). They also use an interdependent girder, column-and-beam system to support the floor and roof assemblies (Figure 6.17). Examples of this type of construction include commercial, mercantile, and industrial buildings. Other examples of this type of construction are reinforced and unreinforced masonry buildings of one to six stories in height.

Figure 6.16 A tilt-up concrete building. *Courtesy of Keith Flood.*

Figure 6.17 Typical concrete columns in tilt-up construction. *Courtesy of Keith Flood.*

HEAVY FLOOR

Structures of this type use cast-in-place concrete construction consisting of flat-slab-panel, waffle, or two-way concrete-slab assemblies. Pretensioned or posttensioned cable systems or reinforcing steel rebar are common components that increase structural integrity (Figure 6.18). The vertical structural supports include integrated concrete columns or concrete-enclosed steel frames that carry the load of all floor and roof assemblies. This classification also includes heavy timber construction that may or may not use steel rods for reinforcement (Figure 6.19). Examples of heavy floor construction include office buildings, schools, apartment buildings, hospitals, and multilevel parking garages. Heights may vary from single-story to high-rise structures.

Figure 6.18 Rebar increases structural integrity.

Figure 6.19 Heavy timber construction. *Courtesy of Keith Flood.*

PRECAST CONCRETE

Structures of this type use modular precast concrete components for floors, walls, columns, and other subcomponents that are assembled on site (Figure 6.20). Individual concrete components use embedded steel reinforcing rods and/or welded wire mesh for structural integrity, and they may have steel-beam-and-column or concrete framing systems for structural assembly and building enclosure. These structures rely on either single or multipoint connections for floor and wall assemblies, so they are a safety concern during collapse rescue operations. Examples of this type of con-

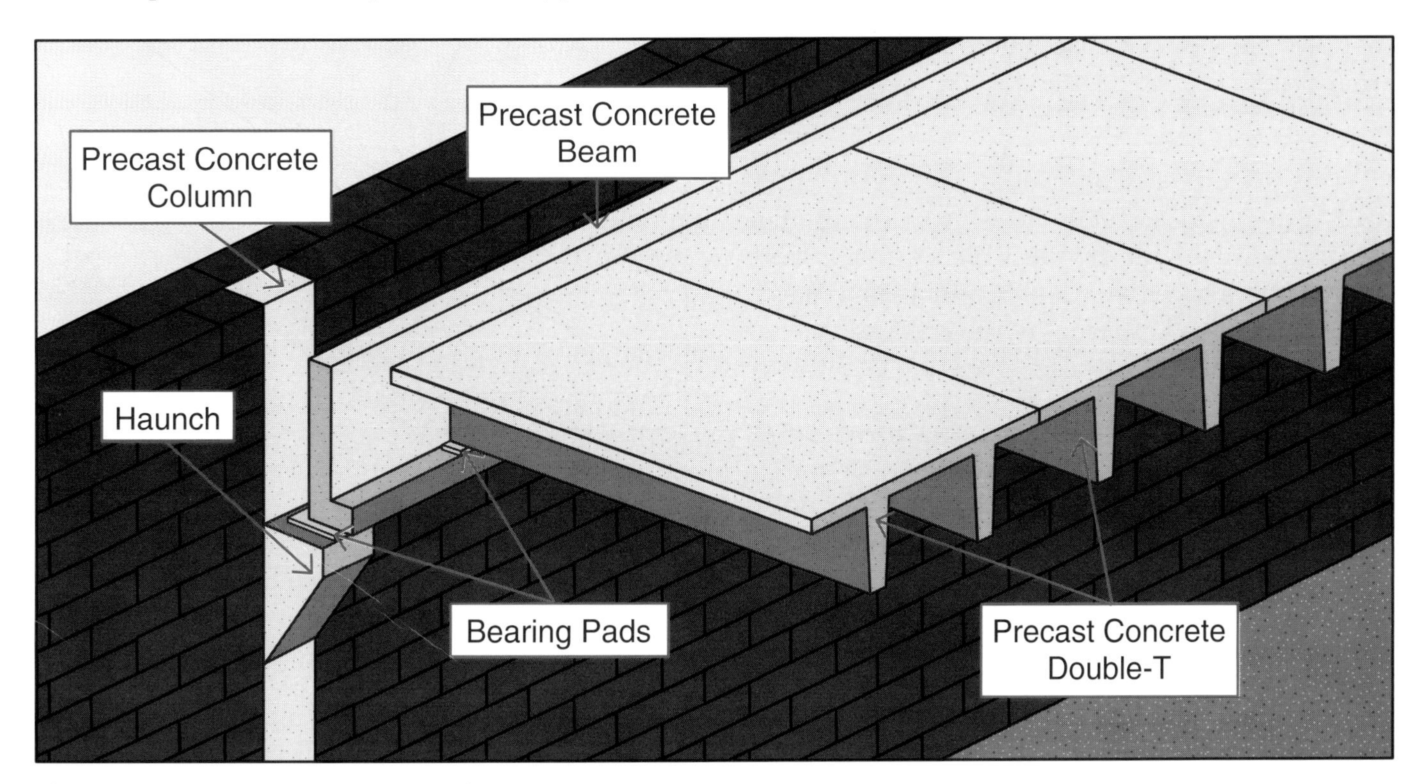

Figure 6.20 Typical precast concrete construction.

struction include commercial and mercantile buildings, office buildings, sports arenas and stadiums, and multilevel parking garages.

Types Of Collapse

Structures collapse in predictable patterns. Knowing and recognizing these patterns can help rescuers make more informed decisions about the likelihood of finding viable victims in the rubble and about the need for shoring and tunneling (see Rescue Skills section later in this chapter). The four most common patterns of structural collapse are pancake, V-shaped, lean-to, and cantilever.

PANCAKE COLLAPSE

This pattern of collapse is possible in any building, but it is most likely in buildings that rely on wood or masonry bearing walls to support the floors and roof. This type of collapse is also likely in older buildings that use steel tie-rods to keep the weight of the roof from pushing the sidewalls out at the top. However, any simultaneous failure of two opposing exterior walls can result in the upper floors and the roof collapsing on top of each other such as in a stack of pancakes — thus, the name of this collapse pattern (Figure 6.21). The pancake collapse is the pattern least likely to contain voids in which live victims may be found, but it must be assumed that there are live victims in the rubble until it is proven otherwise.

Figure 6.21 Pancake collapse.

V-SHAPED COLLAPSE

This pattern of collapse occurs when the outer walls remain intact and the floor(s) and/or roof structure fail in the middle (Figure 6.22). This pattern offers a good chance of habitable void spaces being created on both sides of the collapse.

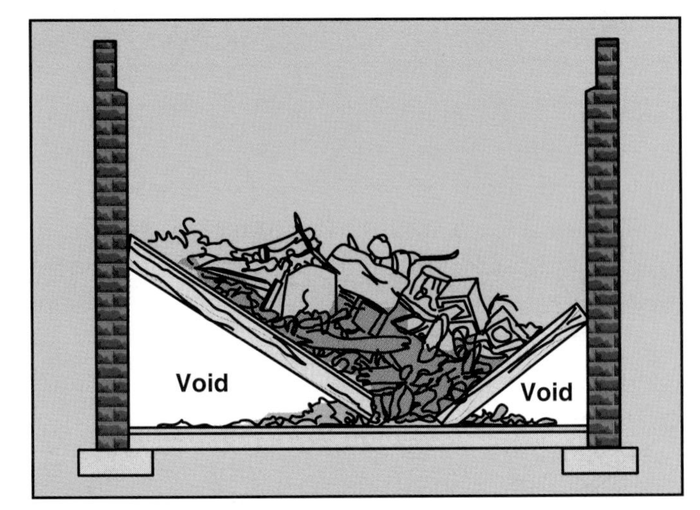

Figure 6.22 V-shaped collapse.

LEAN-TO COLLAPSE

This pattern of collapse occurs when one outer wall fails while the opposite wall remains intact. The side of the roof assembly that was supported by the failed wall drops to the floor forming a triangular void beneath it (Figure 6.23).

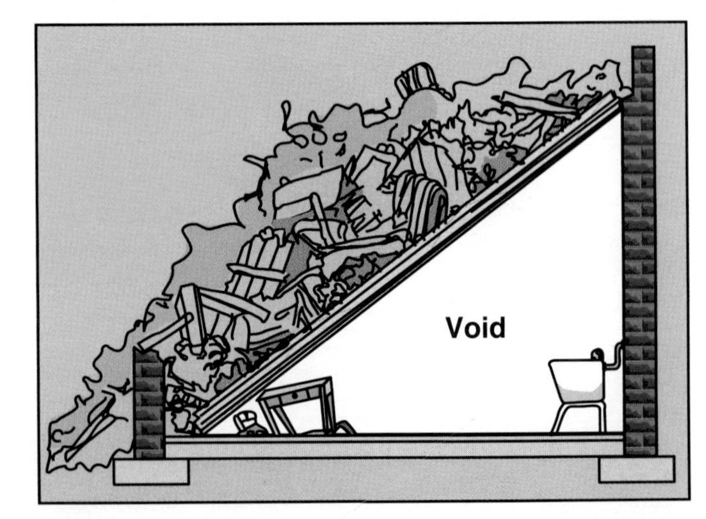

Figure 6.23 Lean-to collapse.

CANTILEVER COLLAPSE

This pattern of collapse occurs when one sidewall of a multistory building collapses leaving the floors attached to and supported by the remaining sidewall (Figure 6.24). This pattern also offers a good chance of habitable voids being formed under the supported ends of the floors. This collapse pattern is perhaps the least stable of all the patterns and is the most vulnerable to secondary collapse.

Figure 6.24 Cantilever collapse.

HAZARDS

There are many actual and potential hazards involved in structural collapse rescue, and they may take any of a wide variety of forms. However, most of the hazards associated with this type of operation fall into one or both of two categories: environmental hazards and physical hazards.

Environmental Hazards

Before rescuers can begin to search the rubble of a collapsed structure for victims, they may have to contend with a number of environmental problems — those that are in and around the collapse. Many of the secondary hazards — those that were created by the collapse or that developed after it — are environmental in nature. Most of the potential environmental hazards involve damaged utilities, atmospheric contamination, hazardous materials contamination, darkness, temperature extremes, noise, fire hazards, or adverse weather.

DAMAGED UTILITIES

Damaged utilities can create life-threatening conditions for rescuers and trapped victims alike. Live electrical wires that are downed on or near the rubble pile are some of the most common and most obvious secondary hazards. However, their full potential is not always so obvious. Wires buried in the rubble can energize electrical conduits, plumbing, and metal structural members with which they come in contact. Broken water pipes can increase the likelihood of electrical shock by wetting the areas where rescuers must work. Broken water pipes can also threaten victims trapped in low areas with drowning and can increase the likelihood of secondary collapse by adding the weight of the water to the building. Broken gas pipes can add the possibility of fire or explosion to the other problems at a structural collapse incident.

ATMOSPHERIC CONTAMINATION

One atmospheric contaminant present at almost every structural collapse is dust. Of particular concern is concrete dust because it is highly alkaline. Asbestos is another major concern because it is a long-term carcinogen. Unless it is dampened down by rainfall or by hose streams, dust stirred up by the collapse can remain suspended in the air for hours, especially in voids and other confined spaces. As already mentioned, broken gas pipes can add a flammable and potentially explosive contaminant to the atmosphere. Depending on the occupancy of the collapsed structure, pure oxygen or other medical gases or a variety of harmful industrial gases may also contaminate the atmosphere in and around the collapse site. Airborne biological hazards may be present in collapsed medical facilities or in any structure where victims' remains cannot be removed within the first day or two.

HAZARDOUS MATERIALS

Many common industrial and commercial chemicals are quite safe when stored and used under normal conditions. But when their containers are damaged in a structural collapse and the chemicals are allowed to leak, they can make an already bad situation many times worse (Figure 6.25). When flammables, corrosives, oxidizers, or

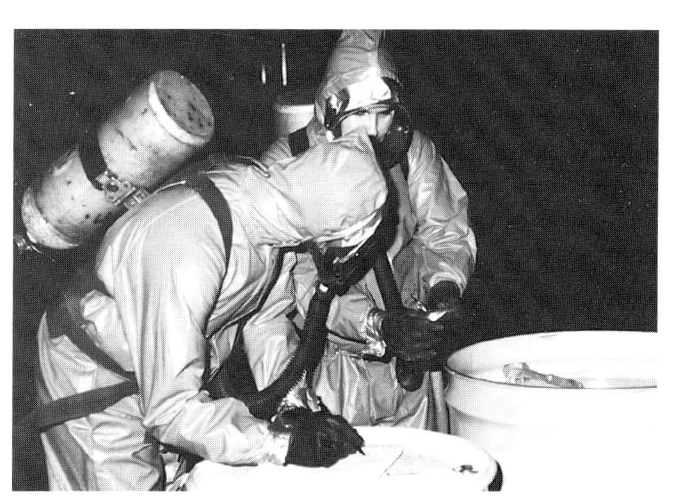

Figure 6.25 Leaking hazardous materials must be mitigated.

toxic substances are released in a structural collapse, the situation can be fatal for victims trapped in the rubble and for rescuers unless recommended procedures are followed.

DARKNESS

Any emergency scene is more hazardous at night. Even in the daytime, voids and other areas deep within the rubble may be in total darkness. Downed electrical wires, unstable debris, and other hazards that might be noticed in full light may go unseen in darkness. In dim light, what may appear to be a wet spot on the ground may actually be a hole into which a rescuer might fall.

TEMPERATURE EXTREMES

Extremes of temperature, whether high or low, can create hazardous conditions for rescuers and victims alike. In very hot weather, rescuers performing heavy manual labor are vulnerable to heat exhaustion or even heatstroke, and rescuers and victims can suffer from dehydration (Figure 6.26).

Figure 6.26 A rescuer in REHAB. *Courtesy of Ron Jeffers.*

Likewise, extreme cold can hamper rescue efforts and subject rescuers and victims to hypothermia, frostbite, and other cold-related conditions. Wet surfaces can become extremely slippery when the moisture freezes.

NOISE

High ambient noise levels can create several hazards at structural collapse sites. Not only do high noise levels add to the stress of everyone involved in the operation, noise can significantly reduce the ability of rescuers to communicate clearly with each other (Figure 6.27). In addition, high ambient noise can obscure important sounds that might otherwise be heard. For example, if there is a lot of noise being produced by power tools, generators, air compressors, and heavy equipment, rescuers may not hear a faint call for help from a victim trapped in the rubble, the creaking sound of debris shifting, or the hiss of escaping gas.

Figure 6.27 Noise can make communication difficult.

FIRE HAZARDS

The physical damage to buildings when they collapse often significantly increases the fire hazard. The forces involved in the collapse change the shape and form of wooden structural members making them more susceptible to ignition. Natural gas pipes may be broken and therefore may add readily ignitable fuel to the scene. Electrical wiring and equipment may be damaged, increasing the sources of ignition. In addition, the friction caused by massive structural members sliding on the surface of exposed wood can create enough heat to cause ignition.

ADVERSE WEATHER

In addition to the temperature extremes already mentioned, adverse weather conditions can increase other environmental hazards at the scene. Rainfall can increase the slip hazard by wetting exposed surfaces. Wind can add significant stresses to weakened walls increasing the threat of secondary collapse.

Physical Hazards

In structural collapse incidents, physical hazards are those associated with working in and around piles of heavy, irregularly shaped pieces of rubble that may suddenly shift or fall without warning. The primary physical hazards are those related to secondary collapse, working in unstable debris, working in confined spaces (some of them below grade), working around exposed wiring and rebar, and dealing with heights.

SECONDARY COLLAPSE

Perhaps the most dangerous and most feared of all hazards in structural collapse incidents is secondary collapse. Even though rescuers are aware of this danger and take steps to minimize it, the possibility of the structure suddenly collapsing further is often present. Secondary collapse can be caused or triggered by earthquake aftershocks; secondary explosions; vibration from helicopters, trucks, trains, other heavy equipment operating at or near the scene; rescuers moving debris; or high winds.

UNSTABLE DEBRIS

Unstable pieces of debris may dislodge and fall on rescuers or victims at any time during structural collapse search and rescue operations. The same mechanisms that cause or trigger secondary collapse can cause pieces of rubble or other debris to suddenly shift and fall. When rescuers move one piece of debris, it can dislodge other unstable pieces and cause them to fall.

CONFINED SPACES

Rescuers may have to enter and work in void spaces left by the collapse of the structure. Getting to and gaining access into these spaces is often hazardous. The spaces themselves may be dark, wet, contaminated, and surrounded by unstable debris. In short, they can be a hostile environment within a hostile environment.

BELOW-GRADE SPACES

Because the basements or subbasements of buildings are designed to support the weight of the building and because they are underground and are supported laterally by the surrounding earth, they may suffer less damage than the rest of the building in a collapse. These spaces may contain voids in which occupants can survive the initial collapse. However, because these spaces are below grade, leaking liquids and heavier-than-air gases can flow and collect in them. As time passes after the collapse of the structure above them, these spaces can become less and less habitable for victims and more and more hazardous for rescuers because of accumulated contaminants.

HEIGHTS

In cantilever collapses, the sections of floors remaining in their original position may be especially vulnerable to secondary collapse. However, victims may be stranded on upper floors, and even if they survived the initial collapse uninjured, they may be left without stairs or other means of self-rescue. At the very least, rescue personnel must gain access to these elevated locations to determine whether there are victims in these areas in need of assistance. Working in these situations can be very hazardous because the places that must be searched may be several stories above the ground. If rescuers find victims, they must spend even more time in these hazardous locations in order to safely remove the victims.

EQUIPMENT

In many ways, the equipment needed for structural collapse rescue is no different from that which firefighters use in other emergency operations. However, there are some differences, and some of the ways in which familiar equipment is used may be different. Following is a discussion of the personal protective equipment (PPE) and other equipment most commonly used in structural collapse rescue.

Personal Protective Equipment

Structural fire fighting turnout gear may be more appropriate in structural collapse rescues than in any other type of rescue except those on the fireground. Structural turnouts are certainly appropriate in structural collapse rescues if there is a fire in the structure or the imminent threat of one starting. Protective clothing suitable for the environment of the rescue area should be worn. When working with injured victims or with any other possible biological hazards, rescuers should wear the protective gear recommended in the universal precautions. In the majority of structural collapse rescue operations, rescuers wear the gear described in the following sections.

HELMETS

In situations where structural gear is not needed, the type of helmet most often worn in structural collapse rescue operations is one similar to the wildland firefighter's helmet (Figure 6.28). These helmets have a chin strap, are lighter in weight, allow better peripheral vision, and are less cumbersome in tight places than structural fire fighting helmets.

EYE PROTECTION

Eye protection is needed in most structural collapse rescue situations because there are numerous opportunities for liquids to splash into rescuers' eyes or for flying debris from operating power tools to strike them in the eye. Eye protection is also a part of the universal precautions. Just to keep dust out of their eyes is reason enough for rescuers to wear goggles or safety glasses (Figure 6.29).

Figure 6.28 A typical rescue helmet.

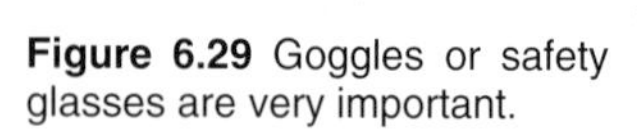

Figure 6.29 Goggles or safety glasses are very important.

OUTERWEAR

Except where protection from actual or potential fires is needed or where extremely wet or cold environmental conditions warrant, structural turnout gear may not be appropriate for structural collapse rescue operations. However, structural turnouts can provide some protection from sharp and jagged debris and from damaged structural members. The outerwear selected should fit snugly enough to avoid getting snagged on or entangled in debris but fit loose enough to allow freedom of movement (Figure 6.30).

FOOTWEAR

Unless the operation must be carried out in standing water or other liquids, rubber boots are usually not needed and are not the best choice for collapse rescue work. Leather boots often provide rescuers a good combination of function and protection. Some agencies provide their personnel with steel-toed boots; others do not. Regardless, selected footwear should be lightweight, have nonslip tread, support the ankles, and be appropriate for the particular environment in which the rescue is to be performed (Figure 6.31).

GLOVES

Heavy-duty work gloves are the most appropriate for structural collapse rescue situations (Figure 6.32). They should allow good freedom of movement and tactile dexterity while providing an acceptable level of protection. As mentioned earlier, when possible biological hazards exist, latex exam gloves should be worn inside of the work gloves.

RESPIRATORY PROTECTION

When rescuers must operate in known or potentially contaminated atmospheres, they must wear supplied-air breathing apparatus. These can take two forms: (1) open- or closed-circuit SCBA (Figure 6.33) or (2) airline equipment with an escape cylinder (Figure 6.34). In atmospheres with dust or other particulates suspended in the air, suitable filter masks may be used if there is ample atmospheric oxygen (Figure 6.35).

Other Equipment

The equipment outlined next is that used to increase the effectiveness and protection of the

Figure 6.31 Typical leather boots worn by rescuers.

Figure 6.30 Rescuers should wear appropriate clothing.

Figure 6.32 Sturdy leather gloves are needed for rescue.

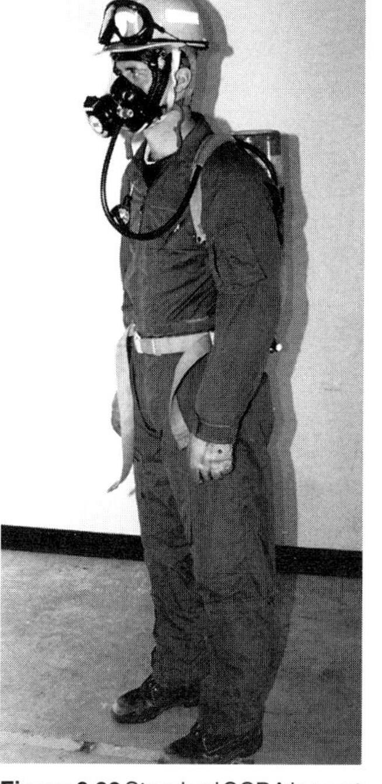

Figure 6.33 Standard SCBA is most often used in rescue.

Figure 6.34 Airline breathing apparatus is used for protracted operations.

Figure 6.35 Filter masks may be used in some circumstances. *Courtesy of Phoenix (AZ) Fire Department.*

rescue teams, as opposed to the PPE worn by each team member. This equipment includes monitoring equipment, hand tools, power tools, air-supply equipment, ventilation equipment, lighting equipment, ropes and related equipment, ladders, and heavy equipment.

MONITORING EQUIPMENT

In structural collapse rescue there are three forms of monitoring equipment used. One is the equipment used to monitor the stability of the structure, another form is used to monitor the atmosphere, and the third form is the equipment used to locate hidden survivors.

Structural-stability monitors. Because of the threat of a secondary collapse, rescuers should use some means of monitoring the stability of the structure. Some of the devices that can be used to sense any movement of the building are plumb lines, builder's transits, and carpenter's levels (Figure 6.36).

Atmospheric monitors. The monitoring equipment used to sample and analyze the atmosphere in voids and other confined spaces must be accurately calibrated, direct-reading instruments that are capable of measuring the oxygen concentration within the space. They must also be capable of measuring the concentrations of toxic and flammable gases and not just be capable of detecting their presence (Figure 6.37).

Figure 6.36 A typical builder's transit.

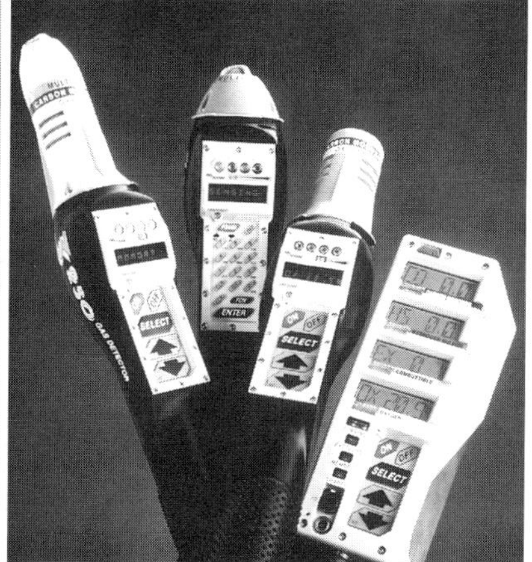

Figure 6.37 A variety of broad-spectrum gas analyzers. *Courtesy of AIM Safety, USA, Inc.*

Victim-locating devices. The equipment used to locate victims in the rubble of a structural collapse can take a variety of forms. Electronic devices used for this purpose include seismic motion detectors, radar motion detectors, acoustic listening devices, fiber-optic cameras, and thermal imaging (infrared) devices. While not considered "devices," specially trained search dogs are also used to locate hidden victims.

HAND TOOLS

In addition to the pry bars, crowbars, and wrecking bars that firefighters use for forcible entry, other hand tools are especially useful in structural collapse incidents. For dealing with rubble and debris, rescuers also need bolt cutters, sledgehammers, and jacks of various types. For constructing and dismantling timber shoring systems, claw hammers and small sledgehammers (single jacks) are also needed (see Rescue Skills section later in this chapter).

POWER TOOLS

The power tools normally used for vehicle extrication (such as hydraulic spreaders and rams; electric and gasoline-driven saws with masonry-, wood-, and/or steel-cutting blades; air chisels; and air bags) are often used in structural collapse incidents (Figure 6.38). Electric, pneumatic, or gasoline-powered jackhammers (concrete breakers) and hydraulic or pneumatic shoring devices are also frequently needed in these incidents (Figure 6.39). Pneumatic nailers can be used to help construct shoring systems (see Rescue Skills section later in this chapter).

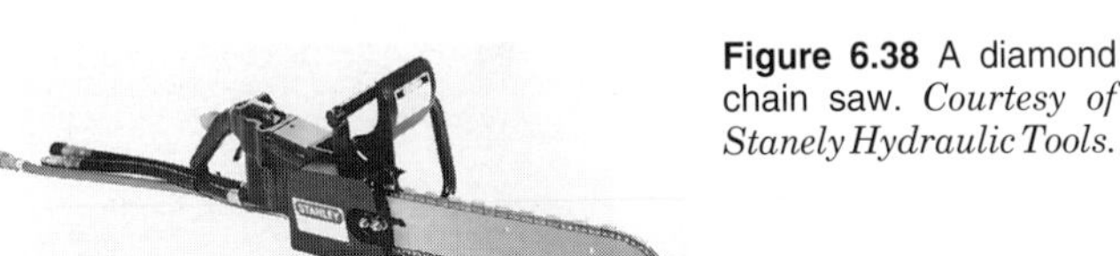

Figure 6.38 A diamond chain saw. *Courtesy of Stanely Hydraulic Tools.*

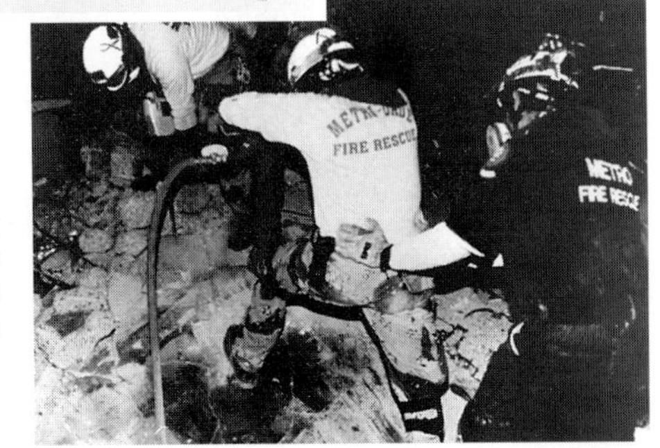

Figure 6.39 A rescuer uses a jackhammer to breach a concrete slab.

AIR-SUPPLY EQUIPMENT

Included in this category is all the equipment needed to provide a safe and dependable source of breathing air for rescuers who must enter contaminated atmospheres and nonbreathing air for pneumatic tools. The equipment may consist of a cache of SCBA cylinders, an air supply unit with a cascade system for refilling empty SCBA cylinders, or an air supply unit with large air tanks and/or an air compressor for supplying airline respirators (Figures 6.40).

Figure 6.40 An adequate air supply is essential.

VENTILATION EQUIPMENT

To enhance the level of safety for rescuers working in voids and other confined spaces in which the atmosphere is contaminated and to increase the chances of survival for trapped victims, the interior of these spaces must be adequately and continuously ventilated by mechanical means. This is usually accomplished with a blower equipped with ducting located outside the contaminated areas to channel the fresh air into the spaces (Figure 6.41). When gasoline-powered blowers are used, they should be positioned downwind of the rubble to reduce the likelihood of exhaust from the engine contaminating the scene.

LIGHTING EQUIPMENT

The type and amount of lighting equipment needed for structural collapse rescues can vary tremendously; but because the victims are often trapped deep within the rubble, some form of artificial lighting is usually needed. Individual flashlights or hand lanterns may be sufficient (Figure 6.42). In many cases, however, some higher level of lighting is required even during the day (Figure 6.43). When portable generators are used, the same precautions are taken regarding exhaust from the generator's engine contaminating the area.

Other types of lighting devices are also available that can make interior search and rescue operations safer. One is a flexible, transparent polyvinyl chloride (PVC) cord that contains tiny clear lights (Figure 6.44). Designed for search and rescue operations, this device is durable and is available in 100-, 200-, and 300-foot (30 m, 60 m, and 90 m) lengths. It is waterproof and can be used in flammable atmospheres. If the cord is deployed as the search team progresses through the structure, it not only provides illumination but makes it easier for the team to find its way out or for others to locate the team inside the structure. Another type of lighting device is a helmet-mounted, high-intensity light (Figure 6.45). As with all other helmet-mounted lights, using this type of light frees both hands of the rescuer for work. However, these lights are rechargeable, and they provide better illumination than the disposable lights that firefighters often strap to their helmets.

Figure 6.41 Fresh air can be supplied through ducts. *Courtesy of Mike McGroarty.*

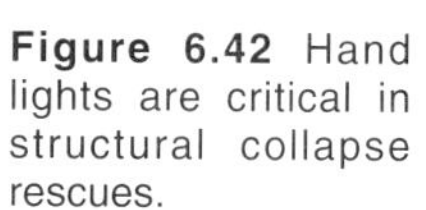

Figure 6.42 Hand lights are critical in structural collapse rescues.

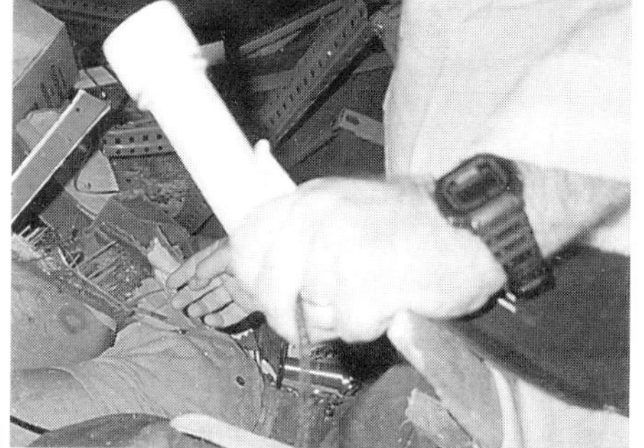

Figure 6.43 Additional lighting is often needed.

Figure 6.44 A rescuer deploys a lighted line. *Courtesy of Flexlite Inc.*

Figure 6.45 One type of helmet light. *Courtesy of NITERIDER LIGHT SYSTEMS.*

RETRIEVAL SYSTEMS

Depending upon the configuration of any confined spaces they must enter, the rescuers may need to be fitted with some form of retrieval system. There must also be an attendant outside of the space(s) monitoring those inside. Retrieval systems consist of a retrieval line (lifeline) attached to a chest harness or to a full-body harness or to wristlets (Figure 6.46). The purpose for using such a system is so that the attendant can pull the rescuer from the space without actually entering the space.

ROPES

Because structural collapse victims may be trapped in areas well above or below grade, rescuers must be trained and equipped to package and extricate them safely and efficiently (Figure 6.47). Any or all of the equipment and skills discussed in Chapter 4, Rope Rescue, may be needed in structural collapse incidents.

LADDERS

All types of fire department ladders have some possible application in structural collapse incidents. Aerial devices may be needed to remove victims from elevated or other difficult-to-reach locations (Figure 6.48). Ground ladders may be needed for climbing up to or down to areas where victims are trapped (Figure 6.49).

HEAVY EQUIPMENT

Other equipment that may be needed at structural collapse incidents is that used in moving massive amounts of rubble and debris. Cranes, booms, dozers, and wheel tractors with front-end loaders are often needed (Figure 6.50).

RESCUE SKILLS

In some cases, making collapsed structures safe for rescuers to enter and free victims from entrapment requires only that rescuers safely and effec-

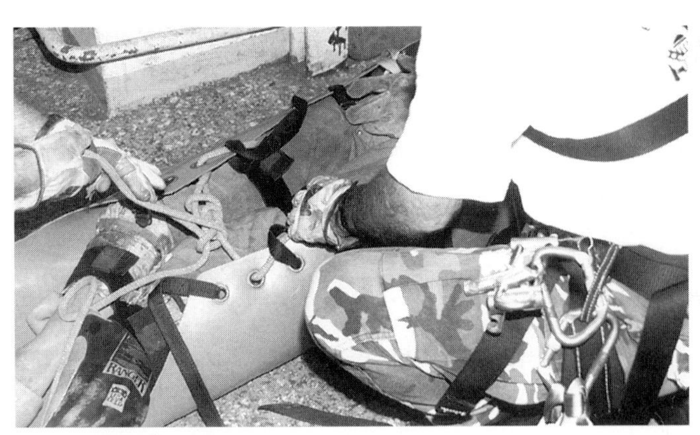

Figure 6.47 Rescuers package a victim for transport.

Figure 6.48 Aerial devices can make access easier and safer.

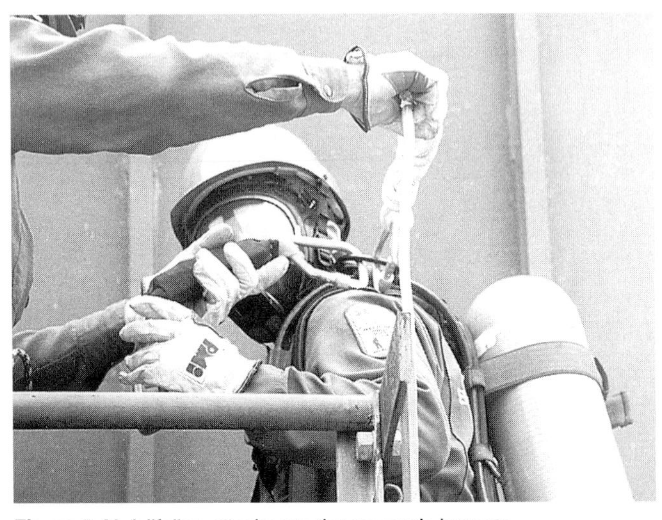

Figure 6.46 A lifeline attaches to the rescuer's harness.

Figure 6.49 Ground ladders are often needed in structural collapse rescue.

Figure 6.50 Heavy equipment is often needed to move rubble.

tively use their PPE and the hand tools and power tools listed earlier. However, other situations may require using technical rescue skills. Rescue personnel need specialized training to apply these technical skills correctly. In structural collapse incidents, the technical rescue skills needed are:

- Using monitoring devices
- Rigging mechanical advantage systems
- Shoring and tunneling

Using Monitoring Devices

As mentioned earlier in this chapter, the monitoring devices used in structural collapse incidents fall into three categories: those used to monitor structural stability, those used to monitor the atmosphere, and those used to locate hidden victims. Following is a discussion of the specific techniques for using these devices.

STRUCTURAL MONITORING

As previously mentioned, there are a variety of ways in which the stability of the building can be monitored. The most common ways are with a plumb line, a transit, or a level.

Plumb lines. In an area sheltered from the wind, a plumb line can be attached to a high point on the structure close to a wall or other vertical structural member with the plumb bob a short distance above the floor. When the plumb bob has stopped moving after attachment, the distance from the point of the bob to the wall or member is carefully measured (Figure 6.51). The measurement is recorded and rechecked at regular intervals. Any difference in measurements is reported to the Safety Officer immediately. If the line is attached near an inside corner, movement of both walls can be monitored from one plumb line.

Transits. A surveyor's transit can be set up on stable ground a short distance from the structure. The transit is then sighted on a mark on a wall of the building (Figure 6.52). As with any form of structural monitoring, any movement detected is reported immediately. If the collapse were the result of an earthquake, the stability of the transit could be disturbed by aftershocks, rendering the subsequent sightings unreliable.

Levels. A construction level — the longer the better — can be taped to a vertical wall or structural member (Figure 6.53). Any movement of the bubble should be reported immediately.

Figure 6.51 A plumb line can be used to monitor the building.

Figure 6.52 A rescuer uses a transit to check for building movement.

Figure 6.53 A builder's level can also be used.

ATMOSPHERIC MONITORING

Atmospheric monitoring devices are available in a wide variety of models and configurations to match the particular settings for their intended use. For structural collapse and other confined space rescues, the substances that must be monitored are oxygen and flammable or toxic gases and vapors. The devices used must be capable of indicating the level of the gas or gases detected; most do so in percentages or parts per million (ppm) or both. Some devices detect and measure oxygen only, while others detect and measure flammable vapors or gases only. Other devices detect toxic gases or vapors only. Many departments now have combination units that detect and measure all three. The devices available also differ in their design. Some devices must be used in the atmosphere to be monitored, while others are equipped with a probe or wand that can be inserted into a confined space from outside the space.

Before atmospheric samples are taken, some instruments must be zeroed upwind of the space in an atmosphere that is free of contaminants. Whenever possible, atmospheric sampling is done from outside of the space being monitored. In all cases, however, the atmosphere is first checked for oxygen because most combustible-gas meters do not give an accurate reading in an oxygen-deficient atmosphere. The space is then checked for flammable/combustible gases or vapors because the threat of fire or explosion is — in most cases — more immediate and more life threatening than exposure to toxic gases or vapors. Finally, the space is checked for those toxic gases or vapors that are most likely to be present in the space.

Figure 6.54 A rescuer checks the NIOSH pocket guide.

The results of these tests are used to determine whether and/or when it is safe for rescuers to enter, what type and level of PPE they need, and the likelihood of finding viable victims inside. For rescuers to safely enter a confined space without supplied-air breathing apparatus, the oxygen level within the space should be between 19.5 percent and 23.5 percent. Flammable gases or vapors should be below 10 percent of the material's lower flammable limit (LFL). Toxic gases or vapors should be below the limit specified in the applicable Material Safety Data Sheets (MSDS) if available or in an appropriate reference such as NFPA 49, *Hazardous Chemicals Data*, or the National Institute for Occupational Safety and Health *(NIOSH) Pocket Guide to Chemical Hazards* (Figure 6.54). Atmospheres with toxic gases or vapors that prove to be beyond these limits may require that rescuers wear supplied-air respirators, that the spaces be mechanically ventilated, or both.

As long as a space continues to be occupied by rescuers or victims, the atmosphere within the space should be monitored *continuously*. In order to do this, however, the device or its sampling probe may have to be removed from the space from time to time and recalibrated in clear air.

LOCATING HIDDEN VICTIMS

As mentioned earlier in this chapter, the devices used to locate victims hidden in the rubble of a structural collapse may take several forms. Rescuers should become familiar with the operation of all such devices available within their response district. The following paragraphs discuss how the most common methods and devices should be used.

Hailing. Before using any of the devices described in this section, rescuers should use the oldest method of locating hidden victims — hailing. Rescuers should repeatedly call out during primary and secondary searches and pause quietly for a few moments after each call to listen for a response. Rescuers should shout into voids and inaccessible areas as they make their way through the rubble. Shouting *"Rescue squad, is anyone in there?"* or a similar call may produce a response from victims who are conscious but unable to free themselves.

Motion detectors. Motion detectors are available with either of two operating principles: seismic or radar. Seismic motion detectors are electronic devices that are extremely sensitive to movement. Whenever a victim trapped in the rubble of a structural collapse moves, otherwise imperceptible vibrations may be transmitted through the solid material. With the detectors placed strategically throughout the rubble pile, the unit picks up these vibrations and initiates an audible signal (Figure 6.55). Short-distance radar units are highly portable but can generate a signal that penetrates any nonmetallic building material (Figure 6.56). Any movement within the range of the antenna, about 20 feet (6 m), initiates an audible alarm. Using a cable attachment, the antenna can be placed up to 200 feet (62 m) from the unit.

Listening devices. Acoustic listening devices are extremely sensitive, electronically enhanced listening devices for detecting otherwise imperceptible sounds. In some units, sensors placed throughout the rubble pile pick up otherwise inaudible sounds made by trapped victims (Figure 6.57). In other designs, a tiny microphone in the end of a probe can be inserted into voids through openings too small for a rescuer to enter without moving rubble and debris. However, the noise produced by power tools and other equipment can make using these listening devices very difficult.

Search cameras. Industrial-quality video cameras (camcorders), often equipped with fiber-optic lenses, can be used to look inside voids and other confined spaces when rubble and debris make it too difficult to enter (Figure 6.58). Even with very sensitive camcorders and videotape, additional lighting may have to be provided to get acceptable results. This adds another possible source of ignition unless explosion-proof lighting is used.

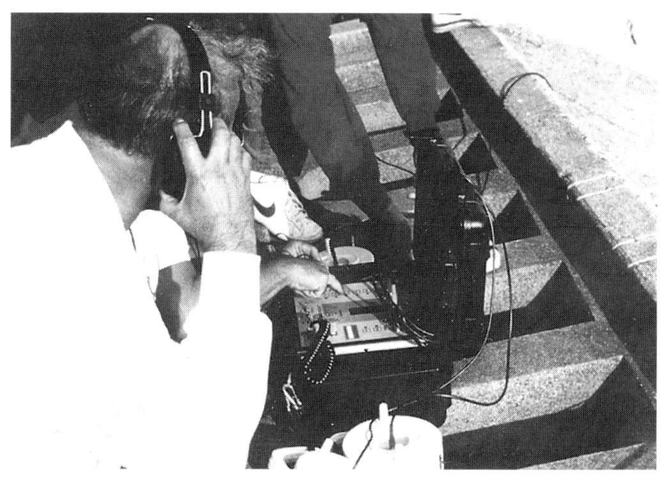

Figure 6.55 A seismic motion detector is monitored. *Courtesy of Mike McGroarty.*

Figure 6.56 A short-distance radar unit. *Courtesy of Hughes Missile Systems.*

Figure 6.57 A typical listening device. *Courtesy of DELSAR, Inc.*

Figure 6.58 A rescuer uses a fiber-optic search camera. *Courtesy of Mike McGroarty.*

Thermal imagers. These infrared devices are similar to video and photographic cameras except that their image is produced by differences in temperature instead of differences in light.

Search dogs. Of course, search dogs are not inanimate devices, but because their primary function is to locate hidden victims, they are also included in this section. These highly specialized dogs are different from those used to track people in the wilderness or those used in law enforcement. Some search dogs are trained to recognize the difference between live victims and cadavers. Others may alert their handlers to any victims, alive or dead. Unless trained to do so, rescuers should not attempt to handle search dogs but should coordinate the activities of those who are trained as handlers (Figure 6.59).

Figure 6.59 Search dogs can be very effective.

Rigging Mechanical Advantage Systems

Mechanical advantage systems are needed in structural collapse incidents primarily to either raise or lower packaged victims to grade. The skills needed are the same as those discussed in Chapter 4, Rope Rescue.

Shoring And Tunneling

Freeing trapped victims of a structural collapse may involve shoring unstable or potentially unstable rubble and/or tunneling through the rubble and debris to reach the victims. While the materials used and the skills needed for shoring and tunneling are the same, how each is accomplished is different.

SHORING

In this context, *shoring* is a general term used to describe any of a variety of means by which unstable structures or parts of structures can be stabilized. It is the process of preventing the sudden or unexpected movement of objects that are too large to be moved in a timely manner and that may pose a threat to victims and/or rescuers. Shoring is not intended to move heavy objects but is just intended to stabilize them. Stabilizing objects with shoring may involve applying air bags, applying cribbing, using jacks, using pneumatic shores, constructing a system of wooden braces, or using a combination of these methods. Deciding the most appropriate type of shoring in a given situation is a matter of judgment based on training and experience.

Air bags. While air bags are often used to lift heavy objects, they also are used to temporarily stabilize an object until more permanent shoring can be applied (Figure 6.60). They should be positioned on a relatively clean surface that is firm enough to withstand the pressure exerted when the bag is inflated against the object. On soft ground, a piece of heavy plywood placed beneath the bag spreads the load and protects the bag from punctures. If the vertical distance between the horizontal surface and the bottom of the object to be stabilized is too great for one bag, the bag can be placed atop cribbing with a solid top tier, or a multicell bag can be used.

Cribbing. Applying cribbing is one of the best methods of stabilizing horizontal objects because cribbing is very strong, is readily available on rescue apparatus, and can be cut from material found on the scene. Plus, it can be applied quickly. Cribbing can be effectively applied to vertical openings equal to a maximum of three times the length of the cribbing (Figure 6.61). Cribbing stacked higher than this distance becomes less stable as the height increases. When the top tier of cribbing does not quite make contact with the bottom of the object to be supported, the remaining space must be filled with shims or wedges (Figure 6.62). If rescuers must work under a heavy horizontal object, such as a collapsed wall, cribbing should be applied on both sides of the entry point and not more than 4 feet (1.2 m) apart (Figure 6.63). Rescuers can then work

Figure 6.60 Heavy objects can be lifted or stabilized with air bags. *Courtesy of Joel Woods.*

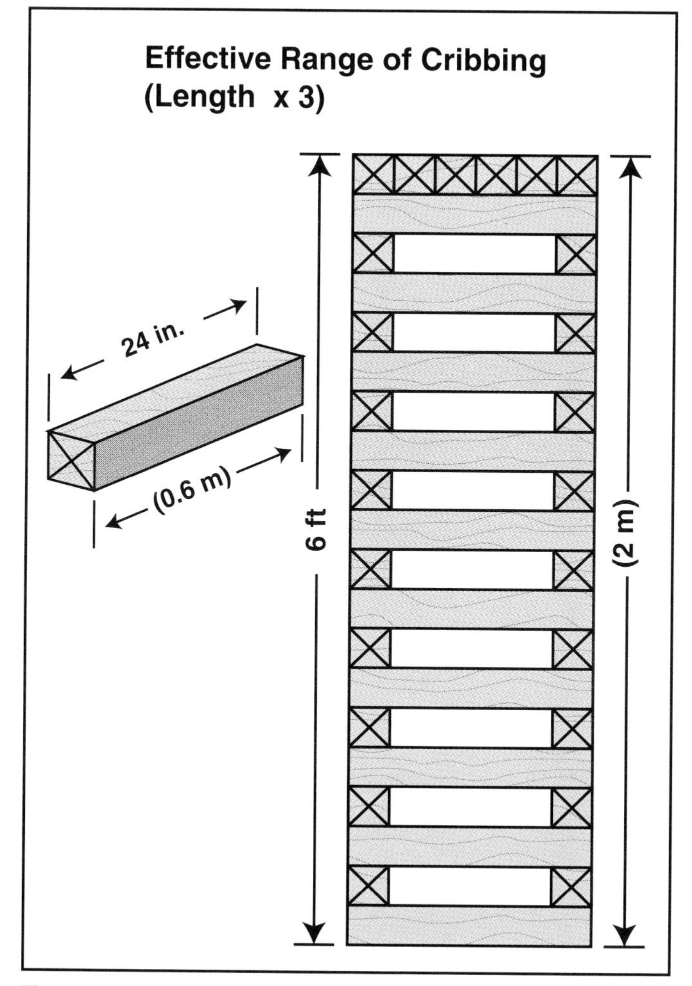

Figure 6.61 Cribbing can be safely stacked three times as high as the pieces are long.

Figure 6.62 Cribbing should make full contact.

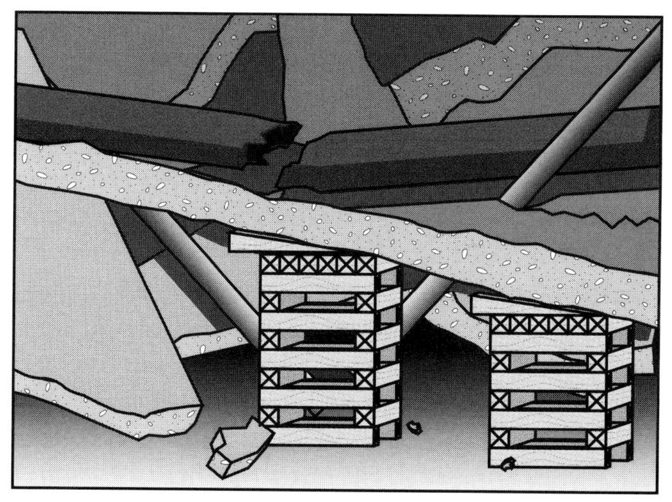

Figure 6.63 Both sides of the entry point should be cribbed.

between the cribs and within an arm's reach beyond them. If the victims are still out of reach, additional pairs of cribs can be constructed forming a safe corridor to the victims.

Jacks. Jacks can also be used to stabilize horizontal objects. The types most often used in structural collapse rescue are screw jacks and hydraulic jacks. Regardless of the type used, they generally work best when the vertical distance between the horizontal surface and the object to be stabilized is less than 2 feet (0.6 m). However, jacks can be used in bigger openings if they are used in combination with cribbing. When screw jacks are used, the threaded shaft should be rotated until the cap makes full contact with the object and is snug against it. No attempt should be made to lift the object with the screw jack.

Pneumatic shores. Pneumatic shores having a minimum load capacity rating of 10 tons (9 tonne) may be used for stabilizing heavy horizontal objects.

Wooden shoring systems. Constructing a system of wooden shores may be the most appropriate and safest way to support heavy horizontal or vertical objects. Wooden shoring is constructed in four basic configurations: *dead shores*, *raker shores*, *flying shores*, and *braces or struts*. These systems are usually constructed of heavy timbers (at least 4 x 4 inch) found on site or brought in. Timbers with a square cross section are better for shoring because they are stronger than those with a rectangular cross section of equal area. All struts or braces should be as short as possible because the shorter they are, the stronger they are. Regardless of the configuration needed, all wooden shoring systems should be laid out and constructed on a flat surface away from the point of application. They should then be installed where needed using as little hammering as possible to minimize the shock delivered to the object being supported. Pneumatic nailers are sometimes used with the system in place because they deliver less shock to the material being nailed. Adjustments to the shoring should be made slowly using a pry bar and wedges or with a screw collar.

Figure 6.64 A dead shore system in place.

Figure 6.65 A raker shore in place. *Courtesy of Mike McGroarty.*

Dead shores. Also called *vertical shores*, dead shores are the simplest form of shoring system and are the easiest to construct. A dead shore consists of a soleplate, vertical strut(s), a header, and wedges or screw clamps (Figure 6.64). While the soleplate may be of any size lumber that will spread the load over a large area, it should have a nominal thickness of not less than 2 inches (50 mm) and should be at least as wide as the struts. Each strut should be cut 1½ to 2 inches (38 mm to 50 mm) shorter than the vertical distance between the soleplate and the header if wedges are used, or 10 inches (250 mm) shorter if screw clamps are used. This space allows wedges to be driven in from opposite sides or clamps to be applied at the bottom of the strut to tighten it. After the struts are securely in place, diagonal braces and/or gusset plates must be installed for additional stability.

Raker shores. A raker shore consists of a wall plate, a raker, and a soleplate. They are most often used to brace freestanding walls that are unstable because they are deformed or out of plumb (Figure 6.65). Raker shores are strongest when the raker is installed at an angle of 45 degrees or less to the horizontal, but it can be installed at any angle up to 60 degrees. If a raker must be installed at an angle greater than 45 degrees, the wall plate may have to be attached to the wall to prevent it from slipping upward when tension is applied. Attaching a wall plate can be a dangerous operation because it may involve nailing it in place or drilling into the wall to insert masonry screws. Either method can deliver a considerable amount of shock to an already unstable wall. If there is a gap between the wall plate and the wall because the wall is deformed, the space should be filled with wooden shims to ensure full contact between the wall plate and the wall (Figure 6.66). The soleplate should be embedded in the horizontal surface at about 90 degrees to the angle at which the raker will be installed (Figure 6.67). The raker should be cut to extend from the wall plate down to the soleplate. Once the top of the raker is against the cleat on the wall plate and the bottom is against the soleplate, a pry bar or a strut and wedges can be used to slowly force the raker into position. A cleat is then nailed onto the soleplate to hold the raker in place. Once the shore is in place, a brace between the raker and the wall plate can be installed for additional stability.

NOTE: When struts or braces are cut for rakers or any other shoring system, the ends should be cut so that they make full contact with the header or wall plate (Figure 6.68).

Flying shores. As the name implies, flying shores are those installed above the ground or other stable horizontal surface. Flying shores consist of horizontal beams, struts, wall plates, and wedges. Flying shores are installed to stabilize two opposing walls. They are configured very much like a raker shore except that a horizontal beam is added in place of the horizontal surface involved in a raker shore. First, a cleat must be nailed to each wall

Figure 6.66 The space between a wall plate and the wall must be filled to allow full contact.

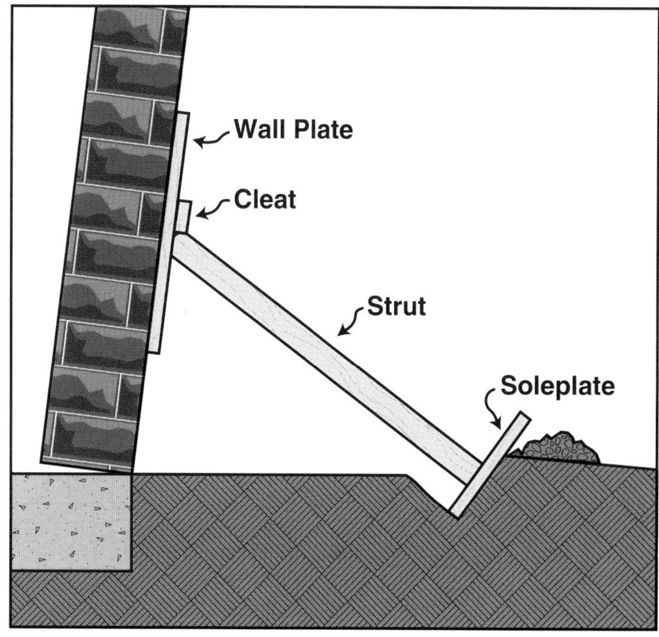

Figure 6.67 The soleplate should be at 90 degrees to the raker.

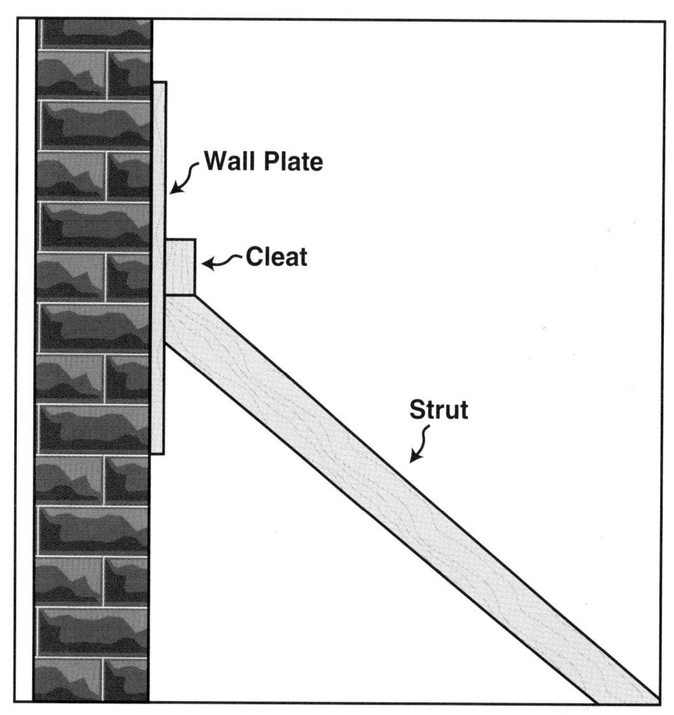

Figure 6.68 The ends of struts/braces should make full contact.

plate and the wall plates attached to the walls, one opposite the other. The beam will rest on these cleats. The beam should be cut slightly shorter that the span between the two walls to allow space for the wedges to be inserted at one end. The wedges should be driven in to tighten the beam between the wall plates. One of the wall plates must also have a cleat nailed to it some distance above the beam. A raker strut must be cut to fit between the upper cleat and another cleat nailed to the top of the beam near the other end. Wedges are driven between the end of the raker strut and the cleat on the beam to tighten the strut (Figure 6.69). A variation of this configuration involves installing a raker between each wall plate and the middle of the beam (Figure 6.70).

Struts/Braces. Vertical or horizontal struts or braces may be used to shore weakened window and doorway openings (Figure 6.71). Depending upon how the openings need to be shored, the shoring system may involve horizontal struts between opposing wall plates or involve vertical struts between a soleplate and a header. In either case, enough room must be left between the struts to allow rescuers and victims to pass through the openings. Hydraulic or pneumatic shores may also be used to shore window or doorway openings.

Tiebacks. When a wall is weakened too high up to make shoring practical and the wall might endanger rescuers or victims below, tiebacks can sometimes be used to stabilize the wall. In this system, the weakened portion of one outside wall is supported by being tied to the opposite outside wall (Figure 6.72). Although commonly referred to as *cable tiebacks*, utility rope, cable, or 2- by 4-inch lumber may be used to tie the walls together.

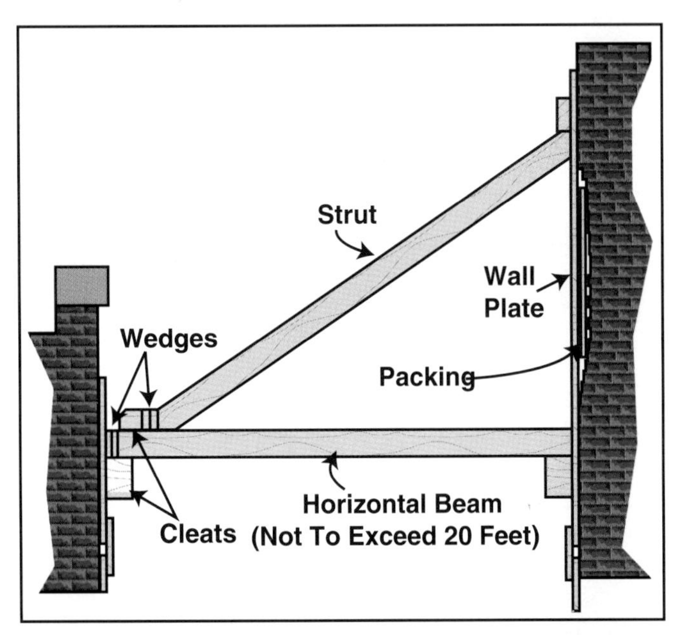

Figure 6.69 Wedges are used to tighten the raker strut in a flying shore.

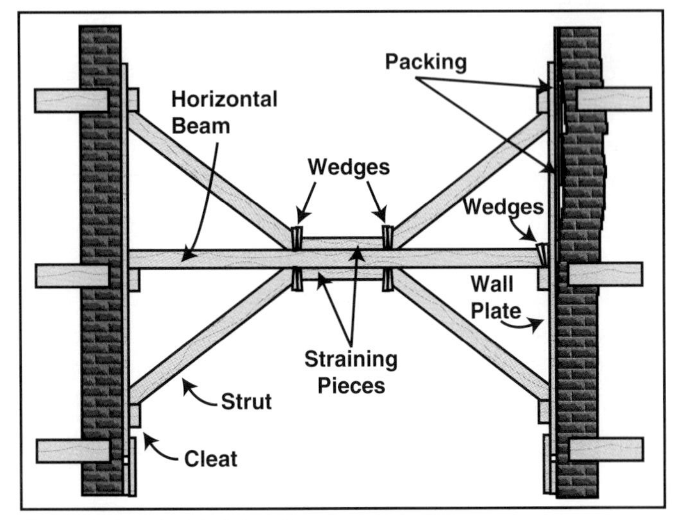

Figure 6.70 A double flying shore.

Figure 6.71 Struts/braces are used to stabilize weakened openings. *Courtesy of Mike McGroarty.*

Figure 6.72 Cable tiebacks can stabilize weakened walls.

TUNNELING

Tunneling may involve shoring large pieces of overhanging rubble, but shoring is not its main function. Tunneling primarily involves removing smaller rubble and debris to create a path to a victim whose location is known. Because it is a slow and dangerous process, tunneling should be used only when all other means of reaching a victim have proven ineffective.

If a victim is known to be under tons of rubble and debris and time does not allow for working down to the victim by removing layers of debris from above, tunneling through the debris may be the only option. Rescuers must be very careful when they begin tunneling because when a piece of debris is moved, there is a chance that it will start a chain reaction of falling debris. That could undo all of the work done up to that point and/or bury the rescuers under tons of debris. There are two ways in which rescuers can protect themselves from this contingency: selective debris removal and constructing a protective enclosure within the tunnel as they progress.

Tunnel debris removal. As rescuers progress through a pile of rubble and debris, they must be very careful with each piece moved. Generally, if pieces of debris are easily moved, they are not under the weight of other debris and are probably safe to move. However, each piece should be picked up individually and checked for wires or other connections to the rest of the debris. Throwing an item still connected by an electrical cord, for example, might dislodge a *key member* — a piece of debris that tends to bind the pile together. Dislodging a key member can start the sort of chain reaction mentioned earlier. When rescuers find a piece of debris that resists being moved, it is likely to be a key member and should be left alone. Because key members must be left in place undisturbed, the tunnel through the debris will have to go around them. Therefore, the tunnel may not be a straight and level corridor to the victim.

Enclosure construction. The other way that rescuers can protect themselves from falling debris is by constructing an open-ended wooden enclosure through which they and the victim may pass. Most debris tunnel enclosures are either square or triangular, but they may be configured in any shape that works in the situation at hand (Figure 6.73). These enclosures can be constructed from materials found on the scene, but because of consistency and uniformity, materials brought in for that purpose may be better (Figure 6.74). The enclosure may be constructed in the form of a frame clad with planks or it may be made of heavy plywood.

TACTICAL CONSIDERATIONS

The tactical considerations involved in any structural collapse rescue may be broken down into six distinct phases: scene survey and control, surface victim removal, void access and search, selective debris removal, general debris removal, and termi-

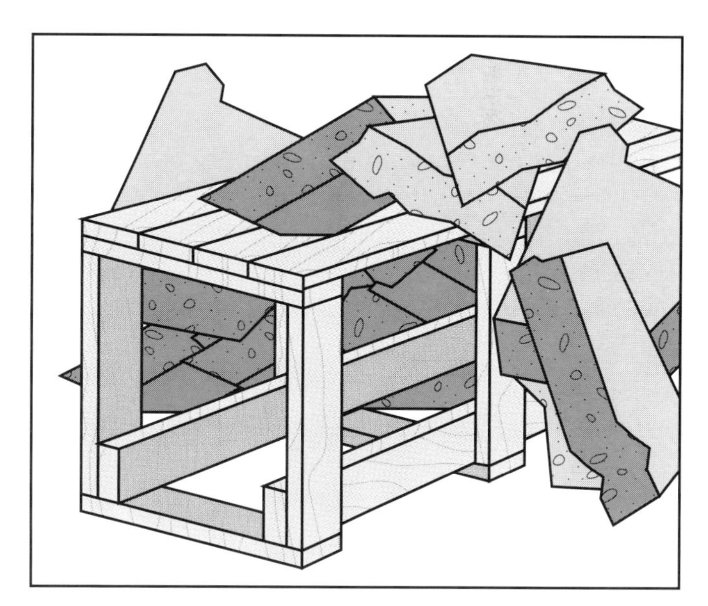

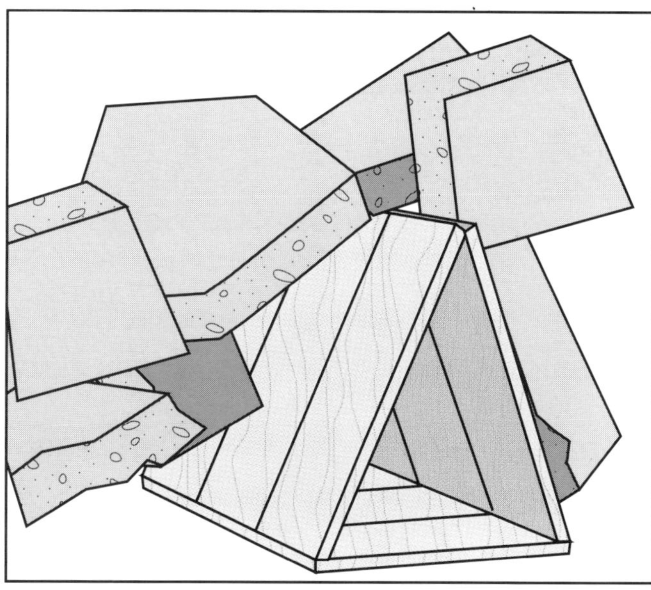

Figure 6.73 Typical debris tunnels.

Figure 6.74 A rescue trailer with shoring/tunneling material. *Courtesy of Mike McGroarty.*

nation. This sequence does not include a specific rescue phase because rescues are performed during any or all of the first five phases.

Phase I: Scene Survey And Control

The first-arriving units must make an initial assessment of the situation and take control of the scene. This will involve conducting primary and secondary assessments.

PRIMARY ASSESSMENT

The primary assessment actually begins with the initial dispatch and continues during response and after arrival. The first-due rescuer must begin to formulate a mental picture of the situation based on the information provided in the dispatch, the time of day, the weather, and the traffic conditions during response to the scene. This mental picture will be further refined and enhanced during the initial size-up.

Information gathering. The process is continued once on the scene by attempting to talk to the reporting party (RP) and/or other witnesses (Figure 6.75). Of critical importance in the primary assessment is gathering any available information about the number, condition, and location of victims:

- Have all occupants been accounted for?
- How many victims are there?
- Is their location known?
- Are they injured or merely trapped?
- Are they conscious, and if so, can they communicate?

Decision making. The answers to these questions help the first-in officer make the first critical decision: Can the units on scene or en route handle the situation, or do additional units need to be called? If more resources are needed, they must be requested immediately to get them on scene as soon as possible (Figure 6.76). If he or she has not done so already, the first-in officer should assume formal command of the incident because the answers to the initial questions also form the basis for the incident action plan that the IC must begin to develop at this point.

Scene control. If the information gathered during the primary assessment confirms that a legitimate rescue emergency exists, the area surrounding the collapsed structure should be cordoned off as described in Chapter 2, Rescue Scene Management (Figure 6.77). Which areas are designated as the hot, warm, and cold zones will depend on the size, nature, and complexity of the collapse situation. The boundaries of the zones may be affected by the presence of secondary hazards (Figure 6.78).

Figure 6.75 The first-in rescuer quizzes the reporting party.

Figure 6.76 Additional resources must be called as soon as possible.

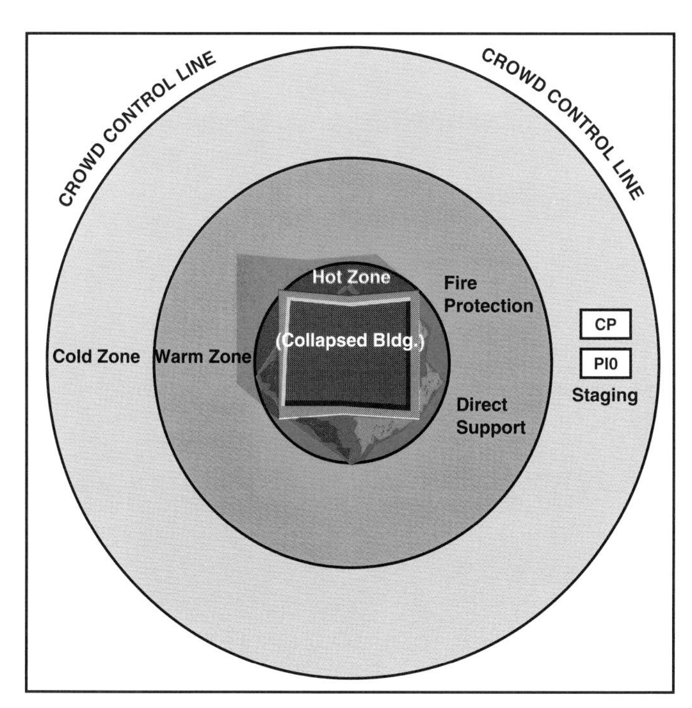

Figure 6.77 Typical control zones.

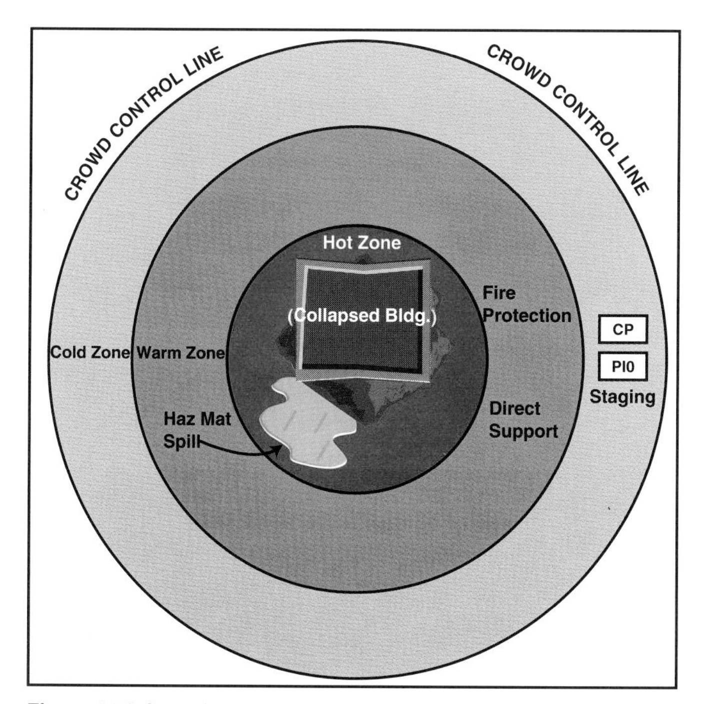

Figure 6.78 Control zones may have to be enlarged.

PERSONNEL ACCOUNTABILITY

The purpose of the accountability system is to ensure that only those who are authorized and properly equipped to enter the structure are allowed to do so and that their location and their status are known as long as they remain inside. As is true of the incident action plan, the degree of formality of the personnel accounting system should reflect the nature, size, and complexity of the particular rescue problem at hand.

Minimum accountability. Some incidents require only one or two rescuers working in the structure at a time, and they do not require respiratory protection. They are close enough to be in constant visual and verbal contact with the team leader. In these cases, the intent of the accountability system can be met with a minimum amount of information being recorded. Merely writing down the entrants' names and their entry and exit times should be sufficient.

Maximum accountability. A more formal system must be employed on larger, more complex incidents requiring more personnel to work inside the structure at once, when rescuers must use respiratory protection, or when they must work out of sight of an attendant. In these cases, the entrants' names and/or other identifiers must be recorded as they enter (Figure 6.79). Their time of entry, SCBA gauge readings, and their *projected* exit times must also be written down. Those using airline equipment can remain in the space for longer periods of time than those wearing open-circuit SCBA, but the same rules of accountability apply. Each entrant's time of exit must be recorded as they leave the structure. As each entrant's projected exit time arrives, the attendant must check the list to ensure that he or she has exited. If anyone has not exited by the time on the list, immediate action must be taken to locate them and escort them out of the structure.

Figure 6.79 Rescuers give personal identifiers to Accountability Officer as they enter the scene. *Courtesy of Clemens Industries, Inc.*

SECONDARY ASSESSMENT

The secondary assessment involves some reconnaissance of the scene to gather information about the structure, its condition, and its contents. All information gathered during both the primary and secondary assessments help determine the mode of operation.

Type of structure. The type of structure involved can indicate some of the problems that must be dealt with. For example, industrial buildings may contain hazardous materials or high-voltage electrical equipment. Residential structures, office buildings, public assemblies, shopping centers, or schools may be heavily occupied depending on when the collapse occurred. Structures that were heavily occupied at the time of the collapse can greatly increase the need for ambulances at the scene. The size of the structure and the type of construction can indicate how much and what types of equipment will be needed to move the rubble and debris (Figure 6.80).

Figure 6.80 Heavy equipment may be needed to move rubble.

Condition of structure. The condition of the structure must also be assessed. Some of the questions that must be answered are as follows:

- What is the extent of damage?
- Is the collapse partial or total?
- Will shoring and tunneling be necessary?
- Will rubble have to be moved to allow access to victims?
- Are there secondary hazards in the rubble?

As rescuers assess the condition of each collapsed structure, they should mark it with spray paint, lumber chalk, or lumber crayon to indicate its status. Before they enter the structure they should make a 2- by 2-foot (0.6 m by 0.6 m) square next to the main entrance. If the square cannot be made immediately adjacent to the entrance, they should make it as close as possible and mark an arrow next to the box pointing toward the entrance point. To the right of the square they should put the date, time of entry, and unit identifier. After their survey of the structure is completed, they should add their assessment of hazardous materials conditions within the structure (Figure 6.81). This mark would indicate that the structure is accessible and safe for search and rescue operations, that damage is minor and there is little danger of secondary collapse, but there is some natural gas hazard.

If the structure is significantly damaged, a slash should be added across the box (Figure 6.82). This would indicate that some are relatively safe, but other areas may need shoring, bracing, or removal of unstable debris.

If the structure is not safe for search and rescue operations, two slashes should be added across the box (Figure 6.83). This would indicate that the structure may be subject to sudden additional collapse. Selective, limited search operations may be done at significant risk. If rescue operations are undertaken, areas of safe refuge must be created and rapid escape routes identified.

The results of these surveys provide the IC with more information on which to base decisions about what additional resources to request. Heavy equipment for moving slabs of concrete or jackhammers for penetrating them may be needed. Timbers and other materials needed for shoring and tunneling

Figure 6.81 A typical structural hazards marking.

Figure 6.82 A mark indicating significant structural damage.

Figure 6.83 A mark indicating that the structure is not safe for search and rescue.

may have to be requested. Portable pumps may be needed to keep voids from filling with water.

Contents of structure. In addition to information about the level of occupancy and the structure's condition, information about the inanimate contents of the structure needs to be gathered. The presence of heavy machinery or merchandise may indicate a need for shoring materials or for heavy equipment to move the items. If documents such as MSDS are available, they can provide critical information about the nature and quantity of the contents. This information may indicate the possible condition of the victims and the type and level of protection needed by the rescuers (Figure 6.84).

Figure 6.84 Some situations may require special protective clothing.

Mode of operation. Finally, all of the information gathered during the primary and secondary assessments confirms the nature and extent of the rescue problem and helps the IC finalize the incident action plan. The information also helps the IC make one of the most important decisions affecting the action plan: whether it is reasonable to think that the victims are alive and that the operation should be conducted as a rescue or that victims could not be expected to have survived and that the operation should be conducted as a body recovery. However, the IC should not make this decision based on the amount of structural damage alone; many survivors have been rescued from buildings that were completely demolished.

During this initial phase, all of the things necessary to make the rescue operation as safe as possible must be done. This includes finalizing the incident action plan, gathering the necessary resources, monitoring and managing the atmosphere in and around the collapse site, and implementing an appropriate personnel accountability system.

Phase II: Surface Victim Removal

The actions necessary to complete this phase will usually be taken by the first-arriving units and/or by civilian bystanders. This phase involves the removal of any victims who have made their way out of the rubble and those still in the rubble but who are only lightly buried by the debris. These rescues can normally be made using only the tools and equipment carried on most engines and trucks.

The majority of survivors of building collapses are recovered in this phase. These survivors include those who were removed by civilians prior to the arrival of the rescue squad, those who made it out of the building on their own, and those who were not necessarily in the building at the time of the collapse but were injured by falling debris.

OCCUPANT/VICTIM TRACKING

As part of this initial rescue phase, someone should be assigned to identify all occupants of the building and to keep track of them (Figure 6.85). Some may leave the scene and simply go home, some may be transported to various hospitals, some may remain in the building awaiting rescue, and others may be unaccounted for. Before any

Figure 6.85 The names of uninjured occupants should be recorded.

occupants of the collapsed structure are allowed to leave the scene on their own, they should be interviewed to obtain the following information:

- Where were they at the time of the collapse?
- By what means/route did they escape?
- Was anyone else with them or near them?
- Were others known to be in the structure?
- Who were they — name, sex, age?
- What were these others wearing?
- Where were they last seen?
- How would the room/area be described?
- What were the contents and finishes in the room/area?
- What was the room/area used for?
- Was there any warning before the collapse?

It is essential for rescuers to remember that they did not cause the collapse, that they are not responsible for the victims being in that situation, and that they are not obligated to sacrifice themselves in a heroic attempt to save the victims and especially not in an attempt to recover a body or bodies. In fact, it is irresponsible and unprofessional for rescuers to take unnecessary risks that might result in their being incapacitated by an injury and therefore unable to perform the job for which they have been trained. It is not the function of the fire/rescue service to add victims to the situation. Having said this, we can now discuss the safe and proper ways in which rescuers should deal with structural collapse rescues.

Phase III: Void Access And Search

The next largest group to survive a building collapse, as many as 30 to 40 percent, are rescued during this third rescue phase. Voids can be located by surveying the rubble pile and identifying the primary type(s) of collapse: pancake, V-shape, lean-to, or cantilever (See Types of Collapse section earlier in this chapter). Each type of collapse tends to leave voids in specific patterns.

Void exploration involves entering accessible spaces, calling out (hailing) to possible victims, and waiting for a reply. All voids should be searched visually, or search dogs can be sent in if they are available (Figure 6.86). As rescuers search a building, they should mark their progress and findings with a standard marking system.

Figure 6.86 Search dogs may be used in this phase.

Void exploration is different from trenching or tunneling because rescuers do not dig through the rubble and debris. They work their way through the pile without disturbing it and attempt to find existing openings. Rescuers will most often enter voids through existing openings, but they may have to cut through walls to gain access to some voids (Figure 6.87). A minimum of shoring is done during this phase because it is intended to be a quick primary search to locate victims and to rescue those who can be removed with little difficulty. Victims stand a better chance of survival if they can be removed promptly. Those who are more seriously trapped will have to be rescued in the next phase: selective debris removal.

Phase IV: Selective Debris Removal

After the first three phases have been completed and the remaining victims have been located, the IC may implement the selective debris

Figure 6.87 A rescuer uses a diamond chain saw to breach a void wall. *Courtesy of Stanley Hydraulic Tools.*

Figure 6.88 Emergency shoring installed to stabilize rubble. *Courtesy of Mike McGroarty.*

removal phase of the operation. As the name implies, it is the selective removal of debris from the rubble pile in order to reach a specific location. It should never be done without a clear objective — to reach a known victim in a known location — and should never be done if all live victims have been rescued. Selective debris removal may involve trenching, tunneling, or breaching walls or roofs. This is an exceptionally dangerous phase of the operation because rescuers will be disturbing and removing portions of the debris pile, and that could cause a secondary collapse. Only those who have been trained to perform such work should be allowed to do so.

Rescuers must be able to recognize the difference between load-bearing and nonload-bearing elements of the rubble pile. They will have to remove debris in a way that will not precipitate a secondary collapse, and they may have to construct shoring systems to make the work scene safe for rescuers (Figure 6.88). If conditions develop that make the scene too dangerous for rescuers to work, debris removal should be suspended until those dangers are removed or stabilized. Because the point of this phase is to rescue live victims, heavy machinery must not be used during this phase of the operation. The person assigned to track victims should be consulted before this phase of the operation is terminated.

Phase V: General Debris Removal

Once all live victims have been removed from the structure, the next phase — general debris removal — can begin. The point of this phase is to restore the scene to a safe condition so that restoration or demolition can be done in relative safety. This phase should be conducted as a recovery operation, exposing rescuers to as little risk as possible. If all occupants have been accounted for, the debris removal can proceed at a normal pace. If any remains have still not been recovered, the process must be done more slowly. Heavy equipment can be used to remove all rubble from the pile and spread it out for examination. Once this process has been done, the debris can be picked up and hauled away.

Phase VI: Termination

The termination phase of a structural collapse rescue involves an obvious element such as retrieving pieces of equipment used in the operation. But it also involves less obvious elements such as investigating the cause(s) of the incident, releasing the structure to those responsible for it, and conducting critical incident stress debriefings (CISD) with members of the rescue teams.

EQUIPMENT RETRIEVAL

Depending on the size, complexity, and length of time involved in the operation, the job of retrieving all of the various pieces of equipment used may be very easy or it may be very difficult and time-consuming. Under some circumstances, it can also be quite dangerous.

Identifying/collecting. The process of identifying and collecting pieces of equipment assigned to the various pieces of apparatus on scene is much easier if each piece of equipment is clearly marked (Figure 6.89). However, it may be necessary for the driver/operators of rescue units and other pieces of apparatus on scene to conduct an inventory of their equipment prior to leaving the scene (Figure 6.90).

Figure 6.89 Equipment that is marked is easier to retrieve.

Figure 6.90 Apparatus operator may have to inventory the equipment on scene.

If the operation was large enough to require the establishment of a Demobilization Unit, it will coordinate the recovery of loaned items, such as portable radios, and the documenting of lost or damaged pieces of apparatus and equipment (Figure 6.91).

Abandonment. In some cases, the environment within the collapsed structure may be too hazardous to justify sending rescue personnel back in to retrieve pieces of equipment — even expensive ones. Rather than putting rescue personnel at risk, it is sometimes advisable to simply abandon the equipment in place. It may be retrievable after the structure has been completely demolished, or the cost of replacing the abandoned equipment may be recovered from the owner of the collapsed structure.

Figure 6.91 A rescue unit in the Demobilization Unit.

INVESTIGATION

All structural collapse rescues should be investigated at some level. At the very minimum, a departmental investigation should be conducted for purposes of reviewing and critiquing the operation. However, if an employee was injured in the incident, the incident will be investigated by the Occupational Safety and Health Administration (OSHA) and perhaps by other entities such as the employer's insurance carrier. Obviously, if the collapse was the result of a crime, such as a bombing, law enforcement agencies will also investigate the incident.

RELEASE OF CONTROL

Once rescuers respond to the scene of a structural collapse, they assume control of the structure and the immediate surrounding area. Within certain limits, they can deny access to anyone, including the owner of the structure. Legitimate members of the news media have certain constitutionally protected rights of access, but the interpretation of these rights varies from state to state and from country to country. Rescuers should be guided by local protocols.

The process of releasing control of the structure back to the owner or other responsible party is sometimes not as straightforward as it might seem. The owner or responsible party should be escorted on a tour of the structure, or as close to it as possible consistent with safety, and should be given an explanation of any remaining hazards. If the structure is still too hazardous to leave unattended, the owner may be required to post a security guard, erect a security fence around the hazard, or both. Before the structure is released, the department may require the owner to sign a written release that describes the hazards and stipulates the conditions the owner must meet.

CRITICAL INCIDENT STRESS DEBRIEFING

Because the injuries suffered by the victims of structural collapse incidents can sometimes be extremely gruesome and horrific, the members of the rescue teams and any others who had to deal directly with the victims should be *required* to participate in a critical incident stress debriefing process. Because individuals react to and deal with extreme stress in different ways — some more successfully than others — and because the effects of unresolved stresses tend to accumulate, participation in this type of process should not be optional.

The process should actually start *before* rescuers enter the scene if it is known that conditions exist there that are likely to produce psychological or emotional stress for the rescuers involved. This is done through a prebriefing process wherein the rescuers who are about to enter the scene are told what to expect so that they can prepare themselves (Figure 6.92).

If rescuers will be required to work more than one shift in these conditions, they should go through a minor debriefing, sometimes called "defusing," at the end of each shift. They should also participate in the full debriefing process within 72 hours of completing their work on the incident.

Figure 6.92 Rescue personnel are briefed before entering the scene.

Chapter 6 Review

Directions

The following activities are designed to help you comprehend and apply the information in Chapter 6 of **Fire Service Rescue**, Sixth Edition. To receive the maximum learning experience from these activities, it is recommended that you use the following procedure:

1. Read the chapter, underlining or highlighting important terms, topics, and subject matter. Study the photographs and illustrations, and read the captions with each.
2. Review the list of vocabulary words to ensure that you know the chapter-related meaning of each. If you are unsure of the meaning of a vocabulary word, look up the word in the glossary or a dictionary, and then study its context in the chapter.
3. On a separate sheet of paper, complete all assigned or selected application and review activities before checking your answers.
4. After you have finished, check your answers against those on the pages referenced in parentheses.
5. Correct any incorrect answers, and review material that was answered incorrectly.

Vocabulary

Be sure that you know the chapter-related meanings of the following words:

- ambient *(180)*
- carcinogen *(179)*
- conduit *(179)*
- contaminant *(179)*
- corrosive *(179)*
- flammable *(179)*
- hydraulic *(184)*
- nominal *(192)*
- oxidizer *(179)*
- pneumatic *(184)*
- seismic *(184)*
- toxic *(188)*
- viable *(178)*
- void *(178*

Application Of Knowledge

1. Examine your department's pre-incident structural collapse plans. Working with other rescue personnel, update these plans as necessary. *(Local protocol)*
2. Choose from the following structural collapse rescue procedures those appropriate to your department and equipment, or ask your training officer to choose appropriate procedures. Mentally rehearse the procedures, or practice the chosen procedures under your training officer's supervision.
   - Don proper PPE for a structural collapse rescue. *(181-183)*
   - Set up ventilation and lighting equipment for a structural collapse rescue. *(184, 185)*
   - Set up and interpret structural stability monitors. *(183 & 187)*
   - Operate and interpret atmospheric monitors. *(183 & 188)*
   - Use victim-locating devices to find structural collapse "victims." *(184, 188-190)*
   - Shore a section of a collapsed structure with air bags. *(190)*
   - Shore a section of a collapsed structure with cribbing. *(190, 191)*
   - Shore a section of a collapsed structure with jacks and/or pneumatic shores. *(191)*
   - Shore a section of a collapsed structure with a wooden shoring system. *(192-194)*
   - Tunnel to a trapped "victim." *(195)*
   - Construct an enclosure to protect victims and workers from tunnel debris. *(195)*
   - Terminate a structural collapse operation. *(201-203)*

Review Activities

1. Identify the following abbreviations and acronyms:
   - PVC *(185)*
   - ppm *(188)*
   - FEMA *(172)*
   - NIOSH *(188)*
   - NFPA 1470 *(172)*
   - NFPA 49 *(188)*
   - MSDS *(188)*
2. Identify the following:
   - Mutual aid *(173)*
   - Secondary hazard *(173)*
   - Pretensioned/posttensioned cable systems *(176)*
   - Integrated concrete column *(176)*
   - Rebar *(176)*
   - Airborne biological hazard *(179)*
   - Frostbite *(180)*

- Retrieval line *(185)*
- Explosion-proof lighting *(189)*
- Gusset plate *(192)*
- Cleat *(193)*
- Key member *(195)*

3. Compare and contrast the following pairs:
   - Heat exhaustion vs. heatstroke *(180)*
   - Hypothermia vs. hyperthermia *(180)*
   - Shoring vs. tunneling *(190 & 195)*
   - Minimum vs. maximum personnel accountability *(197)*
4. List the four major areas that fire/rescue personnel should familiarize themselves with in order to prepare for a major structural collapse within their response district. *(172)*
5. Name the two primary resource documents that help rescue personnel plan for a structural collapse within their response districts. *(172)*
6. List the four elements of pre-incident planning for structural collapse incidents. *(172)*
7. List the seven areas identified by a hazard survey. *(172)*
8. Name the structures that should be addressed first when fire/rescue personnel develop operational plans based on identified target hazards. *(172, 173)*
9. List the questions that fire/rescue personnel should ask when visiting structures to develop operational plans. *(173)*
10. Explain the process for identifying resources after the various target hazards have been studied and site-specific plans have been developed. *(174)*
11. List possible considerations when making general plans for dealing with large numbers of mutual aid resources. *(174, 175)*
12. Briefly describe and list one example for each of the following types of construction:
    - Light frame *(176)*
    - Heavy wall *(176)*
    - Heavy floor *(176)*
    - Precast concrete *(177)*
13. List and briefly describe the four most common patterns of structural collapse. *(178)*
14. Provide examples of ways in which the following environmental hazards may affect rescue operations:
    - Damaged utilities *(179)*
    - Atmospheric contamination *(179)*
    - Hazardous materials contamination *(179, 180)*
    - Darkness *(180)*
    - Temperature extremes *(180)*
    - Noise *(180)*
    - Fire hazards *(180)*
    - Adverse weather *(181)*
15. Provide examples of the ways in which he following physical hazards associated with structural collapse incidents may affect rescue operations: *(181)*
    - Secondary collapse
    - Unstable debris
    - Confined spaces
    - Below-grade spaces
    - Heights
16. Explain the protective benefits of the following personal protective equipment specific to performing structural collapse rescue operations: *(182)*
    - Helmet
    - Eye protection
    - Outerwear
    - Footwear
    - Gloves
    - Respiratory protection
17. Describe the structural collapse rescue uses of each of the following pieces of monitoring equipment:
    - Structural-stability monitors *(183)*
    - Atmospheric monitors *(183)*
    - Victim-locating devices *(184)*
18. Within the following categories, provide specific examples of tools and equipment and their particular uses for structural collapse rescue.
    - Hand tools *(184)*
    - Power tools *(184)*
    - Air-supply *(184)*
    - Ventilation equipment *(184)*
    - Lighting equipment *(184, 185)*
    - Ropes *(186)*
    - Ladders *(186)*
    - Heavy equipment *(186)*
19. Describe how each of the following structural monitoring devices may be used during a structural collapse rescue operation: *(187)*
    - Plumb line
    - Surveyor's transit
    - Construction level
20. List guidelines for atmospheric monitoring at the site of a structural collapse. *(188)*
21. Compare and contrast the uses of the following methods/ devices used to locate victims hidden in the rubble of a structural collapse:
    - Hailing *(189)*

- Motion detectors *(189)*
- Listening devices *(189)*
- Search cameras *(189)*
- Thermal imagers *(190)*
- Search dogs *(190)*

22. Outline guidelines for shoring using each of the following methods:
    - Air bags *(190)*
    - Cribbing *(190, 191)*
    - Jacks *(190)*
    - Pneumatic shores *(190)*
    - Wooden shoring system *(192)*
23. Compare and contrast the following wooden shoring systems:
    - Dead shores *(192)*
    - Raker shores *(192)*
    - Flying shores *(193)*
    - Braces/struts *(194)*
    - Tie-backs *(194)*
24. State the main function of tunneling. *(195)*
25. Describe hazards and safety procedures associated with tunneling. *(195)*
26. List the six phases involved in structural collapse rescue. *(195, 196)*
27. Using the list in Activity 26, outline the procedures/tactical considerations involved in structural collapse rescue. *(196-203)*
28. Describe the spray paint marks/methods used to identify the condition and hazardous materials status of assessed structures. *(198)*
29. List the information that should be obtained through occupant interviews during the second phase of structural collapse rescue. *(200)*
30. Identify the key points in Phases III - VI of structural collapse rescue. *(200, 201)*
31. Explain when it is best to abandon rather than retrieve a piece of equipment. *(202)*
32. List the four basic steps in releasing control of the scene back to the owner or other responsible party. *(202)*
33. Discuss critical incident stress, prebriefings, and "defusings." Why should participation in a full critical incident stress debriefing within 72 hours of the incident be mandatory for all involved rescuers? *(202, 203)*

Questions And Answers

7

Elevator Rescue

Chapter 7
Elevator Rescue

INTRODUCTION

The modern elevator is an essential part of our ability to fully occupy tall structures. Without elevators, occupying the upper floors of high-rise buildings would be impractical (Figure 7.1). Because of their importance and the obvious need for safety and reliability, elevators have developed into one of the safest and most reliable of all modes of transportation. Because of the need for safety, elevator design, construction, and operation are stringently controlled and monitored by all levels of government. Most elevator regulations are based on ASME A17.1, *Safety Code for Elevators*, published by the American Society of Mechanical Engineers.

Although elevators are extremely safe because of the rigid standards to which they must conform, they are still mechanical devices and do occasionally malfunction. Elevator failures may result from a loss of electrical power, malfunctioning controls, or mechanical breakdowns. When one of these failures traps passengers within an elevator, the fire department is usually called to rescue them.

This chapter discusses the operation of elevators, the design of elevator cars and hoistways (vertical shafts in which elevator cars travel), accessing and controlling elevator cars in rescue incidents, elevator safety devices, methods of removing passengers from disabled elevator cars, and preplanning for elevator incidents. However, simply reading this material will not make firefighters proficient in elevator rescue. Hands-on training and opportunities to apply the procedures discussed in this chapter in regular periodic exercises are absolutely essential.

Figure 7.1 Without elevators, buildings this tall would not be practical. *Courtesy of New York (NY) Fire Department.*

TYPES OF ELEVATORS

Almost all elevators in use today are either hydraulic or electric, and the characteristics of both types are examined. In addition, the different types of electric elevators are discussed.

Hydraulic Elevators

The operating principle of hydraulic elevators involves a fluid being forced under pressure into a cylinder containing a piston or ram. As the fluid is pumped in, the ram rises and the attached elevator car moves upward. As the fluid drains out, the car is lowered by gravity.

The assembly consists of an electric motor that powers a hydraulic pump that pumps hydraulic fluid from a reservoir into the cylinder. When the elevator is called to an upper floor, fluid is pumped into the cylinder until the ram raises the elevator car to the desired floor. If the elevator is subsequently summoned to a lower floor, the fluid is allowed to drain out of the cylinder and back into the reservoir until the car is at the correct floor. Hydraulic elevators do not have brakes as do traction elevators (see the following section, Electric Elevators); cars are slowed and stopped by controlling the flow of hydraulic fluid back into the reservoir.

The elevator car is attached to the top of the ram, so the ram must be long enough to reach the highest floor served by that elevator. This means that the cylinder must also extend an equal distance into the ground. This puts the practical upper limit for hydraulic elevators at about six stories (Figure 7.2).

Figure 7.2 Hydraulic elevators are limited to six stories.

For most hydraulic elevators, the machine room is located near the elevator pit (the area extending from the lowest landing floor down to the bottom of the hoistway). There may be situations, however, that require the machine room to be located some distance away. As a practical matter, this only requires some extra piping to transport the hydraulic fluid, but it can make the machine room difficult for rescuers to locate unless they have done adequate pre-incident familiarization and planning.

There are other types of hydraulic elevators, but they are extremely rare in North America. One type popular in Europe uses a telescoping ram. In another type, the car is suspended by cables attached to a hydraulically operated lifting mechanism.

Electric Elevators

DRUM

Older style electric elevators employ a large drum on which the hoisting cable is wound. The drum is located in a motor room directly over the hoistway. Like hydraulic elevators, drum-type elevators have practical height limitations because of the size of drum that would be required for tall buildings (Figure 7.3). This type of elevator is obsolete and is found only in very old structures; this type may still be in use as freight elevators.

Figure 7.3 A typical cable drum elevator.

TRACTION

The most common type of elevators in modern buildings taller than six stories is the traction type because they are very fast and do not have the height limitations of either hydraulic or drum-type elevators. Since they are so numerous, they are also the type with which firefighters must deal in most elevator rescue situations in buildings over six stories in height. As the name implies, traction elevators use a counterweight in the same way a traction device on a hospital bed uses a counterweight. Hoist cables attached to the elevator car run up and over the drive pulley at the top of the hoistway and down the back wall of the hoistway to connect to a movable counterweight. The combined weight of the car, cables, and counterweight provide the traction needed by the drive pulley. With the aid of compensating cables, the weight of the elevator car, the lifting cables, and the counterweight offset so that the drive motor has to lift only the weight of the passengers. The drive equipment for traction elevators is contained in a machine room, usually located directly above the elevator hoistway.

Even though counterweights reduce the amount of energy needed to raise these elevators, the heights to which they operate may require them to have as much as a 500-volt power supply. The drive motors may be either direct current (DC) or alternating current (AC) types. Obviously, firefighters must be extremely careful when a rescue operation requires them to work in the vicinity of such high-voltage equipment.

Traction elevators have a braking system that operates during both normal operation and malfunctions. The system employs a brake drum located on the shaft of the drive motor (Figure 7.4). Under normal conditions, the spring-operated brake shoes are held away from the drum by electromagnets. In the event of a power failure, the electromagnets release, and the brake shoes are forced against the drum. This provides an additional level of safety because any loss of power will result in the brakes being applied and the car being stopped wherever it was when the power failed.

During normal operation, the brakes on traction elevators with AC motors aid directly in stopping the car at the correct floor. On those with DC motors, the brakes play no part in actually stopping the elevator car. The motor stops the car, and then the brakes are applied to hold the car in place.

Figure 7.4 The brake on a typical traction elevator.

ELEVATOR CARS

Elevator cars are assemblies consisting of a safety plank on which the platform rests, a crosshead, and uprights that connect these main structural members together. This framework is enclosed with removable panels. The safety plank forms the bottom of the frame and the base for the floor of the car (platform) and is the point of attachment for the ram in hydraulic systems. The crosshead forms the top support of the frame and includes the cable sheaves in cable-hoist systems. Roller guides — sets of wheels that roll against T-shaped hoistway guide rails — are mounted at each end of the safety plank and the crosshead (Figure 7.5).

Doors

Doors in elevator installations include both car doors and hoistway doors. The two are usually designed to open in the same direction and to open and close simultaneously. This section deals primarily with car doors; hoistway doors are discussed later (see Elevator Hoistway Doors section).

Passenger elevator car doors are powered by an electric motor mounted on the top of the elevator car (Figure 7.6). The car door does not have locks

Figure 7.5 A typical roller guide mounted on the end of the crosshead.

Figure 7.6 A typical door-opening motor. *Courtesy of New York (NY) Fire Department.*

and can be pushed open at any time. However, electric interlocks will not allow a car to move when the car doors are open, and a moving car will immediately stop if the doors are pushed open. When the doors are closed again on most elevator cars, the cars will start to move again. There are some types, however, that will not start moving again until they have been reset.

The doors are designed to open and close automatically when the car stops at the floor to which it was summoned. When the elevator stops at the correct level, the hoistway doors are unlocked by a driving vane attached to the car door (Figure 7.7). As the car door opens, the vane strikes a roller that releases the hoistway door lock. The car doors then push the hoistway doors completely open. When the controller signals the doors to close, a weight forces the hoistway doors closed, the driving vane moves away from the roller, and the hoistway doors are relocked.

Passenger elevator car doors can be of three types: single-slide, two-speed, and center-opening. Freight elevators often use vertical bi-part doors. Firefighters should be familiar with every type of elevator door found in their jurisdiction.

SINGLE-SLIDE

Single-slide doors are the simplest types and consist of one panel that slides horizontally to one side of the doorway opening. Doors of this type tend to function slowly and are not found in high-speed elevator installations (Figure 7.8).

Figure 7.7 One type of driving vane.

Figure 7.8 A typical single-slide elevator door.

TWO-SPEED

Two-speed doors consist of two panels with one behind the other. They move horizontally in the same direction, and both reach the open position simultaneously. As a result, the rear panel, which strikes the jamb, must move faster than the front panel. The motor that opens the doors is connected only to one of the panels. The other panel is moved by a cable — the relating cable — connecting the two panels (Figure 7.9).

CENTER-OPENING

The most common passenger elevator doors are the center-opening type. They consist of two movable panels that meet in the center of the doorway opening (Figure 7.10). The panels move away from

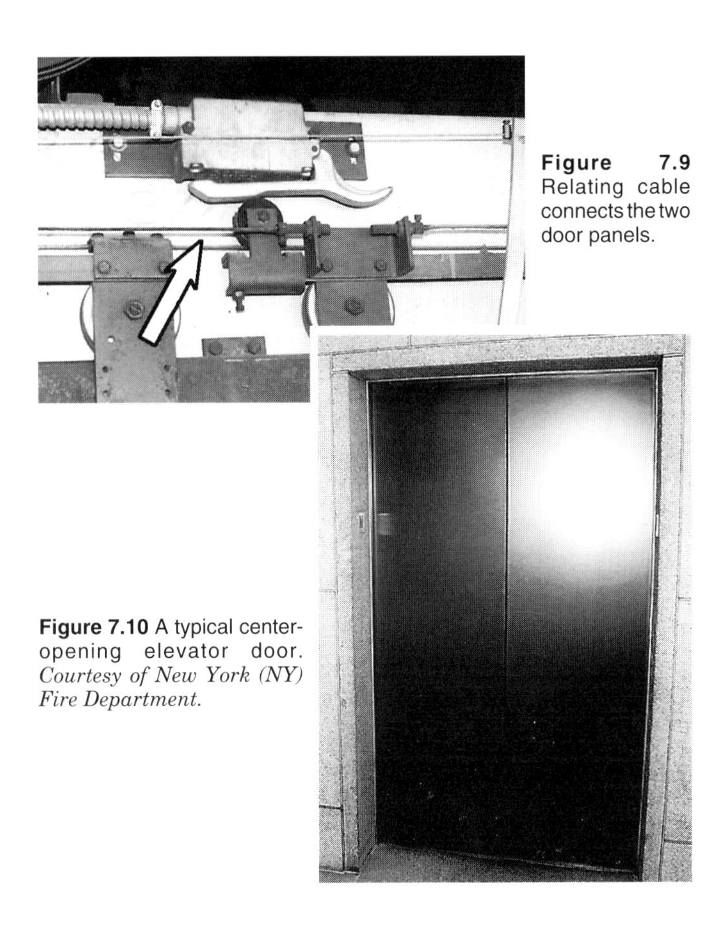

Figure 7.9 Relating cable connects the two door panels.

Figure 7.10 A typical center-opening elevator door. *Courtesy of New York (NY) Fire Department.*

each other when they are opening. As in two-speed doors, a relating cable connects the two panels.

VERTICAL BI-PART

Freight elevator doors are called *vertical bi-part doors*, and they consist of two panels that move vertically. One panel comes down from the top, the other comes up from the bottom, and they meet in the middle (Figure 7.11). Manually operated doors are counterweighted because there is no motor on the car to assist in opening and closing them; others have a motor at each landing.

Figure 7.11 A vertical bi-part door.

Access Panels

In situations where the elevator doors are inaccessible, the emergency exits from the car must be used. These exits consist of either a hinged access hatch through the top of the car or hinged or removable panels on the sides of the car. Because using either of these exits is time-consuming and involves some added risk of injury to the passengers, they should be used only as a last resort.

TOP EXIT

A top exit is provided on all electric traction elevators. On hydraulic elevators, a top exit may or may not be provided depending upon whether the system is equipped with a manual lowering valve. Top exits are required only if the elevator does not have this valve, which permits the lowering of the car in the event of trouble. A top exit is optional otherwise.

Some top exit panels are designed to be opened from inside the car, but all can be opened from the outside, and all open in that direction (Figure 7.12). Some elevator cars are provided with electrical interlocks that prevent car movement while the

Figure 7.12 A top exit from an elevator car. *Courtesy of New York (NY) Fire Department.*

panel is open. However, this feature is not required and will not be found on all models.

SIDE EXIT

In multiple elevator hoistways (more than one elevator in a common shaft), most elevator cars are equipped with side exits to allow passengers to be transferred laterally from a stalled car into a functioning car next to it. Side exit panels can be opened from the outside where a permanent handle is provided. Some panels are locked from the inside and cannot be opened without a special key or handle. Side exits are required to have electrical interlocks to prevent car movement when the panels are open. Side exits may not be provided on cars in hydraulic elevator systems where a manual lowering valve is provided.

Electrical Equipment

A set of elevator controls, normally used by an elevator mechanic for inspection and maintenance functions, is located on top of the elevator car (Figure 7.13). The control box is usually mounted on the crosshead, and it allows the operator to switch operation of the car from automatic to manual, to move the car up and down, and to remove all power from the elevator drive motor. Thus, the operator can ride the car to a desired location and then remove power from the motor (which automatically sets the brakes) for maximum safety.

Figure 7.13 The auxiliary controls atop an elevator car. *Courtesy of New York (NY) Fire Department.*

Another safety feature built into these inspection stations is the safety switch. This switch must be pushed simultaneously with the direction switch for the car to move up or down. This is designed to prevent the car from moving as a result of accidental contact with the direction switch. In addition, both switches are spring loaded so that they must be held down in order for the car to move. If either switch is released, the car will stop.

ELEVATOR HOISTWAYS

An *elevator hoistway* is the vertical shaft in which the elevator car travels and includes the elevator pit (Figure 7.14). The pit extends from the lowest floor landing to the bottom of the hoistway. Hoistways are constructed of fire-resistive materials and are equipped with fire-rated doors. Hoistways are fully enclosed and designed to reduce the likelihood of fire entering the shaft.

Types Of Hoistways

There are three common types of hoistways in use: single, multiple, and blind.

SINGLE

Shafts that contain only one elevator car are known as *single hoistways*. They are normally found in private installations and in smaller buildings with a low number of occupants.

MULTIPLE

Installations with more than one elevator in a common shaft are known as *multiple hoistways*. A large building may contain more than one multiple hoistway, but each hoistway is limited to no more than four elevators. The elevator cars within a given hoistway are usually not separated by any sort of wall or partition.

BLIND

Blind hoistways, whether single or multiple, are used for express elevators that serve only upper floors of tall buildings. There will be no entrances to the shaft on floors between the main entrance and the lowest floor served. In single-car hoistways, however, access doors are provided for rescue purposes (Figure 7.15).

Elevator Hoistway Doors

Hoistway doors are rated assemblies that work in conjunction with the car doors and, with the exception of freight elevators, are dependent upon the car doors for their power. Hoistway doors are of the same types as those described earlier for elevator cars, but with one additional type: the swinging door.

Swinging doors are like regular doors in that they are hinged at one side, and they swing outward from the hoistway (Figure 7.16). Like most other

Figure 7.14 A typical cable-elevator hoistway.

Figure 7.15 An access door into a blind hoistway.

Figure 7.16 A hoistway door that swings open. *Courtesy of Mike Wieder.*

hoistway doors, swinging doors are not powered, so they are equipped with a handle for manual operation. With this type of hoistway door, the elevator car door will usually be of the single-slide type.

Hoistway Door Locks

Hoistway doors are equipped with locks that prevent the doors from opening when an elevator car is not at the landing. This is a protective feature designed to keep people from pushing the hoistway doors open and perhaps falling into the open shaft.

The locking mechanism is mounted on one panel of the hoistway door. The latching device is shaped on one end so that it hooks a stop and prevents the doors from opening in the locked position (Figure 7.17). In the unlocked position, the shaped end of the latching device is pushed away from the stop where it will not catch, and the door can be opened (Figure 7.18). The latch is attached to a roller or drive block, depending upon the style of elevator. When an elevator car comes to a landing, a driving vane mounted on the car door is positioned near the roller or block. If the car is not programmed to stop, the vane passes by the roller without contact. If the car does stop, the vane is pushed against the roller, and the hoistway latch is released. When the car door opens, it pushes the hoistway door open also. If the car is then summoned to another floor, or after a programmed interval, the doors will automatically close and lock.

Figure 7.17 A latching device in the locked position.

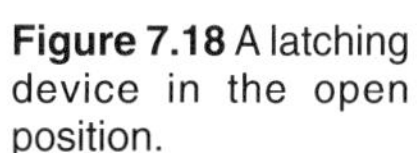

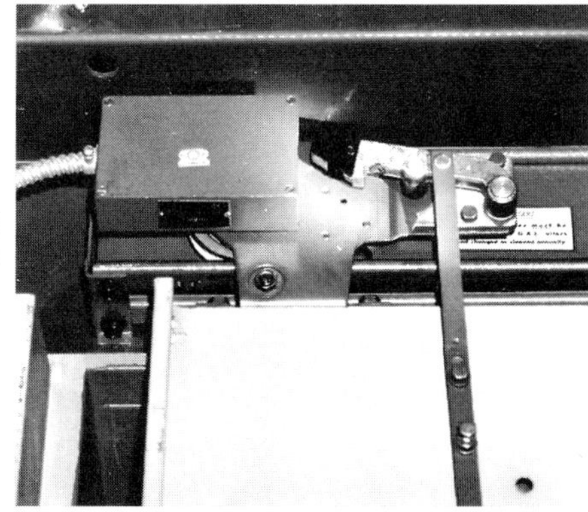

Figure 7.18 A latching device in the open position.

In addition to the mechanical locks, the hoistway doors are equipped with electrical interlocks that must be closed in order for the car to operate. All the doors of a hoistway must be closed, or the elevator will not run. As a result, a moving car will stop if a hoistway door is opened.

ELEVATOR KEYS

Using the proper key is by far the easiest, fastest, and most efficient way to open a hoistway door from the landing side. There are two types of elevator keys: hoistway door keys and elevator control keys. Using the elevator control key is the only way firefighters can safely operate an elevator during an emergency.

Hoistway Door Keys

Since hoistway doors are unlatched only when an elevator car is at the landing, firefighters must have some means of unlatching the door when an elevator is not there. In most cases, special keys are provided with which hoistway doors can be unlatched from the landing side, but this is not universally true. Hoistway doors in all single hoistway installations are now required to be so equipped, but some older installations may not have them. Multiple hoistway installations are only required to have them on the lowest landing. This requirement is primarily so that maintenance personnel can enter the elevator pit because they have no other entrance. Some elevator manufacturers may provide these devices on the highest level landing also, but it is not required.

Several types of keys for unlatching elevator hoistway doors are currently in use. Hoistway keys come in a variety of shapes to match the shape of the locks that are located near the top of the hoistway door. Once the latch is released, the hoistway doors may be pushed open by hand.

One older type key is called the T-shaped key (Figure 7.19). It releases the lock when it is pushed straight into the lock opening. Another common key type is the semicircular or lunar key (Figure 7.20). This type releases the lock when it is inserted into the opening and pulled downward. Both the T-shaped and semicircular keys are rigid, one-piece devices.

Figure 7.19 A T-shaped elevator door key. *Courtesy of New York (NY) Fire Department.*

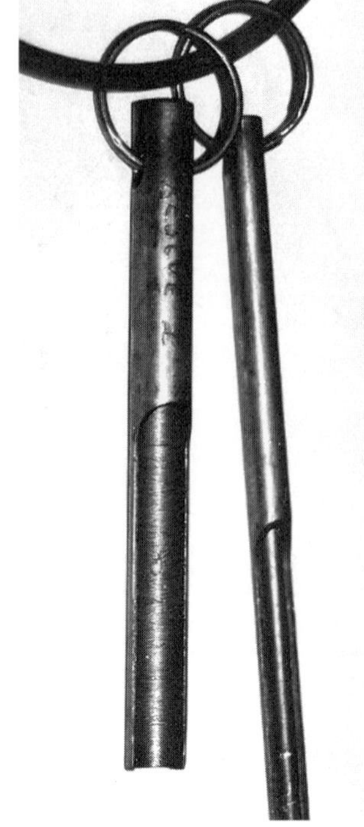

Figure 7.20 A typical lunar key. *Courtesy of New York (NY) Fire Department.*

Another common device for opening hoistway doors, the drop key, is made in two sections of unequal length. The two sections are hinged together (Figure 7.21). When the short section is inserted as far as it will go into the door, the key is then rotated 90 degrees. This allows the short section to drop, forming an "L" shape. Turning the key away from the leading edge of the door will release the hoistway door lock.

Because of problems with vandalism and theft, some building managers remove hoistway door keys from public access areas, so it may be difficult for firefighters to obtain these keys during an incident. Therefore, it is very important for firefighters to conduct pre-incident planning surveys to determine the locations of these keys and how they may be obtained when needed. In most cases, the most efficient method is to have the occupant install a lockbox containing these and other essential keys to the building (Figure 7.22).

In the absence of the manufacturer-provided key, it is sometimes possible to improvise with common tools or other objects if the shape of the lock can be duplicated. This is easiest with locks that require T-shaped or semicircular keys because

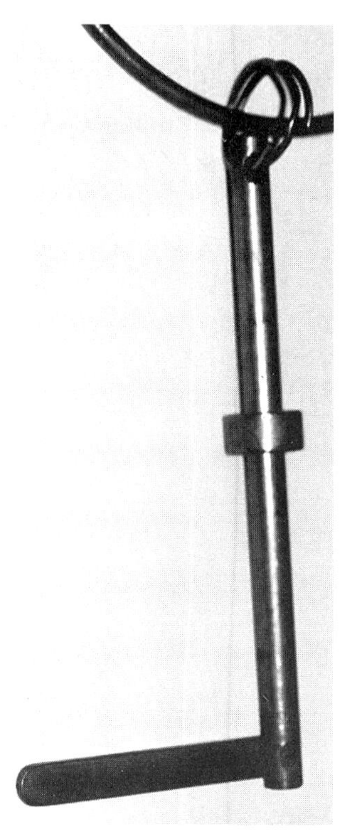

Figure 7.21 One type of drop key. *Courtesy of New York (NY) Fire Department.*

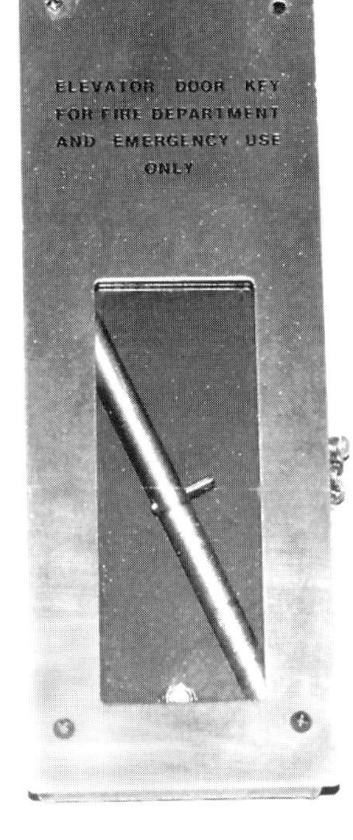

Figure 7.22 A typical lockbox.

the keys are straight. The drop key lock does not lend itself to being opened without the real key because of its drop feature.

Elevator Control Keys

Except in some older installations, there is a key-operated three-position switch located in the main elevator lobby of the Designated Level and within sight of the elevator doors. Most switches are marked "ON," "OFF," and "BYPASS" (Figure 7.23). It will normally be found in the "OFF" position. Turning it to the "ON" position allows firefighters to take control of all elevators controlled by that switch. The "BYPASS" position is used to take control of the elevators after they have been recalled by the activation of the smoke detectors.

In normal operation, elevators are controlled automatically; however, in fires or other emergency situations, it is necessary for firefighters to have complete control of the elevator car. The two methods available for taking control are *Independent Service* and *Emergency Service*, although they may be called by other similar names depending on the elevator manufacturer. Both services are controlled by keys that may or may not be readily available to firefighters. On some models, Independent Service requires a key to unlock the panel that encloses a toggle switch. Emergency Service requires a key to select that mode of operation (Figure 7.24). Once again, pre-incident planning should be done to identify how these keys may be obtained when needed.

Figure 7.23 A three-position switch.

Figure 7.24 Emergency Service is selected with a key. *Courtesy of New York (NY) Fire Department.*

Pre-incident planning is important not only for making provisions for obtaining keys but also for knowing how to use them with different elevators. It is important for firefighters to become familiar with the operation of all elevators within their area of responsibility because of differences in how various brands of elevators are controlled during emergencies and even differences between various models of the same brand. This can be accomplished only through pre-incident planning (Figure 7.25).

Although some older installations may not have been retrofitted, modern elevator installations have a feature known as *Phase I operation*, initiated by

Figure 7.25 A firefighter inspecting an elevator installation.

the building's smoke detectors or the three-position switch in the elevator lobby. When this mode is activated, all operating elevator cars stop wherever they are and immediately return to the Designated Level (usually the main elevator lobby). The car doors open and remain open. If Phase I was initiated by a smoke detector on the Designated Level, all cars are automatically diverted to a preselected Alternate Level.

INDEPENDENT SERVICE

Independent Service is a feature on most elevator installations that allows the car to be switched from automatic to manual control. Independent Service is usually activated by a switch inside the car (Figure 7.26). The switch may be a toggle switch inside a locked panel marked "I.S." or a key-operated three-position switch on the elevator control panel. The switch may also be located in the lobby of the Designated Level. In this mode, the car will respond only to instructions from inside the car and will disregard calls initiated from hall landings; however, the doors will automatically open when the selected floor is reached.

EMERGENCY SERVICE

As in Independent Service, when an elevator car is switched to Emergency Service (also known as *Phase II operation*), the car will respond only to signals initiated from within the car and will disregard those from hall landings. Emergency Service is initiated and controlled by turning the three-position key switch in the lobby of the Designated Level to "ON" (Figure 7.27). Once Phase II operation has been activated, the key is removed from the lobby control panel and is inserted into the three-position switch on the car's control panel and turned to the "ON" position (Figure 7.28).

In this mode the elevator car doors do not open automatically when the selected floor is reached. In most cases, the doors will remain closed until the *door open* button on the control panel is pressed. The pressure on the button must be maintained until the door is fully open. If the button is released before the door is fully open, the door will immediately close. The reverse is also true when using the *door close* button to close the door. Once at the selected floor, firefighters may keep the elevator there by turning the switch to "HOLD" and removing the key (Figure 7.29). This will alert anyone else on that floor that firefighters are on that floor.

In addition, all elevator cars manufactured after 1991 must have a *call cancel* button on the car's control panel (Figure 7.30). This feature, active only in Phase II operation, allows the car to be

Figure 7.26 Independent Service switch inside an elevator car. *Courtesy of New York (NY) Fire Department.*

Figure 7.27 The switch in the lobby must be turned to "ON." *Courtesy of New York (NY) Fire Department.*

Figure 7.28 The switch inside the car must also be turned to "ON." *Courtesy of New York (NY) Fire Department.*

Figure 7.29 The key is removed with the switch on "HOLD."

Figure 7.30 The Call Cancel button. *Courtesy of New York (NY) Fire Department.*

stopped before it reaches the selected floor and to return to the Designated Level.

SAFETY DEVICES

The excellent safety record of elevators can be attributed to strict regulation, rigorous engineering to reduce the likelihood of failure, and numerous safety features designed to limit the effects of those failures that do occur. Maximum passenger protection is maintained by the use of equipment that can safely stop the elevator car in the event of a malfunction.

Some safety devices are common to both types of elevators; others, to traction types only. Both traction and hydraulic elevators are equipped with terminal switches, and both systems may include buffers (see Buffers section) in the elevator pit.

Terminal Switch

The terminal switch is designed to stop the elevator car before it reaches the upper or lower limits of the hoistway (Figure 7.31). When activated, the terminal switch either removes power from the hydraulic pump in that type system, or removes power from the drive motor in traction systems while activating the brake system. This prevents the elevator car from striking the top or bottom of the hoistway in the event of a malfunction of the elevator controls.

Figure 7.31 A typical terminal switch.

Buffers

Buffers are located in the bottom of the elevator pit, and they act as shock absorbers if the terminal switches fail. Buffers consist of either large springs or hydraulic cylinders and pistons (similar to automobile shock absorbers) filled with oil (Figure 7.32). Oil buffers are the most common type and are always used on high-speed elevators. Spring buffers are effective only on slow-moving elevators.

In traction systems, buffers are provided for both the car and the counterweight. If the counterweight bottoms out, traction is lost because of slack developing in the cables, and the car is stopped before it can strike the top of the hoistway. If the car travels beyond the lowest floor landing, the buffers are intended to bring it to an abrupt but safe stop. However, buffers cannot safely stop a free-falling elevator. They are only effective in stopping a car moving at its normal rate of speed. Other devices are provided to stop a free-falling car.

Figure 7.32 One type of elevator buffer.

Since the rate of descent of cars in hydraulic systems is limited by the rate at which the hydraulic fluid is lost from the cylinder, they have no need of the devices described in the following sections. Therefore, these devices will be found only on traction elevators.

Speed-Reducing Switch

Activated by a device known as a *speed governor*, the speed-reducing switch is designed to slow the driving motor when the elevator car begins to travel faster than a predetermined safe speed. Through a cable called a *governor rope*, the speed-reducing switch is connected to the operating lever of the elevator safeties, which are tapered jaws (see Car Safeties section).

Overspeed Switch

Also connected to the speed governor is the overspeed switch. This switch is activated if the speed-reducing switch fails to slow the car sufficiently. Like the terminal switch, the overspeed switch removes the power from the drive motor and applies the braking system.

Car Safeties

Car safeties are tapered jaws that wedge against the guide rails and bring the elevator to a stop. Located at both ends of the safety plank, the safeties are activated when all other speed control devices have failed and the car has begun to travel faster than a predetermined safe speed.

Elevator safeties are designed to stop a free-falling elevator car. Having a car in free-fall is an extremely unlikely possibility because all the hoist cables and all the braking systems would have to fail. Each hoist cable is strong enough to carry the weight of a fully loaded car, and most installations have at least four cables for each car. Traction elevators have at least two emergency braking systems.

ELEVATOR RESCUE OPERATIONS

Calls for assistance involving elevators are either routine incidents or true emergencies. True elevator emergencies are rare, and the vast majority of all elevator-related responses fall into the category of routine incidents. For example, some minor malfunction may cause the elevator car to stop between floors with passengers temporarily confined inside. They would be in no danger at this point, but they might be experiencing some irritation at being inconvenienced and perhaps a little anxiety about their continued safety. In such cases, firefighters should not feel pressured to take some action that might jeopardize their safety or that of the passengers. Rather, they should follow the primary and secondary procedures described in the following sections and wait for an elevator mechanic to arrive.

Elevator emergencies that require immediate intervention by rescuers are those that involve a stalled car with occupants in real jeopardy. These situations may preclude firefighters from following all of the primary and secondary procedures before forcing entry into the car. Some typical elevator emergencies are:

- Occupant(s) in serious need of medical treatment
- Occupants too hysterical to follow instructions
- Fire within the car or hoistway
- Fire on a floor below the stalled car

Whenever possible, it is best to wait for an elevator mechanic to arrive at the scene of a disabled elevator before attempting to remove the passengers. The mechanic can usually repair the equipment or manually move the elevator to a landing and open the doors. There are other reasons for waiting for the mechanic, the most important of which is to ensure passenger safety. Anytime a person has to leave an elevator other than by walking out through the main doors, the possibility of injury is increased. Generally, as long as passengers remain inside a stalled car, they will be safe, even if a little uncomfortable.

Rescuer safety is another reason for waiting for the mechanic. Some of the rescue procedures involve a certain degree of risk for the rescuer, so having the mechanic free the passengers accomplishes the task without endangering anyone.

Though of much less importance, yet another reason for using the mechanic, is that the chance of property damage is reduced. When firefighters must force entry, some damage usually occurs. Using the elevator mechanic allows for the release of the passengers without damaging elevator doors or other components.

Rescue Equipment Required

With a few exceptions, the equipment used in elevator rescue is common and can be found on practically all fire apparatus (Figure 7.33). The following equipment list is not all inclusive but is what has generally been found effective from past experience.

- *Portable radios.* Communication between members of the rescue team and between the team and command is essential. With firefighters spread throughout the building, communication is extremely difficult without radios. (**NOTE:** Because of the mass of some buildings, communication can be difficult even with radios, and other means of communication must be used.)
- *Short extension ladder.* This type of ladder is to be used when the elevator car stalls above or below the landing. The passengers use the ladder to climb either up or down to reach the landing.
- *Folding ladder.* The folding ladder is used to remove passengers through the top exit of the elevator car.
- *Forcible entry tools.* Various forcible entry tools may be needed for prying hoistway doors or forcing entry into machine rooms, etc. The tools commonly carried on the apparatus will usually suffice.

Figure 7.33 An often used piece of equipment in elevator rescue.

- *Safety belts.* Several safety belts are required, both for firefighters working in and around the hoistway and for passengers being removed through top or side exits if this becomes necessary.
- *Lifelines.* Sufficient lifelines must be provided for all passengers and firefighters in the hoistway.
- *Emergency medical equipment.* It is desirable to have emergency medical equipment and personnel trained in its use handy in the event of injuries.
- *Walking plank.* A wooden plank at least 4 feet (1.2 m) long and at least 1 foot (300 mm) wide.

Safety Procedures

Whenever an elevator rescue is conducted, there is some possibility of injury to firefighters or passengers. Working at heights, being close to moving mechanical devices, working in unfamiliar surroundings, and sometimes being near high-voltage electrical equipment expose firefighters to hazards that must be guarded against. In elevator rescues, firefighters may face all of these hazards in combination. The following are some safety suggestions that can reduce the risk of injury during elevator rescue operations:

- Whenever possible, wait for an elevator mechanic to effect the release of the trapped passengers.

- Firefighters should not attempt rescue procedures with which they are unfamiliar.
- Always wear protective equipment for head, eyes, and hands during these operations. Wearing turnout coats is discouraged because their looseness may allow them to become caught in the elevator moving equipment.
- The main power to the car should be shut off before beginning rescue operations (Figure 7.34). This will help prevent unexpected movement of the car.
- A firefighter with a portable radio should be stationed at the main disconnect switch to prevent the restoration of power before the operation is completed.
- The Emergency Stop switch inside the stalled car should always be activated. This provides a backup to the main disconnect (Figure 7.35).
- Rescuers should never attempt to override the elevator's control system in order to drift the car (move it slightly). This should be done only by an elevator mechanic.
- To reduce the possibility of someone falling into the hoistway, an open hoistway door should never be left unattended.
- When removing passengers from a car stalled above the landing, be sure to guard the opening beneath the car.
- All firefighters and passengers entering the hoistway should be secured with lifelines.
- During poling-down operations (See *Poling* in Gaining Access to Elevator Car section), only the operator's head and arms should extend into the hoistway, and another firefighter should maintain a firm grip on the operator during the poling operation. This reduces the likelihood of the operator falling into the hoistway.
- During poling-up operations (See *Poling* in Gaining Access to Elevator Car section), two firefighters should be used to operate the pole. They should be secured by lifelines held by other firefighters.
- Rescuers should maintain constant physical contact with passengers during removal. This helps to minimize passenger anxiety and allows firefighters to maintain control of them (Figure 7.36).

Figure 7.34 A firefighter shuts off the power to the elevator.

Figure 7.35 A rescuer activates the Emergency Stop switch.

Figure 7.36 A rescuer maintains contact with a passenger.

Communicating With Trapped Passengers

The first step in handling elevator incidents is to establish communications with the occupants of the stalled car. This can be done by using the elevator intercom system or by yelling through the doors or into the hoistway (Figure 7.37). The passengers should be reassured that they are safe and that work on getting them out has begun. Rescuers should ask how many people are in the car and if anyone inside is ill or injured. They should also determine if the lights inside the car are on and what floor the car is near.

Figure 7.37 An emergency phone can be used to calm the occupants.

It is important that constant communication be maintained with the occupants. This will aid in keeping their spirits up and in preventing panic. They will be calmer and more rational if they have some understanding of what is happening and are reassured that everything possible is being done.

Coordinating With Building Maintenance Personnel

Whenever possible, rescuers should coordinate operations with building maintenance personnel because they are more familiar with the building. They can open locked doors, answer questions, operate equipment, and furnish tools (Figure 7.38). The staff may also include an elevator mechanic.

If forcible entry is required and time allows, the building manager should be notified of what is to be done and why it is necessary. Damage should be kept to a minimum and done so that it can be repaired.

Incident Determination

Firefighters arriving at the scene of a disabled elevator must make some initial determinations before attempting rescue efforts. The determinations to be made are:

- That an occupied elevator car is disabled. Empty elevators require no actions by firefighters.

Figure 7.38 The rescue team leader consults with the building engineer.

- The exact location of the stalled car.
- The urgency of passenger removal.

DETERMINING CAR LOCATION

It will sometimes be necessary to locate the stalled car to determine if it is occupied or empty. The most common method of determining car location is to check the position indicator (Figure 7.39). Position indicators are usually located in the first floor elevator lobby, and their purpose is to show the floor at which the elevator car is located. When cars are stalled between floors, the indicator will show the nearest floor. However, in certain situations these indicators may not always be accurate. For example, in a high-speed elevator the indicator may read up to three floors difference from the actual location.

If the indicator is inoperative, there are other means by which the elevator position can be deter-

Figure 7.39 The indicator can help locate a stalled car.

mined. After making certain that the main power is turned off, firefighters may open the hoistway door at the lowest level and visually observe the car.

When hoistway doors on other levels have provisions for being opened by a key, the location of the car can also be determined by looking into the hoistway at some other point (Figure 7.40). Firefighters can count the number of doors between them and the stalled car to determine its location. Each door represents a floor. This is sometimes difficult when a car is stalled near the top of a very tall building.

Figure 7.40 A rescuer looks into an open shaft to locate the car.

An elevator car's location may also be determined by checking the selector located in the overhead machine room (Figure 7.41). The selector is a device that instructs the elevator car to respond to signals from various floors. The position of the pointer on the selector indicates what floor the elevator car is stalled near. However, the positions may not always be labeled to correspond with individual floors, so firefighters may need help from someone more familiar with that particular system.

Figure 7.41 The selector may reveal the car's location. *Courtesy of New York (NY) Fire Department.*

When a car is stalled near the top of a very tall building, the overhead machine room can provide another means of determining the car's location. There is usually an opening between the top of the shaft and the machine room to allow air movement as the car travels up and down in the shaft. This opening is often covered by a grate, but it will allow firefighters to visually locate the car by counting the doors between the car and the top of the shaft (Figure 7.42). Firefighters must exercise extreme caution when using this method because the grate over the opening may be missing, weakened, or out of position. In addition, this room contains high-voltage electrical equipment and perhaps unguarded mechanical devices that can represent serious safety hazards to unwary firefighters.

DETERMINING URGENCY OF PASSENGER REMOVAL

It cannot be overemphasized that the best way to get passengers out of a stalled elevator car is to wait for an elevator mechanic to move the car to the nearest landing so they can simply walk out. However, some situations make it impossible to wait. If it appears that life will be endangered by waiting, rescue efforts must be initiated immediately. This can result either from a passenger's deteriorating medical condition or from the car being threatened by a spreading fire. However, if passengers must be removed by other than normal means, the methods used must be well planned and responsible. Anytime the passengers leave an elevator except by walking out the main doors, they are at some risk, so other means of exiting should be used only as a last resort.

Figure 7.42 The car may be seen through an opening in the floor.

Removal Of Passengers

Once the decision has been made to effect the release of the passengers using rescue techniques, adequate preparation must be done before the operation begins. Preparation is essential to the safety of both the passengers and the firefighters. To ensure that all instructions are clearly understood and that the operation is conducted with speed and efficiency, an incident command system should be set up to organize and control the incident. Radio communication must be maintained between command and all operating units involved in the incident.

PRIMARY REMOVAL PROCEDURES

As mentioned earlier, the first step at any elevator incident is to establish contact with the occupants of the stalled car (Figure 7.43). This will aid in sizing up the incident, in determining the urgency of the situation, in determining the location of the car, and in reconstructing the history of the incident. Establishing communication with the occupants will not only help to reassure them and reduce their level of anxiety, but it may also allow them to help solve their own dilemma.

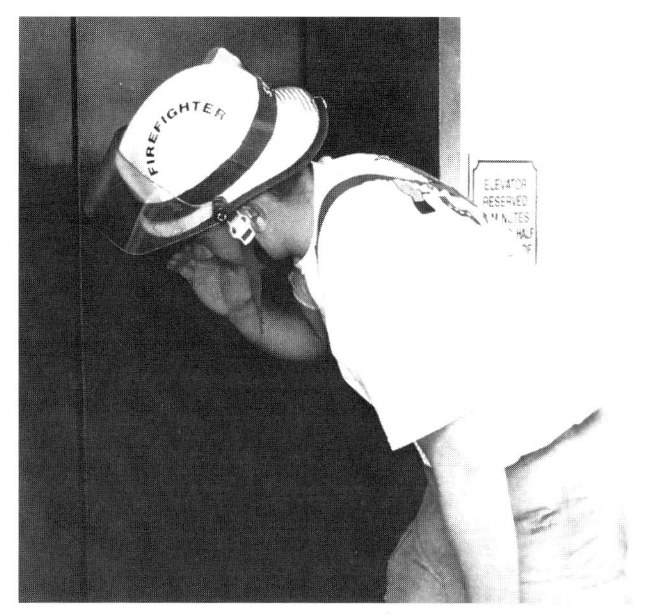

Figure 7.43 A rescuer shouts to trapped passengers.

If no emergency exists, the primary means of removal is to attempt to restore movement to the stalled car so that the passengers can simply walk out of the car. Depending upon the type and design of the elevator, the following actions should be taken:

- Warn occupants to expect sudden movement of the car.
- Have occupants try the Emergency Stop switch (it may have been activated inadvertently, causing the car to stop) (Figure 7.44).
- Have occupants push the Door Open button.
- Have occupants manually push the car door toward the closed position because it may be slightly ajar (Figure 7.45).

Figure 7.44 A passenger tries the Emergency Stop switch.

Figure 7.45 Occupants may be able to close a partially open door.

- Have occupants push several floor selection buttons.
- Firefighters should attempt to take control of the car using the Independent Service or the Emergency Service mode from the three-position switch in the main elevator lobby.
- Check the hoistway doors on every floor to make sure they are closed. Open doors will keep the car from moving.
- Check the main power supply switch in the elevator machine room. Movement can be restored to some types of elevators by momentarily switching the power off and then back on again.

> **WARNING**
>
> **If the elevator has blown a fuse or is otherwise in need of repair, mechanical or electrical repairs must be left to an elevator mechanic. Firefighters should never attempt such repairs themselves.**

Figure 7.46 The main power switch is locked-out. *Courtesy of New York (NY) Fire Department.*

SECONDARY REMOVAL PROCEDURES

If following the primary removal procedures has not restored movement to the car, the following secondary removal procedures should be implemented:

- Call for an elevator mechanic.
- Have a firefighter in the machine room ensure that the power to the affected elevator is off; if it can't be locked-out, a firefighter should be left there to make sure the power is not turned back on prematurely (Figure 7.46).
- Direct passengers to attempt to push the car door open, and if the car is at a floor landing, direct them to manipulate the hoistway door opening mechanism.
- Firefighters should open a hoistway door near the stalled car, using one of the special hoistway door keys if available. They should then use the appropriate poling technique to open the hoistway door adjacent to the stalled car (see *Poling* in Gaining Access to Elevator Car section). This may be done from a landing or from another adjacent car.
- In hydraulic elevator installations, firefighters can use the manual lowering valve to drift a stalled car to a lower landing (Figure 7.47). The doors can then be opened and the passengers released.

Figure 7.47 The lowering valve on a hydraulic elevator.

IMMOBILIZING THE ELEVATOR CAR

One important initial step in elevator rescue is ensuring that the elevator car does not move during the extrication activities. The chance of this happening is virtually eliminated by opening the main disconnect switch to the system and by activating the Emergency Stop switch in the elevator car.

The main disconnect is located in the machine room. One switch is provided for each elevator. When the switch is opened, the power is removed from the drive motor and the braking system. The motor is then inoperable, and the brakes are set. Someone should be assigned to stand by the main disconnect to be certain that power is not restored during the rescue.

Another means of immobilizing the car is to activate the Emergency Stop switch on the car control panel. This also interrupts the power to the drive motor and acts as a backup safety feature to the main disconnect. Passengers should be instructed to activate the switch, and firefighters should check it when they enter the car.

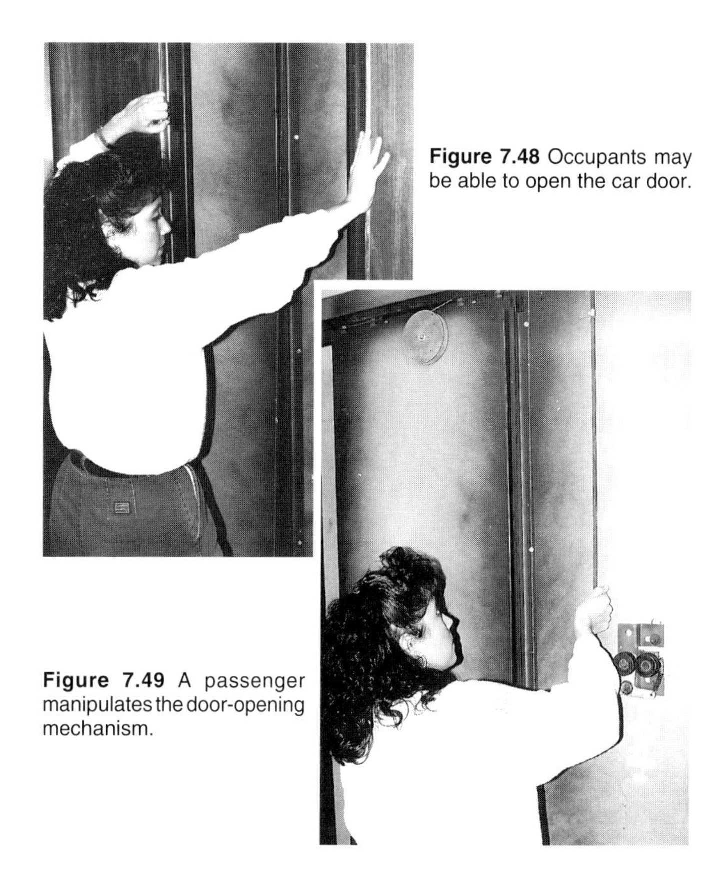

Figure 7.48 Occupants may be able to open the car door.

Figure 7.49 A passenger manipulates the door-opening mechanism.

GAINING ACCESS TO ELEVATOR CAR

Access to the elevator car may become difficult if a hoistway door does not have an accessible locking mechanism or if a key is not available. In these cases, alternate methods must be used.

Passengers open car door. If the elevator is stalled near or within the landing zone, the passengers can be instructed to push open the elevator car door (Figure 7.48). If the car is close enough to the landing to unlock the hoistway door, it will also open. If the car is not close enough to the landing, opening the car door will expose only the locked hoistway doors. If the locking mechanism is exposed, the passengers can be instructed to release it (Figure 7.49). The hoistway doors can then be pushed open.

Poling. In cases where the elevator is not close enough to the landing for the previous procedures to be used, it is possible to open the hoistway door by poling from a landing or from an adjacent elevator in a multiple hoistway. Poling from another elevator is done by first placing the adjacent car at the landing nearest the stalled car. A slender pole or slat is then slipped between the car and wall of the functional elevator (Figure 7.50). The pole is used to reach the hoistway door in front of the stalled car. If a pike pole will not fit into the opening, a thinner pole or slat with a hook attached can be used. The hook is used to pull or push the roller to release the locking mechanism. Moving the rollers away from the leading edge of the door will release the lock.

In cases where poling laterally is impossible, such as in a single hoistway, poling can also be done from a landing above or below the stalled car (Figure 7.51). Poling up or down may have to be used if not all of the hoistway doors are designed to be opened with the special key. If only the doors on either the lowest level or the highest level are so equipped, it may be necessary to open the nearest hoistway door with the special key and then pole up or down, one floor at a time, until the level of the stalled car can be reached. A firefighter should be stationed at the level nearest the stalled car to push the hoistway door open when the lock is released.

Opening hoistway door from inside. In multiple hoistway installations, an alternate method of releasing the hoistway door is also available. It involves entering the stalled car from an

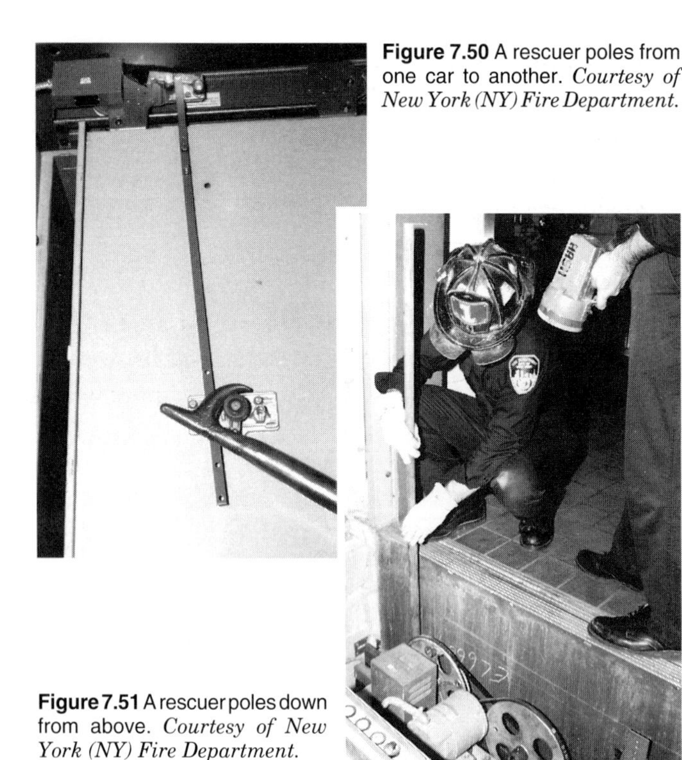

Figure 7.50 A rescuer poles from one car to another. *Courtesy of New York (NY) Fire Department.*

Figure 7.51 A rescuer poles down from above. *Courtesy of New York (NY) Fire Department.*

adjacent one and opening the door from inside the stalled car. This method may have to be used when building design precludes the application of poling techniques.

Aligning the rescue elevator with the stalled elevator may be difficult because of the firefighters' inability to see the stalled elevator from within the rescue car. Initially, the rescue elevator should be moved as close to the stalled car as possible on judgment alone. Then using a key or handle, the side exit of the rescue car should be opened to determine its position relative to the stalled car. The rescue car can then be moved until the two are even. However, the side panel has electrical contacts that prevent the car from moving when the panel is open, so it may have to be replaced before the car will move. In some cases, the contacts can be closed manually, but this is not true of all installations. Each brand and type of elevator may be different in the way Independent Service and Emergency Service modes function. Consequently, moving the elevator up or down slightly in order to line it up with the stalled car may be difficult. Some installations have an Inspector switch that can be used to start and stop the car, while others do not (Figure 7.52). This is another example of why it is important for firefighters to do pre-incident planning to gather the necessary information in advance.

Once the cars are aligned, rescuers can then unlock the side exit of the stalled car from the outside. A handle is located on the outside of the panel. Before the rescuer leaves the rescue elevator, the Emergency Stop switch in both cars should be activated and a short plank should be placed across the gap between the two cars. Once inside the stalled car, the rescuer should double-check the Emergency Stop switch to make sure it has been activated. Passengers can then be provided with a lifeline and assisted into the rescue elevator. The plank can then be removed and the side exit panels on both cars replaced. The freed passengers can then be taken to the nearest landing or to the main elevator lobby.

Forcible entry. A less desirable method of opening the hoistway door is by prying it open with forcible entry tools. The door may just crinkle and bend without releasing the lock, so this method should be attempted only after all else has failed.

In prying hoistway doors open, force should be applied at the top, near the lock, and in the direction that the door opens (Figure 7.53). When successful, the locking mechanism or the relating cable that connects the two doors together will

Figure 7.52 Some cars are equipped with an Inspector switch. *Courtesy of New York (NY) Fire Department.*

Figure 7.53 Under some conditions, entry may have to be forced.

break. If the relating cable breaks, only one panel will be free. The panel containing the lock will have to be released manually.

Passenger Egress

Depending upon the location of the elevator car in relation to the hoistway door, the method of assisting passenger egress will vary. The techniques to be presented include passenger egress through the doors and also through the top and side exits.

> **WARNING**
>
> **Whenever passengers are being removed from a car stalled below the landing, it is imperative that firefighters ensure that the car cannot suddenly drop.**

WITHIN LANDING ZONE

When an elevator car stalls within the landing zone, passenger removal is very easy because the vane on the car will have released the locking mechanism on the hoistway door. The car door can be pushed open by the passengers, which will open the hoistway door. If the hoistway door will not open because of something in the door track, the door will open once the obstruction is removed. A firefighter should then enter the car to make sure the Emergency Stop switch is activated. Because the car may be located a few inches above or below the level of the landing, passengers should be assisted from the car to prevent them from tripping as they walk out (Figure 7.54).

SLIGHTLY BELOW LANDING

When the elevator is stalled below the landing level, the passengers will obviously have to climb up to reach the landing. After installing shoring, or otherwise ensuring that the car cannot suddenly drop, a firefighter descends a ladder placed down into the car, and activates the Emergency Stop switch. The firefighter then heels the ladder as the passengers climb up and out of the car (Figure 7.55). Another firefighter at the landing aids the passengers in stepping from the ladder while maintaining a secure hold on them at all times.

SLIGHTLY ABOVE LANDING

When the elevator car is stalled above the landing, the passengers must climb down in order to reach it. Before anything else is done, a ladder is placed horizontally across the opening between the bottom of the car and the landing to prevent anyone from falling into the open hoistway (Figure 7.56). Again, a firefighter climbs a ladder into the car and activates the Emergency Stop switch. The firefighter then assists the passengers onto the ladder while another on the landing heels the ladder and assists them down.

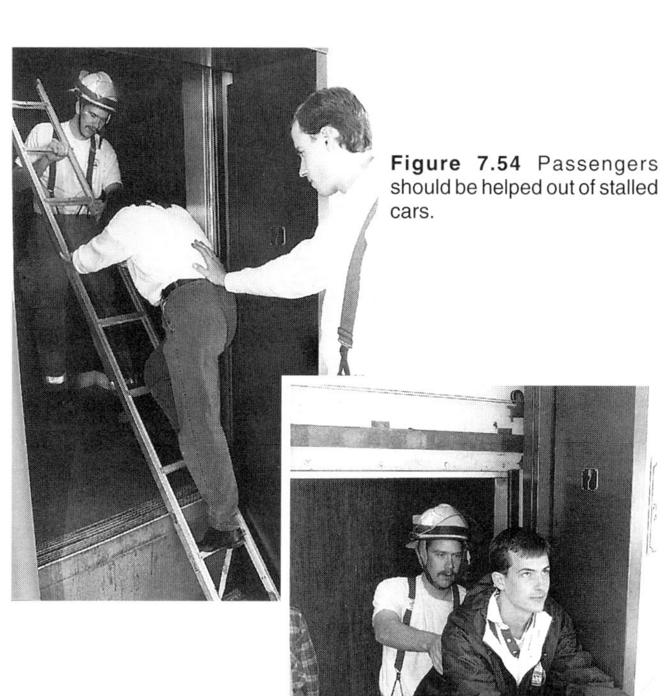

Figure 7.54 Passengers should be helped out of stalled cars.

Figure 7.55 A passenger is assisted up a ladder.

Figure 7.56 A ladder placed horizontally across the opening.

Emergency Rescue Operations

If passengers are in a life-threatening situation, it may be necessary to take extraordinary action to remove them. These situations constitute the few true elevator emergencies. Situations falling into this category are as follows:

- Passenger in the throes of a life-threatening medical condition
- Passenger in uncontrollable hysteria
- Fire or smoke threatening the elevator car

Any or all of these conditions may preclude waiting for an elevator mechanic to arrive or attempting the primary or secondary removal procedures. If the car will respond to the Emergency Service switch, that method should of course be used to bring the car to the main elevator lobby. However, the many possible scenarios make it very difficult to specify any one best way to handle elevator emergencies. Depending on the circumstances, any or all of the following means may be needed:

- Using top hatch
- Using side exit panels
- Forcing hoistway doors
- Breaching hoistway walls

RESCUE THROUGH TOP OF CAR

Removal of passengers through the top exit may be the best option if rescue through the doors is too difficult or impossible. In the case of a single hoistway installation, using the top exit is the only alternative to using the doors to remove passengers.

Access to the top of the stalled car can be gained by opening the hoistway door above and laddering down to it. Firefighters, secured by lifelines, can then climb down to the top of the car (Figure 7.57).

CAUTION: Interrupting power to the elevator hoist motor does not eliminate power to the top of the car. Firefighters must be aware that the fan, lights, etc., are still energized.

Once the exit panel is removed, a ladder should be placed down into the car. A firefighter should descend the ladder and check the Emergency Stop switch. If none of the passengers is in need of immediate medical attention, the removal process can begin. Before passengers are allowed to climb out, they should be secured by a lifeline (Figure 7.58). As they climb, they should be assisted by the firefighter in the car and helped through the opening by another on top of the car. A firefighter located on the landing can then assist each passenger onto the landing and remove the lifeline. If the top of the elevator is not within one normal step of the landing, a ladder should be used to get the passengers onto the landing. Additional firefighters will be needed to stabilize the ladder.

RESCUE THROUGH SIDE EXIT

Side exit removal of passengers is an effective method of rescue if the elevator installation is of the multiple-hoistway type. Methods previously discussed in the Gaining Access to Elevator Car section can be used. A serviceable elevator car is positioned so that it is even with the stalled car. The firefighters open the side panel of the rescue car, reach across, and open the side panel of the stalled car. One firefighter enters the stalled car and checks the Emergency Stop switch. Though the gap between the cars is usually only about 30 inches (762 mm) wide, a walking plank should

Figure 7.57 A rescuer descends a ladder to a stalled elevator.

Figure 7.58 A passenger is fitted with a lifeline. *Courtesy of New York (NY) Fire Department.*

be used to span the gap. A pike pole can be used as an improvised handrail. While passengers are crossing from one car to the other, firefighters should maintain physical contact with them at all times. All passengers should have a lifeline attached before they walk between the cars.

BLIND HOISTWAY REMOVALS

Removing passengers from a car stalled in a blind hoistway presents problems that may seem insurmountable. The rescue may involve damaging the building and could be extremely hazardous. If the installation is of the single hoistway type, the only way to remove the passengers may be by breaking through the hoistway wall. This may require special equipment and expertise not readily available. Because this is so time-consuming and laborious, it is recommended that rescue be delayed until the arrival of an elevator mechanic, unless lives would be jeopardized by waiting.

NOTE: Codes in many jurisdictions require emergency access doors into blind hoistways at intervals of every three or four floors. This will usually preclude having to breach the hoistway wall.

In a blind multiple hoistway, rescues can be performed from an adjacent car (see Rescue Through Side Exit section), except in the case of a complete power outage to the building. If there is power to the elevators, the stalled car can be located by positioning a firefighter in the top exit of the rescue car to scan the hoistway as the rescue car moves.

In case of a complete power outage in a single hoistway installation, locating the car may be extremely difficult. Firefighters may have to use a portable light to see approximately where the car is located and then conduct a floor-by-floor search after the passengers have been instructed to open the car door and tap on the hoistway wall. These instructions may be given over the elevator intercom (if operational) or by using a bullhorn to speak into the hoistway.

Emergency removals from blind-shaft elevators can be performed by accessing the top of the elevator car from the closest emergency access door (which may be 20 feet [6.1 m] or more above the top of the car) or by breaching the shaft wall. In newer buildings, the hoistway wall is often made of gypsum wallboard over metal studs, which is easily breached with standard forcible entry tools. In older buildings, the shaft may be enclosed with walls of brick, cement block, or poured concrete. Brick and cement block walls can be breached with some difficulty using hand or power tools (Figure 7.59). Poured concrete walls are much more difficult to breach, but they are often only on three sides of the shaft, with the fourth side of a more easily penetrated material. This more vulnerable wall is usually the one in which the doors are located as well.

Blind-shaft elevators are usually found in the tallest buildings, which can make locating the stalled car very difficult. Since blind-shaft elevators are at least in part express elevators, they may not have floor indicators except in the machine room. For example, if an elevator serves only floors 20-30, it is blind from the main lobby through the nineteenth floor, and the hallway indicator will show "X" (for express) at any point between those floors. Opening the hoistway door in the lobby may be useful if the shaft is not obscured by smoke (Figure 7.60). Even though the emergency access

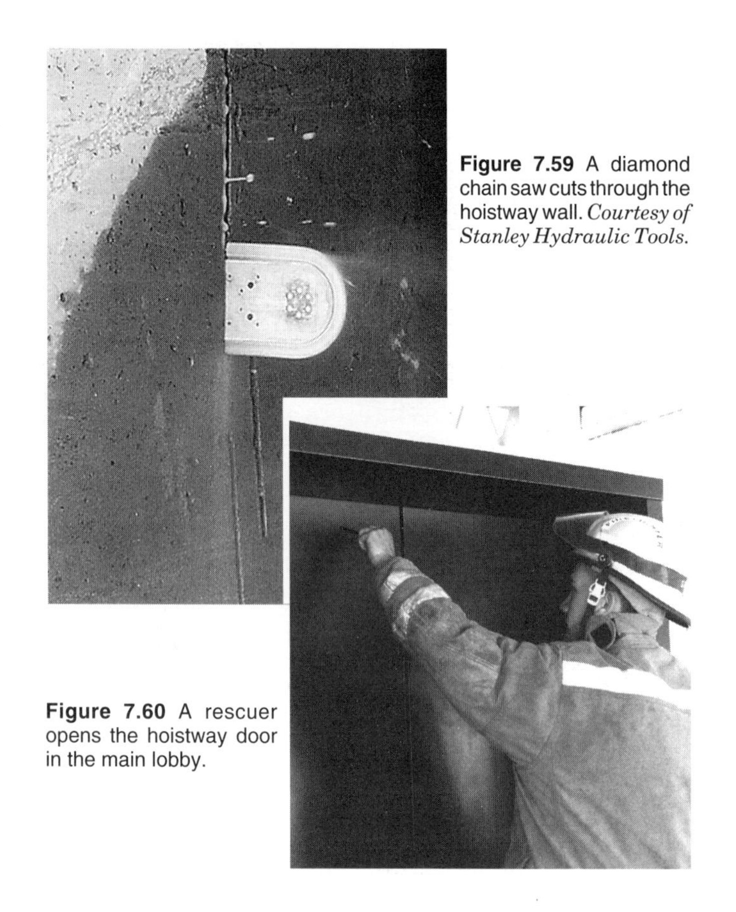

Figure 7.59 A diamond chain saw cuts through the hoistway wall. *Courtesy of Stanley Hydraulic Tools.*

Figure 7.60 A rescuer opens the hoistway door in the main lobby.

doors may be two, three, or four floors apart, counting the doors between the lobby level and the stalled car will tell rescuers the approximate location of the car.

Locating the hoistway wall on the floor nearest the stalled car may also be difficult. Unless there is an emergency access door on that floor, the hoistway wall is likely to look just like any other wall from the hallway side. One way is to orient the hoistway door in the lobby to something that also appears on every floor such as a standpipe connection or hose cabinet. The position of the object in relation to the elevator door should be noted, and the distance between them paced off. Rescuers can then use the same references on the floor nearest the car to locate where to breach the wall.

RESCUES INSIDE THE SHAFT

One of the rarest and often most difficult of all elevator rescue situations involves victims who have fallen into the elevator shaft. They may have leaned against a faulty hoistway door and fallen into the pit (and perhaps the door fell as well) or onto the top of the car. When rescuing someone from the pit, power to the elevator must first be shut off, usually in the elevator machine room, or (in a true emergency) by wedging a screwdriver in the Lower Terminal Switch in the elevator pit. Once this has been done, the operation can proceed as any other traumatic injury incident.

If the victim fell onto the top of the car, the best approach is to access the victim from the landing above rather than through the roof hatch. Being mindful of the potential electrical hazards on the roof of the car and the danger of falling into the hoistway, rescuers should attempt to stabilize victims on the car top before moving them.

Victims may also be wedged between the car or the counterweights and the wall of the shaft (Figure 7.61). While this may result from some elevator malfunction when people are working in and around the elevator shaft, the majority of these incidents result from a very dangerous game popular with some adolescents called *elevator surfing*, in which they ride on top of the car as it moves up and down in its normal operation. In multiple-hoistway installations, they may attempt to jump from one car to another as the cars pass traveling in opposite directions (Figure 7.62).

Regardless of how victims come to be wedged in the shaft, the rescuers' initial actions should be as follows:

- Activate Emergency Stop switch.
- Shut off power to the involved car.
- Call for an elevator mechanic.
- Secure victims in place with lifelines.

Figure 7.61 A victim may be wedged between a car and the wall.

Figure 7.62 A dangerous game called *elevator surfing.*

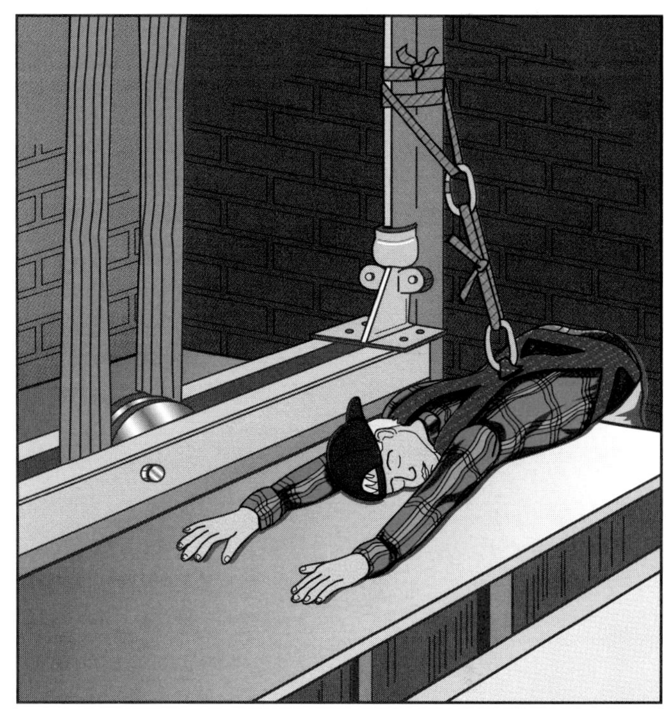

Figure 7.63 Victim should be secured to a solid object.

Lifelines applied to victims and tied off to a guide rail or other fixed object above them will usually prevent them from slipping farther down in the shaft (Figure 7.63). Rescuers must then create a safe working platform around victims. This may involve improvising some temporary scaffolding from which to access the victims. It may also be necessary to breach the hoistway wall to gain better access.

Rescuers should not rely on the elevator's emergency braking system. Before anyone is allowed to work beneath the car, it should be secured in place with ropes, timbers, jacks, or whatever combination will work under the circumstances. However, the means by which the car is secured should be easily manipulated because the car or the counterweight will almost certainly have to be moved slightly (by the elevator mechanic) to free the victim. If necessary, it is also possible to shift the car sideways using air bags or hydraulic spreaders. Care should be taken not to move the car far enough sideways to force it off the guide rails because this may release the emergency brakes. As a last resort, it may be necessary to cut the guide rails to free the victim. This can be done with a rotary saw or a cutting torch. In either case, a charged hoseline should be at hand in case litter in the pit or grease on elevator parts are ignited by the sparks.

PRE-INCIDENT PLANNING

Pre-incident planning is essential to the timely and efficient handling of elevator incidents. Firefighters need to be familiar with the building, with the elevator installation, and with how to obtain the necessary keys for each elevator in their response district. It is critical that the information be kept up to date and that any changes in the facility or the elevator installation be noted. Unless firefighters have this information before an incident occurs and have an opportunity to study it and devise appropriate action plans, valuable time can be lost while attempting to do this at the scene during an incident. Fire departments should develop an elevator data sheet similar to the following example:

ELEVATOR DATA SHEET

Name and Address of Location: ______________________________

____

Type of Installation:

_____ Passenger

_____ Freight

_____ Hydraulic

_____ Traction

Location of Hoistways: ______________________________

____

Number of Cars: ______________________________

Manufacturer/Address/Phone Number: ______________________________

____

Emergency Phone Number: ______________________________

Elevator Service Number: ______________________________

Location of Hoistway Door Key: ______________________________

____

Location of Hoistway Door Key Holes: ______________________________

____

Location of Elevator Machine Room: ______________________________

____

Independent Service: ___ Yes ___ No Operation ______________________________

Emergency Fire Operation: ___ Yes ___ No Operation ______________________________

Emergency Stop Switch Procedure:

____

Intercom Operation and Location:

____

Recommended Door-Opening Technique:

_____ Poling

_____ Hoistway

_____ Door Key

_____ Other

Other critical information such as blind hoistways, hazards, etc.:

____

____

Chapter 7 Review

Directions

The following activities are designed to help you comprehend and apply the information in Chapter 7 of **Fire Service Rescue**, Sixth Edition. To receive the maximum learning experience from these activities, it is recommended that you use the following procedure:

1. Read the chapter, underlining or highlighting important terms, topics, and subject matter. Study the photographs and illustrations, and read the captions with each.
2. Review the list of vocabulary words to ensure that you know the chapter-related meaning of each. If you are unsure of the meaning of a vocabulary word, look up the word in the glossary or a dictionary, and then study its context in the chapter.
3. On a separate sheet of paper, complete all assigned or selected application and review activities before checking your answers.
4. After you have finished, check your answers against those on the pages referenced in parentheses.
5. Correct any incorrect answers, and review material that was answered incorrectly.

Vocabulary

Be sure that you know the chapter-related meanings of the following words:

- conjunction *(214)*
- drift *(222)*
- lateral *(214)*
- obsolete *(210)*
- pole *(222)*

Application Of Knowledge

1. Complete an elevator data sheet for at least one elevator in your jurisdiction. *(234)*
2. Review your department's pre-incident plans for elevator rescue. *(Local protocol)*

Review Activities

1. Identify each of the following:
   - ASME A17.1 *(209)*
   - Pit *(210)*
   - Shaft/hoistway *(209)*
   - Car *(209)*
2. Identify the following parts of a hydraulic elevator: *(210)*
   - Electric motor
   - Hydraulic fluid
   - Hydraulic pump
   - Piston/ram
   - Cylinder
   - Reservoir
3. Describe the general operation of hydraulic elevators, including the way in which they are slowed and stopped. *(210)*
4. Explain why the practical upper limit for hydraulic elevators is about six stories. *(210)*
5. Compare and contrast the operation of drum and traction electric elevators. *(210, 211)*
6. Sketch a simple traction-type electric elevator and hoistway in which you locate and identify the following: *(211)*
   - Counterweight
   - Hoist cables
   - Drive pulley
   - Drive motor
   - Brake drum
7. Contrast brake operation on AC and DC traction-type electric elevators. *(211)*
8. List and identify the functions of the basic parts of an elevator car assembly. *(211, 212)*
9. Describe how elevator car doors and hoistway doors open and close simultaneously and automatically. *(212)*
10. Compare and contrast the operation of the four types of elevator car doors:
    - Single-slide *(212)*
    - Two-speed *(212)*
    - Center-opening *(212)*
    - Vertical bi-part *(213)*
11. Distinguish among functions, location, and method of opening top exit and side exit access panels. *(213, 214)*

12. Explain why side exit and top exit access panels may not be provided on some hydraulic elevator cars. *(214)*
13. Identify the usual location and safety features of the elevator controls normally used by elevator mechanics for inspection and maintenance. *(214)*
14. Compare and contrast the three common types of hoistways: single, multiple, and blind. *(214)*
15. Describe the operation of the locking mechanism on elevator hoistway doors. *(215)*
16. Describe the operation of each of the following hoistway door keys: *(216)*
    - T-shaped
    - Lunar (semicircular)
    - Drop
17. State the functions of the key-operated three-position switch located in the main elevator lobby of the designated level. *(217)*
18. Explain how rescue personnel activate independent elevator service and emergency elevator service. *(217)*
19. Describe what happens when Phase I rescue mode is activated. *(217)*
20. Distinguish between independent service and emergency service (Phase II rescue operation). *(217, 218)*
21. State the Phase II purposes of the *door open, door close,* and *call cancel* buttons on the elevator car's control panel. *(218)*
22. Identify the locations and functions of the following elevator safety devices:
    - Terminal switch *(219)*
    - Buffers *(219)*
    - Speed-reducing switch *(220)*
    - Overspeed switch *(220)*
    - Car safeties *(220)*
23. List possible elevator emergencies that preclude firefighters/rescuers following all primary and secondary rescue procedures before forcing entry into the elevator car. *(220)*
24. List four reasons why rescuers should wait for an elevator mechanic to arrive at the scene of a disabled elevator if passenger safety is not threatened. *(220, 221)*
25. Describe uses for the following rescue equipment used in elevator rescues: *(221)*
    - Portable radios
    - Short extension ladder
    - Folding ladder
    - Forcible entry tools
    - Safety belts
    - Lifelines
    - Emergency medical equipment
    - Walking plank
26. List hazards specific to elevator rescue. *(221)*
27. List safety guidelines that can reduce the risk of injury during elevator rescue operations. *(221, 222)*
28. Describe information that should be sent and received during the initial contact with trapped passengers. *(223)*
29. List the initial determinations that firefighters arriving at the scene of a disabled elevator must make. *(223)*
30. Describe ways in which rescuers can determine car location in the hoistway. *(223, 224)*
31. List the actions that should be taken during primary passenger removal procedures. *(225, 226)*
32. List the actions that should be taken during secondary passenger removal procedures. *(226)*
33. State methods of immobilizing an elevator car to effect rescue. *(227)*
34. Describe each of the following methods of gaining access to an elevator car if a hoistway door does not have an accessible locking mechanism or if a key is not available. Be sure to explain the conditions that dictate when to use each method.
    - Passengers opening car door *(227)*
    - Poling *(227)*
    - Opening hoistway door from inside *(227)*
    - Forcing entry *(228)*
35. Describe how to assist passengers from an elevator car stopped in the following positions: *(229)*
    - Within landing zone
    - Slightly below landing
    - Slightly above landing
36. Describe each of the following rescue methods and tell when each is appropriate.
    - Rescue through top of car *(230)*
    - Rescue through side exit panels *(230, 231)*
    - Rescue from a blind hoistway *(231, 232)*
    - Rescues inside the shaft *(232, 233)*
37. Explain what is meant by *elevator surfing*. *(232, 233)*

Questions And Notes

8

Confined Space Rescue

Chapter 8
Confined Space Rescue

INTRODUCTION

The history of people working in tanks and other confined spaces is filled with stories of workers losing consciousness inside the spaces where they were working. All too often, coworkers who entered the spaces attempting to rescue their fallen comrades were themselves overcome and only succeeded in complicating the problem by adding more victims. In Northern California, two winery workers died in a large wine tank after one worker who had entered to clean the tank lost consciousness and another entered, without SCBA, to try to rescue the first one. In Oklahoma, a construction worker died when he entered a large sewage trunk to retrieve a package of cigarettes that had fallen from his pocket into the open access way. In Pennsylvania, three emergency responders died when they entered an abandoned well, without SCBA, while attempting to rescue a child who had fallen in. In all of these cases, and in countless others on record, fatalities occurred because basic confined space safety procedures were not followed. These safety procedures are just as important for rescuers as for the workers themselves.

Because these scenarios were not unusual, several different organizations in the U.S. and Canada developed safety standards in the late 1980s and early 1990s to protect workers who are required by their employers to enter confined spaces. In the U.S., standards were developed by the National Institute for Occupational Safety and Health (NIOSH), by the American National Standards Institute (ANSI), and by the Occupational Safety and Health Administration (OSHA); in Canada, by the Division of Occupational Health and Safety. While this chapter is not intended as a detailed review of these standards, it relates their requirements to confined space rescues, and it uses some of their terminology and definitions. This chapter reviews currently accepted techniques for rescues from all sorts of confined spaces, except trenches, mines, and caves, which are covered in other chapters of this manual.

BACKGROUND

Regulations

Of the many confined space standards developed in the U.S., the most frequently applied are the Occupational Safety and Health Administration (OSHA) regulations contained in Title 29 of the Code of Federal Regulations (CFR). In the U.S., work of any kind within confined spaces is regulated under 29 CFR 1910.146. Within Section 146, two types of confined spaces are identified: (1) confined spaces for which no entry permit is required for workers to enter and (2) those for which an entry permit is required.

Also applicable in some confined space rescues is NFPA 1983, *Standard on Fire Service Life Safety Rope, Harness, and Hardware*. This standard provides minimum performance requirements for the life safety rope, harness, and hardware that rescuers use to support themselves and victims during actual or simulated rope rescue operations.

Confined Space Defined

All of the various confined space standards generally agree on how to define a confined space. Regardless of the specific wording of the standard, a confined space has certain characteristics. Any confined space:

- Is large enough and so configured that an employee can bodily enter and perform assigned work
- Has limited or restricted means for entry or exit
- Is not designed for continuous employee occupancy

However, some confined spaces are considered to be more hazardous to employees working inside than are others. These are termed *permit-required confined spaces*. In addition to meeting *all* the criteria for a confined space, permit-required spaces have *one or more* of the following characteristics:

- Contains or has a potential to contain a hazardous atmosphere
- Contains a material that has the potential for engulfing an entrant
- Has an internal configuration such that an entrant could be trapped or asphyxiated by inwardly converging walls or by a floor that slopes downward to a smaller cross section
- Contains any other recognized serious safety hazard

The regulations specify that before workers are allowed to enter a permit-required space, an entry permit must be prepared by the employer. This is important to rescuers because the required entry permit must be *written,* and it must be *posted* at the entry point. The permit must list more than a dozen essential items of information about the space, the work to be done within the space, and those who will do it (Figure 8.1). Obviously, the information contained in the permit can be of critical importance to rescuers because, among other things, it describes some of the hazards of which the rescuers should be aware, and this description may indicate the level of protection that rescuers need in order to safely enter.

Types Of Confined Spaces

The OSHA regulations contained in 29 CFR 1910.146 apply only to general industry and exclude shipyard employment, construction, and agriculture because other sections in 29 CFR 1910 cover those areas of employment. However, because workers enter all types of confined spaces to clean or repair them or to work on machinery inside the spaces, rescuers must be prepared to function in all types of confined spaces, regardless of the occupational setting. The following sections describe the most common types of confined spaces to which rescuers may be called.

TANKS/VESSELS

Tanks come in a variety of sizes and shapes and may be constructed of wood, fiberglass, or masonry, but most are made of aluminum or steel (Figure 8.2). Some tanks have the additional hazards associated with being elevated high in the air (Figure 8.3). In this context, the term *vessel* can refer to any confined space, but it usually refers to tanks involved in transporting products on land or water. Vessels may be tank trucks, railroad tank cars, barges, or ships with confined cargo spaces.

While the interior of the tanks may contain some sort of product, such as water, wine, vegetable oil, petroleum products, or grain, they are often empty of product when rescuers respond to calls involving such tanks. However, the absence of product does not necessarily mean an absence of hazards within the tanks. Even "empty" tanks can contain hazardous atmospheres as described later in this chapter.

Figure 8.2 One form of confined space.

Figure 8.3 An elevated confined space.

ENTRY PERMIT

PERMIT VALID FOR 8 HOURS ONLY. ALL PERMIT COPIES REMAIN AT SITE UNTIL JOB COMPLETED

DATE:_________ SITE LOCATION/DESCRIPTION ____________________

PURPOSE OF ENTRY ____________________

____

SUPERVISOR(S) in charge of crews   Type of Crew Phone #

____

____

COMMUNICATION PROCEDURES ____________________

RESCUE PROCEDURES (PHONE NUMBERS AT BOTTOM) ____________________

*** BOLD DENOTES MINIMUM REQUIREMENTS TO BE COMPLETED AND REVIEWED PRIOR TO ENTRY ***

REQUIREMENTS COMPLETED	DATE	TIME		DATE	TIME
Lock out/De-energize/Try-out	____	____	**Full Body Harness w/"D" ring**	____	____
Line(s) Broken-Capped-Blank	____	____	**Emergency Escape Retrieval Eq**	____	____
Purge-Flush and Vent	____	____	**Lifelines**	____	____
Ventilation	____	____	Fire Extinguishers	____	____
Secure Area (Post and Flag)	____	____	Lighting (Explosive Proof)	____	____
Breathing Apparatus	____	____	Protective Clothing	____	____
Resuscitator-Inhalator	____	____	Respirator(s) (Air Purifying)	____	____
Standby Safety Personnel	____	____	Burning and Welding Permit	____	____

Note: Items that do not apply enter N/A in the blank.

** RECORD CONTINUOUS MONITORING RESULTS EVERY 2 HOURS **

CONTINUOUS MONITORING ** TEST(S) TO BE TAKEN	Permissible Entry Level									
PERCENT OF OXYGEN	**19.5% to 23.5%**	____	____	____	____	____	____	____	____	____
LOWER FLAMMABLE LIMIT	**Under 10%**	____	____	____	____	____	____	____	____	____
CARBON MONOXIDE	**+35 PPM**	____	____	____	____	____	____	____	____	____
Aromatic Hydrocarbon	+1 PPM * 5 PPM	____	____	____	____	____	____	____	____	____
Hydrogen Cyanide	(Skin) * 4 PPM	____	____	____	____	____	____	____	____	____
Hydrogen Sulfide	+10 PPM * 15 PPM	____	____	____	____	____	____	____	____	____
Sulfur Dioxide	+2 PPM * 5 PPM	____	____	____	____	____	____	____	____	____
Ammonia	* 35 PPM	____	____	____	____	____	____	____	____	____

* Short-term exposure limit: Employee can work in the area up to 15 minutes.

+ 8 hr. Time Weighted Avg.: Employee can work in area 8 hrs (longer with appropriate respiratory protection).

REMARKS: ____________________

GAS TESTER NAME & CHECK #	INSTRUMENT(S) USED	MODEL &/OR TYPE	SERIAL &/OR UNIT #
____	____	____	____
____	____	____	____

SAFETY STANDBY PERSON IS REQUIRED FOR ALL CONFINED SPACE WORK

SAFETY STANDBY PERSON(S)	CHECK #	CONFINED SPACE ENTRANT(S)	CHECK #	CONFINED SPACE ENTRANT(S)	CHECK #
____	____	____	____	____	____
____	____	____	____	____	____

SUPERVISOR AUTHORIZATION - ALL CONDITIONS SATISFIED ____________ DEPARTMENT/PHONE ________

AMBULANCE 2800 FIRE 2900 SAFETY 4901 Gas Coordinator 4529/5387

Figure 8.1 A typical entry permit.

SILOS/ELEVATORS

These enclosures are designed to contain agricultural products, but their nature and the inherent hazards make each of them very different from a rescue standpoint. Silo rescues are likely to involve workers who have been overcome by a hazardous atmosphere, and grain elevator rescues are more apt to involve workers who have been engulfed by flowing grain (Figure 8.4).

STORAGE BINS/HOPPERS

Storage bins and hoppers are similar to silos and elevators in that they normally contain dry products, but bins and hoppers are usually smaller (Figure 8.5). They often contain products such as grain, gravel, sand, dry portland cement, wood chips, or sawdust.

Figure 8.4 A typical grain elevator.

Figure 8.5 Grain bins are confined spaces.

UTILITY VAULTS/PITS

These enclosures are usually of concrete or other masonry construction and are often below ground level (Figure 8.6). They may contain high-voltage electrical equipment, or pumps or valves for controlling gases or liquids in pipelines.

AQUEDUCTS/SEWERS

Pipelines that carry water to cities or industrial complexes or those that carry effluent away from them can be of different sizes and made from a variety of materials. They may be just big enough for one person to fit into or big enough to drive a truck into and may be made of concrete or steel (Figure 8.7).

CISTERNS/WELLS

While cisterns are tanks that usually collect and store rainwater, they are often underground; and both cisterns and wells may contain water or be dry (Figure 8.8). Many of these particular confined spaces are old and abandoned, so they may be made of or lined with a variety of materials — from ungrouted masonry to steel.

Figure 8.6 A common type of utility vault.

Figure 8.7 Sewers and storm drains are confined spaces.

Figure 8.8 An unprotected underground cistern. *Courtesy of Bobby Henry.*

COFFER DAMS

These are watertight barriers installed or constructed to provide a relatively dry environment in which to work. Coffer dams are often used around the forms when concrete must be poured in deep water.

Contamination Of Confined Spaces

Confined spaces can become contaminated in a number of different ways. Rescuers need to be aware of these mechanisms in order to know how to protect themselves and others during rescue operations. The following sections describe the most common ways in which confined spaces may become contaminated.

ABSORPTION/ADSORPTION

This mechanism of contamination occurs when the porous materials of which the enclosure is constructed either absorb contaminants from outside the vessel or adsorb materials formerly contained within it. *Absorption* would usually result from a concrete underground tank allowing the contaminant (most often gasoline or other petroleum product) to pass from a contaminated aquifer through the tank walls and into the tank. *Adsorption* occurs when the product formerly contained within the vessel clings tenaciously to the tank's interior surfaces after the product has been off-loaded. One example of evidence of adsorption is an empty tank that still has product fumes because of some of the product clinging to tank walls.

DESORPTION

The reverse of the absorption/adsorption mechanism, *desorption* occurs when a container within a vessel leaks or leaches the contaminant out of its containment and into the confined space where it is housed.

COMBUSTION

A fire in a confined space can produce some extremely hazardous conditions. As the fire passes through the first and second phases of combustion (incipient and steady state), it consumes all of the oxygen within the space. The fire slowly progresses into the third phase (hot-smoldering) where the fuel can only smolder but continues to produce heat and unburned gases. If the fire continues to be contained within the vessel, it will eventually burn itself out and cool to ambient temperature. The vessel will be heavily contaminated with creosote and other products of combustion. However, by wearing SCBA and impermeable clothing, rescuers can safely enter the vessel. More important for rescuers to remember is that if a hatch cover is opened before the gases within the space have cooled below their ignition temperature (lowest temperature of self-sustaining combustion), a backdraft can occur with devastating force.

BIOLOGICAL ACTIVITY

One of the most common forms of biological activity with which rescuers must be concerned is *fermentation*. In this process, organic matter consumes oxygen and gives off carbon dioxide. In a confined space, the obvious result is a reduced level of oxygen within the vessel. Unless it is properly and continuously ventilated, the atmosphere within a fermentation tank or silo is likely to be below 19.5 percent oxygen, which is oxygen-deficient, so the use of supplied-air breathing apparatus will be required by all who enter.

Another major source of biological contamination in confined spaces is untreated sewage. The biological activity in raw sewage produces hydrogen sulfide gas; in confined spaces such as pipelines and trunks, lethal concentrations are quite likely. Supplied-air respiratory protection is a must for all entrants.

INERTING

Confined spaces may be contaminated by the intentional introduction of an inert gas such as argon or nitrogen into the space. This is often done to purge the space of flammable vapors or gases, but in doing so the level of oxygen within the space is also reduced below 19.5 percent.

INSERTING

The insertion of a product into a tank can contaminate the tank in terms of its suitability for human occupancy. Flammable liquids and many other products that are routinely inserted into tanks and other confined spaces render them uninhabitable until the tanks have been emptied, cleaned, and ventilated. Until the contamination has been mitigated, entrants wear special protective clothing and respiratory protection.

CHEMICAL REACTION

Unless tanks and other vessels are properly cleaned after use, contamination by chemical reaction can occur. This may happen when a chemical that is introduced into the vessel reacts with residue from another chemical that was formerly contained within the space. While this most often involves chemicals in a liquid form, mixing incompatible dry chemicals can produce a similar result.

OXIDATION

The chemical union of oxygen with another material is well known to firefighters because it is the essence of the combustion process. However, in a confined space, even a slow form of the oxidation process, such as rusting, can reduce the level of oxygen within the space, as well as produce toxic by-products.

HAZARDS

Every confined space has certain actual or potential hazards within it. These hazards may be created in a number of different ways and may take a variety forms. Most of the hazards will fall into one of three major categories: atmospheric, physical, or environmental.

Atmospheric Hazards

Atmospheric hazards within confined spaces can be created in a number of different ways and may take different forms. However, the potential for the existence of an atmospheric hazard makes it critically important that the atmosphere within the space be sampled with properly calibrated instruments prior to entry (Figure 8.9).

Figure 8.9 The atmosphere within the vessel must be sampled.

OXYGEN DEFICIENCY

According to 29 CFR 1910.146, the atmosphere within a confined space is considered *oxygen-deficient* whenever the percentage of oxygen drops below 19.5 percent. Oxygen deficiency can result from biological activity such as in fermentation, in a fire or other oxidation process, or in the oxygen being displaced by another gas being intentionally or unintentionally introduced into the space.

OXYGEN ENRICHMENT

Any atmosphere within a confined space that exceeds 23.5 percent oxygen is defined by 29 CFR 1910.146 as an *oxygen-enriched atmosphere*. This can be created by oxygen leaking into the space or by the space inadvertently being purged with oxygen instead of air or an inert gas. Too much oxygen in a confined space, of course, greatly increases the fire hazard.

FLAMMABILITY

According to 29 CFR 1910.146, if the atmosphere within a confined space contains a flammable gas, vapor, or mist in excess of 10 percent of its lower flammable limit (LFL), it is considered hazardous. This will most likely result from residual flammable product remaining in the space, but it can also result from a volatile flammable liquid seeping in from a contaminated aquifer. However, a flammable atmosphere does not have to be related to either flammable gases or flammable liquids. When grain dust or fine dust from woodworking is stirred up and suspended in the air, a very flammable or even explosive atmosphere can be created. The regulations say that the lower flammable limits (LFL) of airborne combustible grain dust can be approximated by a condition in which the dust obscures vision at a distance of 5 feet (1.5 m) or less.

TOXICITY

Because the atmosphere within these spaces is confined, toxic gases and vapors that might otherwise dissipate harmlessly into the atmosphere are maintained at harmful concentrations. Whether the source is a liquid residue that is off-gassing, a biological activity such as in a sewer trunk, or an intentionally introduced substance such as a fumigant, rescuers must be careful to not add themselves to the list of victims. They can avoid becoming victims by following basic confined space safety procedures as outlined later in this chapter.

Physical Hazards

The physical hazards in confined spaces may be created by the lack of structural integrity of the space. They may also be created by hazardous objects within the space.

STRUCTURAL INSTABILITY

The possibility of the space itself collapsing is ever present in some confined spaces, especially immediately after an earthquake, explosion, tornado, flash flood, or other violent event. Spaces under tons of structural debris may collapse at any moment if shaken by an aftershock or secondary explosion. Mud slides and flood waters can exert tremendous pressure on structures and the spaces within, increasing the possibility of collapse. So, in some cases, rescuers must work against the clock to either shore up the space enough to prevent a collapse or to complete rescue operations before the space collapses.

DEBRIS

Loose and unstable debris from a partial collapse or other violent event may fall on rescuers within a confined space. Violent events move things and rearrange the geometry of structures and the spaces within them. This loosens objects and makes them prone to sudden and unexpected movement when disturbed. Debris strewn around the scene can also increase the likelihood of rescuers tripping over it.

ENGULFMENT

Confined spaces that contain large amounts of loose and free-flowing materials, such as grain, gravel, sand, or other dry products, all have the potential to engulf rescuers without warning. Victims can be engulfed when these materials are discharged into the vessel by remotely operated augers, conveyors, or other machinery. These materials can move and behave like liquids when confined. But unlike liquids, they are opaque, and anyone engulfed in them quickly becomes disoriented because there is no light and no way to tell up from down. The weight of these materials pressing in on a victim's upper body can restrict breathing and cause suffocation in a short time.

Environmental Hazards

The environmental hazards of confined spaces are those that are created by conditions within the space but not by the physical or structural condition of the space. Darkness, temperature extremes, and high noise levels can combine to make rescues slower and more difficult and to heighten victims' anxiety and feelings of claustrophobia.

DARKNESS

The absence of natural light within a confined space can make performing even routine tasks much more difficult and time-consuming. With limited artificial light and with rescuers' attention focused exclusively on the victim(s), the darkness can hide certain critical facts such as the water level slowly rising within a space where victims are trapped and in danger of drowning. Darkness within a space also increases the likelihood of rescuers falling over debris or into unprotected openings.

TEMPERATURE EXTREMES

Whether high or low, temperature extremes can threaten both victims and rescuers. Victims trapped in cold, damp spaces for extended periods can suffer from hypothermia or frostbite. Likewise, extremely hot, dry environments can bring on dehydration and other heat-related illnesses.

NOISE

The echo within confined spaces can increase the effects of even normal ambient noise. When the high noise levels created by rescue equipment and personnel are added, the effects of noise can be more than a mere distraction in confined space rescues. The intensified noise can be disorienting for victims and may cause rescuers to miss a faint cry for help or the hiss of escaping gas.

MOISTURE

Moisture is a common phenomenon within confined spaces. Even without broken pipes or cracks in walls that can allow seepage into the space, condensation can add a considerable amount of moisture to a space under the right conditions. Moisture can add unwanted humidity to hot, stifling confined spaces, and it can hasten the onset of hypothermia in cold situations. Wet, slippery floors can slow rescue efforts due to poor footing and can create a serious fall hazard for victims and rescuers alike.

DUST

Mineral dust and other fine airborne particulates are also common to some confined spaces. Structural collapse can stir up large amounts of dust, which can obscure vision and make breathing difficult. In the static atmosphere of a confined space, dust can remain suspended in the air for hours. Other types of dusts, such as grain dust and fine sawdust, can add flammability to the other problems that are created when dust is suspended in the air.

EQUIPMENT

For the most part, the equipment needed for confined space rescues is no different from that which firefighters use in other emergency operations. However, there are some differences, and some of the ways in which familiar equipment is used may be different. Following is a discussion of the personal protective equipment and team gear most commonly used in confined space rescues.

Personal Protective Equipment (PPE)

Structural fire fighting turnout gear is often not appropriate for rescue situations in general or for confined space rescues in particular — unless there is a fire within the space or the imminent threat of one starting. The type of fire fighting personal protective equipment (PPE) that is most appropriate for confined space rescues is similar to that worn by wildland firefighters. In the majority of confined space rescue operations, rescuers wear the following PPE.

HELMETS

The type of helmets most often worn in confined space rescue operations are similar to the wildland firefighters' helmet, but they have chin straps and are designed specifically for rescue work (Figure 8.10). These helmets are lighter in weight than structural fire fighting helmets, they allow better peripheral vision, and they are less cumbersome in tight places.

EYE PROTECTION

Eye protection is a must in confined space rescues. In these operations there are numerous opportunities for liquids or airborne particulates and flying debris to enter rescuers' eyes. Just keeping dust out of their eyes is reason enough for rescuers to wear goggles or safety glasses (Figure 8.11).

Figure 8.10 A typical rescue helmet.

Figure 8.11 A rescuer dons proper eye protection.

OUTERWEAR

Except where protection from actual or potential fires is needed or where extremely wet or cold environmental conditions warrant, structural turnout gear is usually not appropriate for confined space rescues. Coveralls or jumpsuits similar to those worn for wildland fire fighting serve much better (Figure 8.12). They are lighter in weight, allow more freedom of movement, and are generally more comfortable for rescuers than structural gear. Under some circumstances, such as when hazardous materials have contaminated the space, rescuers may need to don special protective clothing (Figure 8.13).

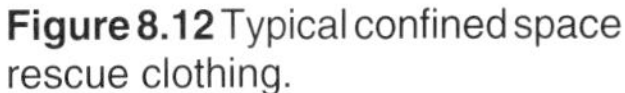

Figure 8.12 Typical confined space rescue clothing.

Figure 8.13 A rescuer in special protective clothing.

FOOTWEAR

Unless the operation must be carried out in standing water or other liquids, rubber boots are usually not needed and are not the best choice for confined space rescues. The leather safety boots that are standard issue in many fire departments provide the best combination of function and protection (Figure 8.14). Whatever footwear is selected, it should be appropriate for the particular environment in which the rescue is to be performed.

Figure 8.14 Typical leather safety boots.

GLOVES

Medium-weight leather gloves similar to those worn by wildland firefighters are the most appropriate for confined space rescues (Figure 8.15). They allow good freedom of movement and tactile dexterity while providing an acceptable level of protection.

Figure 8.15 Sturdy leather gloves are very important.

AIR SUPPLY

When rescuers must operate in known or potentially contaminated atmospheres, they must wear supplied-air breathing apparatus. These can take either of two forms: (1) open- or closed-circuit SCBA or (2) airline equipment with an escape cylinder (Figures 8.16 a - c). Dual-purpose SCBA units allow safe "buddy breathing" as well as the ability to connect to an airline for refill or long-term operations (Figure 8.17).

Team Gear

The equipment outlined next is the gear that is used for protection of the rescue team, as opposed to the PPE worn by each team member. The team gear may include air-supply equipment, ventilation equipment, atmospheric-monitoring equipment, lighting equipment, and retrieval systems.

AIR-SUPPLY EQUIPMENT

Included in this category is all the equipment needed to provide a safe and dependable source of breathing air to individual team members who must enter contaminated atmospheres to effect rescues. The equipment may consist of a cache of SCBA cylinders, an air-supply unit with a cascade system for refilling empty SCBA cylinders, and/or an air-supply unit with both large air tanks and an air compressor for supplying airline respirators (Figures 8.18).

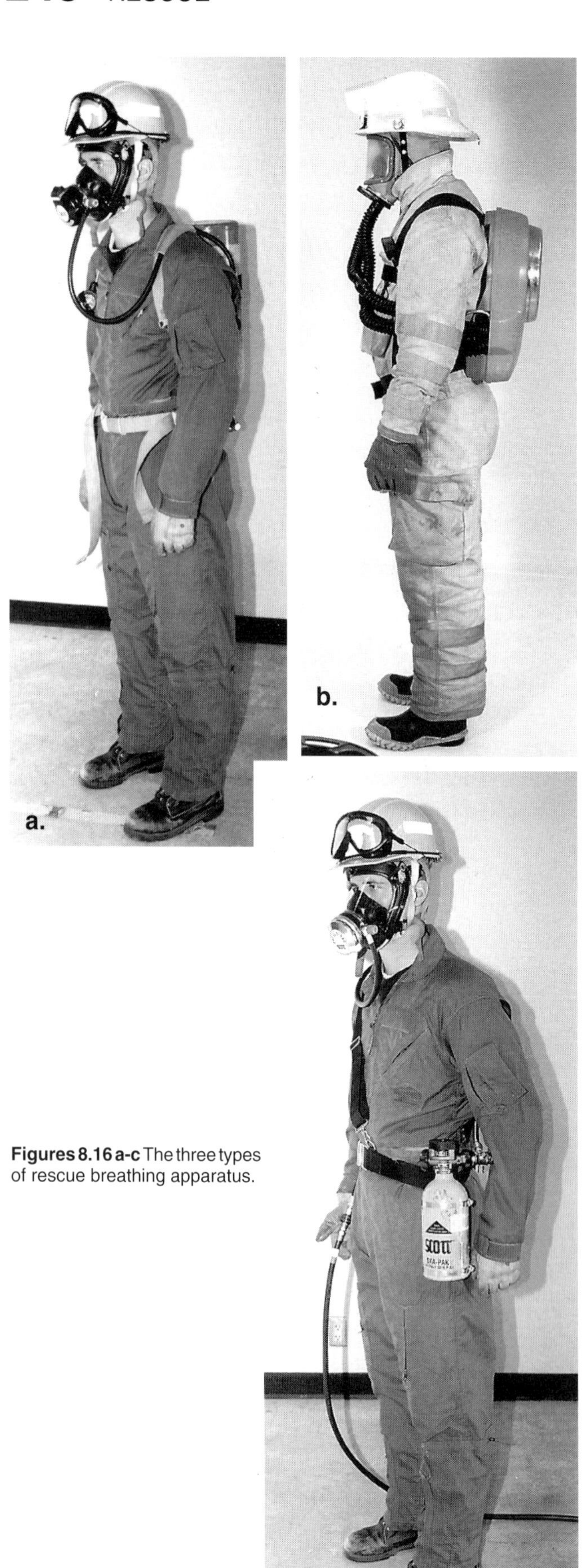

Figures 8.16 a-c The three types of rescue breathing apparatus.

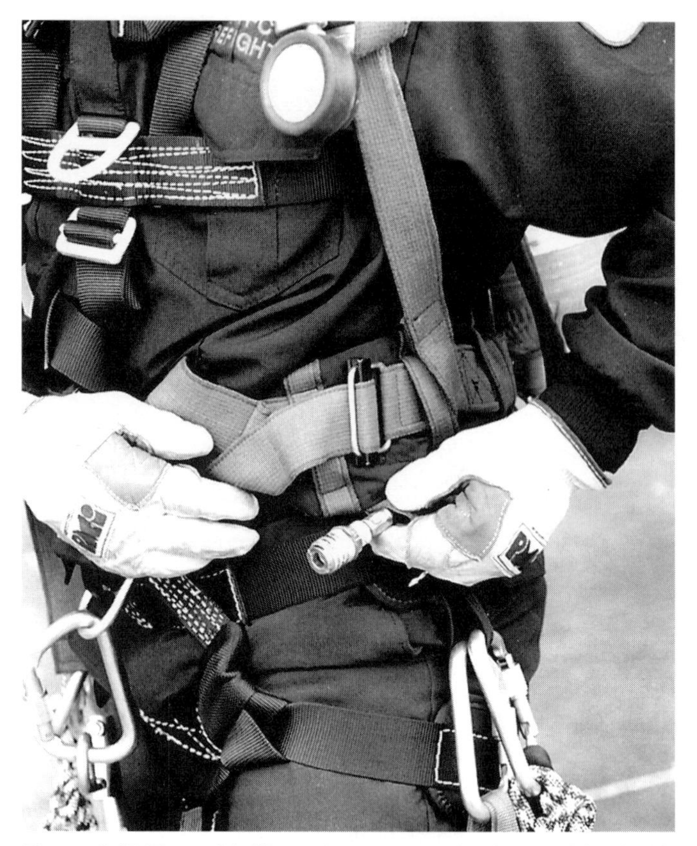

Figure 8.17 The quick-fill attachment may also be used for buddy breathing.

Figure 8.18 A dependable air supply is needed in confined space rescue incidents.

VENTILATION EQUIPMENT

To enhance the level of safety for rescuers working in confined spaces in which the atmosphere is contaminated and to increase the chances of survival for trapped victims, the interior of the space must be adequately and continuously ventilated by mechanical means. Such ventilation is loosely defined in 29 CFR 1910.146, but it is essentially a mechanical means of ventilation (blower) located outside of the contaminated area with some means (ducting) of channeling the fresh air into the space, or an intrinsically safe blower exhausting contaminated air from the space (Figure 8.19).

ATMOSPHERIC-MONITORING EQUIPMENT

The monitoring equipment used to sample and analyze the atmosphere within a confined space must be accurately calibrated direct-reading instruments capable of measuring the oxygen concentration within the space. They must also be capable of *measuring* the concentrations of toxic and flammable gases, not just of detecting their presence (Figure 8.20).

Figure 8.19 A typical mechanical ventilation setup.

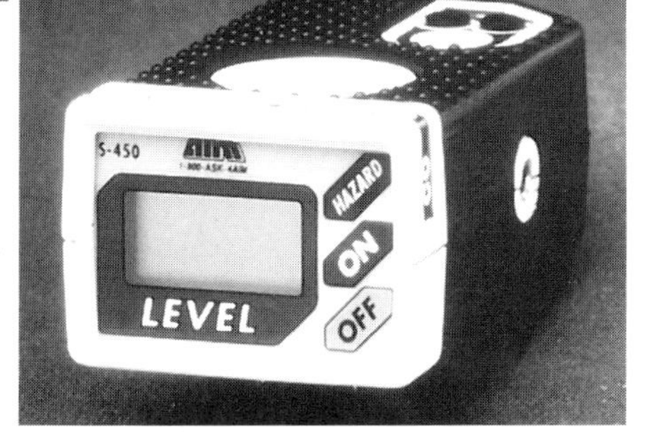

Figure 8.20 A modern multigas analyzer. *Courtesy of AIM Safety, USA, Inc.*

LIGHTING EQUIPMENT

The type and amount of lighting equipment needed for confined space rescues can vary tremendously, but because the access way into a confined space is so small, some form of artificial lighting is almost always needed (Figure 8.21). Individual flashlights or hand lanterns may be sufficient. In the majority of cases, however, some higher level of lighting is required, even during the day. For safety, all forms of lighting used inside the space should be intrinsically safe. When portable generators are used, they must be positioned downwind from the confined space access way to ensure that the exhaust from the gasoline engine powering the generator does not get carried into the confined space.

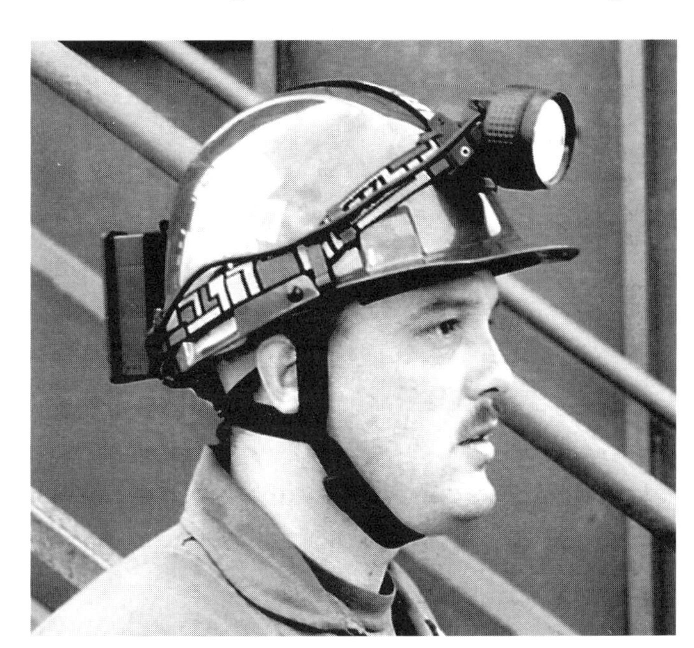

Figure 8.21 A typical helmet light.

RETRIEVAL SYSTEMS

The OSHA regulations require that anyone who enters a permit-required confined space (including rescuers) must be fitted with some form of retrieval system and that there be an attendant outside the space monitoring those inside. Retrieval systems consist of a retrieval line (lifeline) attached to either a chest or full-body harness or to wristlets (Figure 8.22). This allows the attendant to pull the entrant from the space without entering the space. The regulations provide an exception to the retrieval line requirement if the line will actually increase the overall risk by becoming entangled or if the line would not contribute to the rescue of the entrant.

Figure 8.22 A typical retrieval system with air supply.

Figure 8.23 A rescuer talks to a witness.

TACTICAL CONSIDERATIONS

The tactical considerations involved in confined space rescues may be broken down into four distinct phases: assessment on arrival, prerescue operations, rescue operations, and termination.

Phase I: Assessment On Arrival

The assessment of the situation that must be made by the first-arriving unit can be subdivided into a primary and a secondary assessment.

PRIMARY ASSESSMENT

The primary assessment actually begins with the initial dispatch and continues during response and arrival. The first-due officer must begin to formulate a mental picture of the situation based on the information provided in the dispatch, the time of day, the weather, and the traffic conditions during response to the scene.

Information gathering. The process is continued once on scene by attempting to talk to the reporting party (RP) and/or other witnesses (Figure 8.23). Of critical importance in the primary assessment is gathering information about the number, condition, and location of victims:

- How many victims are there?
- Are they injured or merely trapped?
- How long have they been down?
- Are they conscious, and if so, can they communicate?
- Are they all in the same confined space?
- Is there an entry permit available?

Decision making. The answers to these questions will help the first-in officer make the first critical decision: Can the units on scene or en route handle the situation, or do additional units need to be called? If more resources are needed, they must be requested immediately to get them on scene as soon as possible. If he or she has not done so already, the first-in officer should assume formal command of the incident because the answers to the initial questions will also form the basis for the incident action plan that the IC must develop at this point.

The size-up of the situation should be continued using the following checklist:

- Contact victim (if possible).

- Interview witnesses.
- Examine permits.
- Monitor atmosphere within the space.
- Identify hazards.
- Evaluate what has been and is being done.
- Weigh risks vs. benefits of available options.
- Evaluate adequacy of initial response.
- Contact expert assistance.

Scene control. If the information gathered during the primary assessment confirms that a legitimate rescue emergency exists, then the area surrounding the confined space should be cordoned off as described in Chapter 2, Rescue Scene Management. The area within the confined space should be designated the "hot zone," the area immediately outside of the space should be designated the "warm zone," and an area surrounding these two zones should be cordoned off and designated the "cold zone."

SECONDARY ASSESSMENT

The secondary assessment involves some reconnaissance of the scene to gather information about the confined space, its condition, and its contents. All information gathered during both the primary and secondary assessments helps determine the mode of operation.

Type of space. The first thing to be determined is the type of space involved. This can indicate the nature of the problem: A silo or wine tank may be oxygen deficient; a utility vault may contain high-voltage equipment, etc. It is also important to determine what the space is made of. This may indicate the feasibility of breaching the walls if that should become necessary. The RP or others familiar with the space may be able to describe the interior configuration of the space and alert rescuers to potential hazards.

Condition of space. The condition of the space must also be assessed. Is it intact or damaged? Will tunneling be necessary to reach the space, or will it be necessary inside the space? Will debris have to be removed to allow access into the space? Will the space have to be shored up to make it safe to work in during the rescue?

Contents of space. Again, the RP or others may be able to provide information about the contents of the space. If the entry permit or a Material Safety Data Sheet (MSDS) is available, critical information about the contents of the space can be obtained (Figure 8.24). That information can indicate both the likely condition of the victim(s) and the type and level of protection needed by the rescuers.

Mode of operation. Finally, all of the information gathered during the primary and secondary assessments confirms the nature and extent of the rescue problem and helps the IC finalize the incident action plan. The information also helps the IC make one of the most important decisions affecting the action plan: whether it is reasonable to think that the victims are viable and the operation should be conducted as a rescue or that the victims could not be expected to have survived and the operation should be conducted as a body recovery.

Figure 8.24 The available paperwork can provide critical information.

Phase II: Prerescue Operations

During this phase, which may last from a few minutes to several hours, all of the things necessary to make the rescue operation as safe as possible must be done. This phase includes:

- Finalizing the incident action plan.
- Gathering the necessary resources.
- Monitoring and managing the atmosphere inside the space.
- Making the space structurally stable enough to enter.
- Ensuring that there is adequate communications capability to allow the action plan to be carried out safely.

INCIDENT ACTION PLANS

On small, relatively simple rescue operations the incident action plan need not be in writing, but there *must be a plan*. On larger, more complex operations the plan should be in writing and should reflect an incident management system as described in Chapter 2, Rescue Scene Management. In either case, the plan must be finalized and communicated through channels to everyone involved in the operation.

While the original plan should be flexible enough to accommodate a certain amount of the unexpected, *a backup plan should be available* in case something completely unforeseen occurs to invalidate the plan. If the information gathered during the primary and secondary assessments was somehow inaccurate or misleading or if some subsequent occurrence, such as a secondary explosion or a major collapse, changes the situation significantly, a secondary or backup plan should be ready to be implemented. If it suddenly becomes necessary to rescue the rescuers, that plan of operation should be ready.

GATHERING RESOURCES

Resources consist of personnel and equipment, and both are critically important to the success of the operation. If there are too few rescue personnel or if the personnel are insufficiently trained to perform as needed, the best equipment in the world will not get the job done. Likewise, the most highly trained and motivated rescuers will not be able to do what is necessary if they do not have the tools and equipment they need.

The resources gathered at the scene should reflect the incident action plan. But as stated earlier, if both the personnel and equipment in the initial response are insufficient for the rescue problem at hand, the IC must request additional resources as soon after arrival as possible (Figure 8.25). The sooner these resources respond, the sooner they will arrive on scene where they are needed. If the IC is initially unsure about the type and/or amount of equipment that will actually be needed, he or she should call for everything that *might* be needed. Resources that prove to be unnecessary can be returned to quarters either while still en route or after they arrive on scene.

Personnel. Depending upon the nature and extent of the rescue problem, the number of rescue personnel needed will vary. However, even a relatively simple rescue of one victim from a confined space can involve eight to ten rescuers: a two-person entry team, a two-person backup team, an attendant, a safety officer, an incident commander, and two or three support personnel to set up and operate on-scene equipment. This list does not include EMS personnel who may be needed to treat and transport the victim to a medical facility. Obviously, as the number of victims and the complexity of the rescue problem increase, the number of rescue personnel needed increases proportionately.

Equipment. The amount and types of equipment needed will also vary with the nature and

Figure 8.25 Additional resources must be requested as soon as possible.

extent of the rescue problem. The scenario previously described could require a rescue unit, a couple of engine and/or truck companies (depending upon staffing levels), plus a command vehicle and an ambulance. More complex incidents could also require specialized units such as lighting units, air units, communications units, hazardous materials units, and cranes or booms.

ATMOSPHERIC MONITORING

The atmosphere within the confined space should be sampled *from outside the space* before rescuers enter and should be continuously monitored while they remain in the space (Figure 8.26). The information obtained by sampling the atmosphere helps determine the need for mechanical ventilation and the type of respiratory protection required for rescuer safety. If the readings change for the worse after rescuers have entered the space, they should withdraw and reevaluate the situation. It may be prudent to wait until mechanical ventilation has restored a safe atmosphere before reentering.

Oxygen concentration. The atmosphere within the confined space may be either oxygen deficient or oxygen enriched. Neither condition is safe for rescuers, and corrective steps must be taken. If the atmosphere proves to be oxygen deficient, regardless of what created the deficiency, rescuers must wear supplied-air respiratory protection, such as SCBA or airline respirators with escape cylinders, to safely enter the space.

Figure 8.26 The atmosphere within the space should be sampled from outside.

Likewise, regardless of the source, an oxygen-enriched atmosphere greatly increases the flammability or explosive potential of any fuel within the space. Therefore, rescuers must not enter the space until ventilation has decreased the oxygen level within the space below 23.5 percent. A sufficient number of charged hoselines should also be ready in case of ignition.

Flammables. If monitoring equipment indicates the presence of flammable gases or vapors in the atmosphere within the space, rescuers must assume the atmosphere to be flammable or explosive. All nearby sources of ignition should be eliminated, and rescuers must consider the potential risk/benefit involved in entering the space before ventilation has reduced the detectable level of flammable gas or vapor to 10 percent of its lower flammable limit (LFL) or below. Check the entry permit, the MSDS, or either NFPA 49, *Hazardous Chemicals Data*, or NFPA 325M, *Fire Hazard Properties of Flammable Liquids, Gases, and Volatile Solids*, to determine these limits. Obviously, a sufficient quantity of the most appropriate fire protection must be ready in case of ignition.

Toxics. If toxic gases or vapors are detected in the confined space, entry must be reevaluated and perhaps postponed until the atmosphere is mechanically ventilated and the source of the contamination has been identified and eliminated. The atmosphere within the space should be tested continuously until it is found to be within safe limits for the particular contaminant involved. Check the entry permit, the MSDS, or the *NIOSH Pocket Guide to Chemical Hazards* to determine these limits (Figure 8.27).

Examples of highly toxic materials are hydrogen fluoride gas and cadmium vapor. Either may produce immediate, often severe transient effects that may pass without medical attention but that are followed by a sudden, possibly fatal collapse 12 to 72 hours after exposure. After recovering from the initial transient effects, the victim "feels normal" until he or she collapses. Another example is hydrogen sulfide gas. Even though it is easily detected by its characteristic "rotten-egg" odor, in high concentrations it can paralyze a victim's respiratory functions faster than hydrogen cyanide, and it can be fatal in seconds.

Figure 8.27 A rescuer checks the NIOSH pocket guide.

The nature and extent of toxic contamination will also affect the IC's decision to approach the problem as a rescue or as a body-recovery operation. If it is the latter, there is no longer any need for haste, and every effort must be made to reduce the risks to rescue personnel.

VENTILATION

Because the openings in confined spaces are relatively small, natural (nonmechanical) ventilation will almost always be ineffective. This means that some form of mechanical ventilation must be used. Mechanical ventilation takes two forms: positive and negative. The form chosen should be based on the situation and the equipment available on scene. Mechanical ventilation should be used in concert with any prevailing wind, not against it.

Positive ventilation. Also known as *positive-pressure ventilation*, positive ventilation involves creating a slight positive pressure within the space by placing a fan or blower outside the entry opening and blowing fresh air into the space. For this technique to be successful, the blower must be at the correct distance from the entry opening to create a positive seal around the opening, and an exit opening must be created at the opposite end of the space. The exit opening should be from three-quarters to one and one-half times the size of the entry opening.

Negative ventilation. This form of ventilation involves placing an intrinsically safe smoke ejector to exhaust the contaminated air from the space. Unless ducting is used, an opening will also be needed at the opposite end of the space to allow replacement air in. In this case, the part of the exit opening not occupied by the smoke ejector may have to be sealed with a salvage cover or similar material to prevent the contaminants from being drawn back into the space (churning). If ducting is used, replacement air can be drawn in through the entry/exit opening.

PREPARING THE SPACE

Even after the necessary resources have been gathered on scene and the atmosphere within the space has been determined to be safe to enter, the characteristics of the space itself may still have to be manipulated before it is safe for rescue personnel to work inside. This may involve marking potential entrapment areas, identifying and mitigating hazards within the space, shoring up the structure itself, and/or providing intrinsically safe lighting within the space.

Space layout. The interior configuration or layout of the space may represent a hazard to rescuers. Spaces with floors that slope toward chutes, drains, or other outlets may cause rescuers to slide into them, especially if the floors are wet or covered in fine dust. Sloping floors may also cause rescuers to slide into very narrow tapering spaces in which they can become wedged. All such potential hazards should be marked or barricaded to protect rescue personnel working inside the space. If a retrieval line would enhance the safety of rescuers, each should have one on.

Internal hazards. In some cases, the confined space is created by the exterior housing of a large industrial machine. The machine (the confined space) may contain all sorts of pipes, valves, gears, levers, rollers, and all manner of electrical wiring, switches, relays, and controls, sometimes energized with very high voltage. Before rescuers are allowed to work inside a machine, all potential hazards should be neutralized using proper lockout/tagout procedures. Electrical power to the machine may have to be shut off and the switch locked out and tagged out (Figure 8.28). Gears and other moving parts may have to be blocked to prevent any sudden and unexpected movement, and piping may have to be interrupted and the open ends sealed with blankout flanges.

Shoring. As mentioned earlier, explosions and many types of natural calamities can damage and weaken the structure of confined spaces, often causing partial or total collapse. If there is any question about the structural integrity of a confined space, it will have to be shored up to make it safe for rescuers to work in. The emergency shoring techniques described in Chapter 6, Structural Collapse Rescue, will have to be applied as needed.

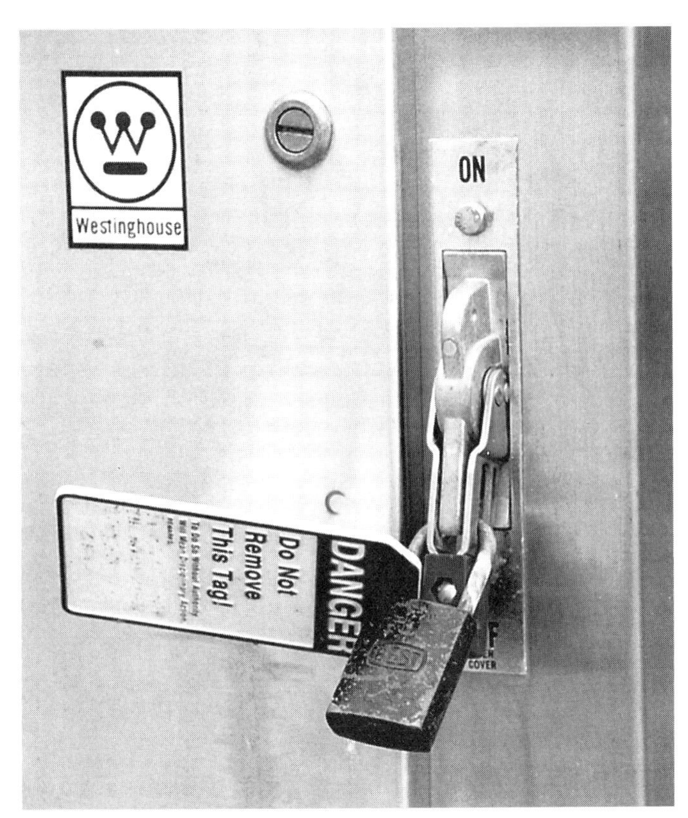

Figure 8.28 A locked-out switch.

Lighting. Most confined spaces have little or no lighting inside. Rescuers must be prepared to provide the lighting they need to do their jobs. Flammable atmospheres will require the use of intrinsically safe lighting equipment. If power is provided by a portable generator, care must be taken not to allow the exhaust from its engine to drift into and contaminate the space.

COMMUNICATIONS

Because confined space rescues can involve everything from small, easily accessed spaces to large, complex spaces with many walls, bulkheads, and other barriers to communication, the form of communication that works best can vary considerably. It may involve direct voice communication, a series of tugs on a lifeline, hard-wired phones, or portable radios. All communications devices used in any potentially flammable atmosphere must be of the intrinsically safe type.

Voice communication. Direct, face-to-face voice communication is preferred whenever the physical arrangement of the confined space allows it. The chances of miscommunication are greatly reduced if those involved can hear each other's voices, even if they cannot see each other.

Lifeline. While the main purpose of a lifeline attached to the harness worn by members of an entry team is to provide a means of pulling them out of the space if necessary, the lifeline can also serve as a primitive means of communication. One or more tugs on a lifeline can communicate a single thought. The acronym OATH is used by some departments to communicate as follows:

- 1 tug = ***O***K
- 2 tugs = ***A***dvance
- 3 tugs = ***T***ake Up (eliminate slack)
- 4 tugs = ***H***elp!

Hard-wired phones. These portable phone systems are very effective over the short distances involved in most rescue situations (Figure 8.29). These systems have proven their value as a communications medium in military operations and in high-rise fires. The main disadvantage is having to lay the phone line from the command post to the operational location.

Figure 8.29 A rescuer uses a hard-wired phone.

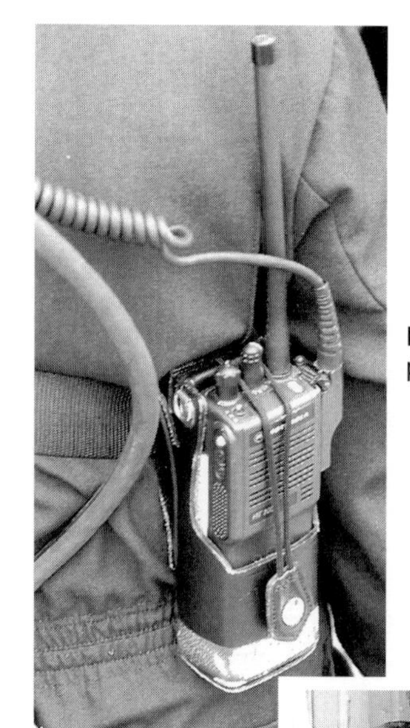

Figure 8.30 A modern multichannel portable radio.

Figure 8.31 Portable radios may be ineffective in some situations.

Portable radios. These versatile units have been the primary means of fireground and emergency-scene communications for many years, and advancing technology makes them even more useful in this role. Modern, multichannel units with scanners allow communication on a number of different frequencies (Figure 8.30). Such flexibility allows each incident in progress at the same time to have its own tactical or incident channel. This obviously reduces the amount of traffic the rescuers must monitor and process, and it greatly reduces the chances of miscommunication. Aside from their cost, the primary disadvantage of modern portable radios is that their signal is sometimes incapable of penetrating the mass of the structure of some confined spaces (Figure 8.31).

ENTRY PERSONNEL

Those personnel who will actually enter the confined space to effect rescues must be assembled, along with those who will work immediately outside the space to directly support them. Anytime there are rescuers inside a confined space, there must also be an attendant at the entry point but outside the space to monitor the safety of those inside. The incident Safety Officer may act as the attendant.

Entry team. Except where the space is too small for more than one rescuer, no one should be allowed to enter a confined space alone. So, in most cases, an entry team will be composed of at least two members. They must be properly dressed and equipped for the conditions inside the space and for the nature of the work they will have to perform. Their gear should be checked by the attendant before they are allowed to enter, and they should be informed of any potential hazards within the space and of any particularly gruesome scenes that are likely to be seen (Figure 8.32).

Backup team. There should also be a fully prepared and equipped backup team ready to enter if the first team gets into trouble. The backup team must be composed of the same number of personnel as the entry team and should have the same level of equipment, training, and expertise.

Phase III: Rescue Operations

Once all of the preparations have been made, the process of actually removing victims from

Figure 8.32 An entry team is briefed before entering a space.

confined spaces can begin. This phase of the operation usually involves entry into the confined space, so the use of a personnel accounting system is essential. In addition, this phase may involve searching for victims, treating them prior to their extrication, and perhaps using retrieval systems.

PERSONNEL ACCOUNTABILITY

The purpose of the accountability system is to ensure that only those who are authorized and properly equipped to enter are allowed to do so and that both their location and their status are known as long as they remain in the confined space. As is true of the incident action plan, the degree of formality of the personnel accounting system should reflect the nature, size, and complexity of the particular rescue problem at hand.

Minimum accountability. Some incidents require only one or two rescuers working in the space at a time, and they do not require respiratory protection. They are close enough to be in constant visual and verbal contact with the attendant. In these cases, the intent of the accountability system can be met with a minimum amount of information being recorded. Merely writing down the entrants' names and their entry and exit times should be sufficient (Figure 8.33).

Maximum accountability. A more formal system must be employed on larger, more complex incidents requiring more personnel to work inside the space at once, when rescuers must use respiratory protection, or when they must work out of sight of the attendant. In these cases, the entrants' names and/or other identifiers must be recorded as they enter (Figure 8.34). Their time of entry, SCBA gauge readings, and their *projected* exit times must also be written down. Those using airline equipment can remain in the space for longer periods of time than those wearing open-circuit SCBA, but the same rules of accountability apply. Each entrant's time of exit must be recorded as they leave the space. As each entrant's projected exit time arrives, the attendant must check the list to ensure that he or she has exited. If anyone has not exited by the time on the list, immediate action must be taken to locate them and escort them out of the confined space.

Figure 8.33 Accountability Officer records entry team data.

Figure 8.34 Accountability Officer collects personal identifiers. *Courtesy of Clemens Industries, Inc.*

ENTRY

Finally, after everything possible has been done to ensure a safe and successful operation, rescue personnel are allowed to enter the confined space to perform rescues. They may enter the confined space singly or in teams of two or more.

One rescuer. As mentioned earlier, if there is little room in which to work within the space, a single rescuer may have to enter alone. This is not desirable and should be done only when absolutely necessary. Except where there is an entanglement hazard, a single rescuer should not enter a confined space unless wearing a harness with a lifeline attached, especially if he or she will be out of the view of the attendant (Figure 8.35). Whenever possible, the attendant should keep up a running conversation with the rescuer to maintain at least verbal contact and as a means of assessing the rescuer's condition and situation.

Figure 8.35 A lifeline is attached to a rescuer's harness.

Two rescuers. Even though members of a rescue team can and should depend on each other, it is still important for all members to wear a retrieval harness with a lifeline attached, except where the potential for entanglement would increase the hazard. Even if the team is composed of only two members, one of the members must act as the team leader; and even though both may have portable radios, only the team leader should communicate with the unit supervisor or the IC (Figure 8.36). All actions taken by the team should be coordinated by the team leader and directed toward achieving the objectives established in the incident action plan.

Figure 8.36 Only the team leader should communicate with Command.

SEARCH

Unless the victim's location is obvious, the rescue team may have to conduct a search of the confined space. The same techniques that are used to search a burning building should be used to search a confined space. The search should be conducted systematically and in a logical sequence. Even though progress may seem inordinately slow, the team should work together as a unit and should avoid splitting up. When searching pockets and isolated areas within the confined space, one member should remain in a fixed location while the other member makes a sweep of the area. The stationary member keeps the other engaged in continuous conversation, just as the attendant does with a single rescuer. The team should occasionally stop its search, call out, and remain quiet for a few moments to listen for sounds or calls from victims. The IC should be kept informed through channels of the team's progress, of anything unusual or unexpected that they find, and especially when they locate the victims.

VICTIM TREATMENT/STABILIZATION

Once the victims have been found, they must be examined and their medical condition evaluated. This is made much easier if the victims are conscious and can describe their symptoms to the team members. If the victims were overcome by a lack of oxygen or by the presence of a toxic gas or vapor, they should immediately be supplied with breathing air or be removed from the space and into fresh air. Entrapment victims may be suffering from dehydration, shock, and the effects of prolonged exposure to heat or cold in addition to any traumatic injuries they may have. These conditions must be stabilized before the victims are manipulated to free them from their situation. Local medical protocols must be followed in stabilizing these victims.

VICTIM REMOVAL

When victims have been medically stabilized and have been freed from whatever was entrapping them, they should be packaged appropriately. In most cases, this involves using some form of basket stretcher or litter. Some of the most functional of these devices are the flexible plastic litters and spinal stabilization devices (Figure 8.37). Because they provide a good combination of rigidity and flexibility, they are especially well suited for confined space rescues.

Once the victims have been packaged, they can be carried or pulled out by the rescue team or pulled from the space by those outside. They can then be turned over to EMS personnel for transportation to the nearest medical facility.

Phase IV: Termination

The termination phase of a confined space rescue involves such obvious elements as accounting for all personnel and retrieving pieces of equipment used in the operation. But it also involves less obvious elements such as investigating the cause(s) of the incident, releasing the space to those responsible for it, and conducting critical incident stress debriefings with members of the rescue teams.

Figure 8.37 A victim appropriately packaged. *Courtesy of SKEDCO, Inc.*

EQUIPMENT RETRIEVAL

Depending on the size, complexity, and length of time involved in the operation, the job of retrieving all of the various pieces of equipment used may be either very easy or very difficult and time-consuming. Under some circumstances, it can also be quite dangerous.

Identifying/collecting. The process of identifying and collecting pieces of equipment assigned to the various pieces of apparatus on scene can be made much easier if each piece of equipment is clearly marked (Figure 8.38). However, it may be necessary for the driver/operators of the rescue unit and other pieces of apparatus on scene to conduct an inventory of their equipment prior to

Figure 8.38 The unit's identifier is clearly visible.

leaving the scene. If the operation was large enough to involve mutual aid units, the Demobilization Unit will coordinate the recovery of loaned items, such as portable radios, and documenting lost or damaged pieces of apparatus and equipment (Figure 8.39).

Abandonment. In some cases, the environment within the confined space is too hazardous to justify sending rescue personnel back in to retrieve pieces of equipment — even expensive ones. Rather than putting rescue personnel at risk, it is sometimes advisable to simply abandon the equipment in place. It may be retrievable after the space has been opened up or dismantled, or the cost of replacing the abandoned equipment may be recovered from the owner of the confined space.

Figure 8.39 A rescue unit is demobilized.

INVESTIGATION

All confined space rescues should be investigated at some level. At the very minimum, a departmental investigation should be conducted for purposes of reviewing and critiquing the operation. However, if an employee was injured in the incident, it will be investigated by OSHA and perhaps by other entities such as the employer's insurance carrier (Figure 8.40). Obviously, if a crime such as breaking and entering was involved, law enforcement agencies will also investigate the incident.

Figure 8.40 The scene may be investigated by OSHA.

RELEASE OF CONTROL

Once rescuers respond to the scene of a confined space rescue, they assume control of the space and the immediate surrounding area. Within certain limits, they can deny access to the space to anyone, including the owner of the space. Legitimate members of the news media have certain constitutionally protected rights of access, but the interpretations of these rights vary from state to state and from country to country. Rescuers need to be aware of local protocols.

However, the process of releasing control of the space back to the owner or other responsible party is sometimes not as straightforward as it might seem. The owner of the space should be escorted on a tour of the space, or as close to it as possible consistent with safety, and should be given an explanation of any remaining hazards. If the space is still open and too hazardous to leave unattended, the owner may be required to post a security guard or erect a security fence around the hazard, or both. The department may require that a written release, describing the hazards and stipulating the conditions the owner must meet, be signed by the owner before the space is released.

CRITICAL INCIDENT STRESS DEBRIEFING (CISD)

Because the injuries suffered by the victims of confined space accidents can sometimes be extremely gruesome and horrific, the members of the rescue teams and any others who had to deal directly with the victims should be *required* to participate in a CISD process. Because individuals react to and deal with extreme stress in different ways — some more successful than others — and because the effects of unresolved stresses tend to accumulate, participation in this type of process should not be optional.

The process should actually start *before* rescuers enter the scene if it is known that conditions exist there that are likely to produce psychological or emotional stress for the rescuers involved. This is done through a prebriefing process wherein the rescuers who are about to enter the scene are told what to expect so that they can prepare themselves (Figure 8.41).

If rescuers will be required to work more than one shift in these conditions, they should go through a minor debriefing, sometimes called "defusing," at the end of each shift. They should also participate in the full debriefing process within 72 hours of completing their work on the incident.

Figure 8.41 A rescue team is prebriefed on what to expect inside.

Chapter 8 Review

Directions

The following activities are designed to help you comprehend and apply the information in Chapter 8 of **Fire Service Rescue**, Sixth Edition. To receive the maximum learning experience from these activities, it is recommended that you use the following procedure:

1. Read the chapter, underlining or highlighting important terms, topics, and subject matter. Study the photographs and illustrations, and read the captions with each.
2. Review the list of vocabulary words to ensure that you know the chapter-related meaning of each. If you are unsure of the meaning of a vocabulary word, look up the word in the glossary or a dictionary, and then study its context in the chapter.
3. On a separate sheet of paper, complete all assigned or selected application and review activities before checking your answers.
4. After you have finished, check your answers against those on the pages referenced in parentheses.
5. Correct any incorrect answers, and review material that was answered incorrectly.

Vocabulary

Be sure that you know the chapter-related meanings of the following words:

- ambient *(243)*
- aquifer *(243)*
- backdraft *(243)*
- blower *(249)*
- churning *(254)*
- dehydration *(245)*
- dissipate *(245)*
- ducting *(249)*
- ejector *(254)*
- fumigant *(245)*
- hypothermia *(245)*
- intrinsically safe *(249)*
- mitigated *(243)*
- off-gassing *(245)*
- oxygen deficient *(251)*
- purge *(243)*

Application Of Knowledge

1. Provide specific examples of the following confined spaces in your jurisdiction: *(Local protocol)*
   - Tanks/vessels
   - Silos/grain elevators
   - Storage bins/hoppers
   - Utility vaults/pits
   - Aqueducts/sewers
   - Cisterns/wells
   - Coffer dams
2. Choose from the following confined space rescue procedures those appropriate to your department and equipment, or ask your training officer to choose appropriate procedures. Mentally rehearse the more complex procedures, or practice these procedures under your training officer's supervision.
   - Monitor confined space atmospheres. *(253)*
   - Perform positive-pressure and negative-pressure ventilation of a confined space. *(254)*
   - Complete a confined space entry permit. *(241)*
   - Prepare a confined space for rescue operations. *(251)*
   - Properly enter a confined space (one-rescuer and two-rescuer entry). *(258)*
   - Use a lifeline to communicate with a rescuer in a confined space. *(255)*
   - Search for and remove a victim from a confined space. *(258)*
   - Terminate a confined space rescue. *(259)*

Review Activities

1. Identify each of the following:
   - ANSI *(239)*
   - CFR *(239)*
   - CISD *(261)*
   - EMS *(252)*
   - IC *(250)*
   - LFL *(244)*
   - MSDS *(251)*
   - NFPA *(239)*
   - NIOSH *(239)*
   - OSHA *(239)*
   - PPE *(246)*
   - RP *(259)*
   - 29 CFR 1910.146 *(239)*
   - NFPA 1983 *(239)*
2. List the characteristics of a confined space. *(240)*
3. List the characteristics of a permit-required confined space. *(240)*
4. Explain what an entry permit is, when it is used, and the general information it must contain. *(240, Figure 8.1)*
5. Identify and describe the hazards associated with each of the following confined spaces:
   - Tanks/vessels *(240)*
   - Silos/grain elevators *(242)*
   - Storage bins/hoppers *(242)*
   - Utility vaults/pits *(242)*
   - Aqueducts/sewers *(242)*

- Cisterns/wells *(242)*
- Coffer dams *(243)*

6. Briefly describe each of the following ways in which a confined space may become contaminated:
   - Adsorption *(243)*
   - Absorption *(243)*
   - Desorption *(243)*
   - Combustion *(243)*
   - Fermentation *(243)*
   - Biological action *(243)*
   - Inerting *(243)*
   - Inserting *(243)*
   - Chemical reaction *(244)*
   - Oxidation *(244)*
7. Describe the causes and rescue problems associated with each of the following atmospheric hazards that may be encountered in confined spaces:
   - Oxygen deficiency *(244)*
   - Oxygen enrichment *(244)*
   - Flammability *(244)*
   - Toxicity *(245)*
8. Describe the causes and rescue problems associated with each of the following physical and environmental hazards that may be encountered in confined spaces:
   - Lack of structural integrity *(245)*
   - Debris *(245)*
   - Engulfment *(245)*
   - Darkness *(245)*
   - Temperature extremes *(245)*
   - Noise *(245)*
   - Moisture *(246)*
   - Dust *(246)*
9. Compare structural fire fighting PPE and confined space PPE in the following areas:
   - Helmet *(246)*
   - Eye protection *(246)*
   - Outerwear *(246)*
   - Footwear *(247)*
   - Gloves *(247)*
   - Air supply *(247)*
10. Briefly describe the following confined space team gear:
    - Air-supply equipment *(247)*
    - Ventilation equipment *(248)*
    - Atmospheric monitoring equipment *(249)*
    - Lighting equipment *(249)*
    - Retrieval systems *(249)*
11. Explain why portable generators should be positioned downwind from a confined space access. *(249)*
12. List information of critical importance that should be gathered during the primary assessment. *(250)*
13. Explain why the first-in officer should assume formal command early in an incident. *(250)*
14. List the size-up items that should be checked off during the primary assessment. *(250, 251)*
15. List and explain the zone designations within a confined space operation. *(251)*
16. Explain the information that should be gathered during the secondary assessment in each of the following areas: *(251)*
    - Type of space
    - Condition of space
    - Contents of space
    - Mode of operation
17. List the general areas included in Phase II Prerescue Operations. *(251, 252)*
18. Describe the course that the IC should take during Phase II to supply adequate personnel and equipment resources. *(252, 253)*
19. Explain the process of monitoring the atmosphere during Phase II operations. *(253)*
20. Identify the hazards associated with the following toxic contaminants that may be found in confined spaces: *(253)*
    - Fluoride gas
    - Cadmium vapor
    - Hydrogen sulfide gas
21. Explain the differences between natural (nonmechanical) ventilation and mechanical ventilation. *(254)*
22. Explain the differences between positive and negative ventilation. *(254)*
23. Describe how the IC may have to manipulate the following when preparing the confined space for rescue:
    - Space layout *(254)*
    - Internal hazards *(255)*
    - Shoring *(255)*
    - Lighting *(255)*
24. Discuss the pros and cons of the following communications systems in relation to confined space rescue:
    - Voice communication *(255)*
    - Lifeline *(255)*
    - Hard-wired phone *(255)*
    - Portable radios *(256)*

25. Identify the roles and responsibilities of the following rescue personnel: *(256)*
    - Entry team
    - Backup team
    - Incident Safety Officer
    - Attendant
26. Explain the difference between minimum and maximum personnel accountability. *(257)*
27. Describe the guidelines for one-rescuer and two-rescuer confined space entry. *(258)*
28. Outline the general confined space rescue procedures for search, victim treatment stabilization, and victim removal. *(258, 259)*
29. Explain the guidelines for identifying/retrieving equipment used in the rescue operation. *(259)*
30. Explain when it is better to abandon rather than retrieve a piece of equipment. *(260)*
31. Describe the variables for releasing control of the scene back to the owner or other responsible party. *(260)*
32. Discuss critical incident stress, prebriefings, and "defusings." Why should participation in a full critical incident stress debriefing within 72 hours of the incident be mandatory for all involved rescuers? *(261)*

Questions And Notes

9

Water And Ice Rescue

Chapter 9
Water And Ice Rescue

INTRODUCTION

Drowning is the second leading cause of accidental death in North America. The majority of drownings occur in open water — lakes and rivers. Less than 10 percent happen in swimming pools. Technically, drowning is death by suffocation (the absence of oxygenated blood). Therefore, the primary objective of water rescue is to restore normal breathing to the victim. Efforts to stabilize a nonbreathing victim and restore normal breathing should not be delayed until the victim is on dry land but should be started while still in the water. As always, the victim's airway, breathing, and circulation (ABCs) must be checked and any required treatment begun immediately.

In addition, victims may have also suffered other life-threatening injuries as a result of the accident that put them in the water. The victims must receive proper pre-hospital care and transportation as soon as possible if their chances for recovery are to be maximized. For more information on emergency medical care, refer to an EMS first responder manual.

Rescuing victims involved in water-related accidents presents special problems for firefighters or other rescue personnel. Water rescue may appear to be less dangerous than fire fighting, but water can be just as deadly as fire. Water rescue includes more than rescuing drowning people. Incidents may also involve ice rescue, dive rescue, vehicle rescue, and body recovery. The number of victims may range from one to several hundred. This chapter discusses the many facets of water rescue as well as proper safety procedures, water hazards, and equipment requirements.

RESCUER SAFETY

A rescue operation cannot be considered successful unless it is conducted safely, so rescue personnel should never enter the water or be out on ice unless they have been trained to do so. As with any other aspect of rescue work, training increases competence, which builds confidence. All rescue personnel should be trained to safely handle the various water-related emergencies mentioned earlier. Departments that have boats should have all personnel trained and certified in both boat operation and boat safety.

THEORY OF AQUATIC RESCUE

Also necessary for an aquatic rescue to be considered a success is that life must be preserved, including the lives of the rescuers. To accomplish this, the rescuer must apply the four components of a successful rescue: (1) *knowledge* of the techniques available, (2) *skill* necessary to perform the techniques, (3) physical *fitness* needed to apply the skills, and (4) *judgment* to decide which techniques to apply and when. Rescuers can only acquire, develop, and maintain these components through training.

The following are classifications of aquatic rescue situations in which the rescuer will need to apply the four components of a successful rescue:

- The victim is not in immediate danger of drowning, but the rescuer must use specialized skills to get the victim out of the water safely; for example, ice rescue and vehicle rescue skills may be needed.
- The victim is struggling to keep from drowning, or the victim has gone underwater, but there is still hope for resuscitation.

- Victim recovery is probably beyond resuscitation, making the situation a body recovery.

HAZARDS IN THE WATER

Rescuers must be aware of the variety of hazards that exist in the aquatic environment, and they must constantly monitor their surroundings during rescue operations. Following is a brief explanation of some common hazards found in swift-water environments.

Currents

River current (rapidly moving water) is created by the *laminar flow* (gravity and the river bottom) and the *helical flow* (against the shore). The laminar flow travels down the center of the body of water, which takes the victim (rescuer) downstream. The helical flow cuts in at the shoreline pulling the victim (rescuer) out into the laminar flow (Figure 9.1). Similar currents and flows can be created by incoming and outgoing ocean tides in coastal estuaries and inlets.

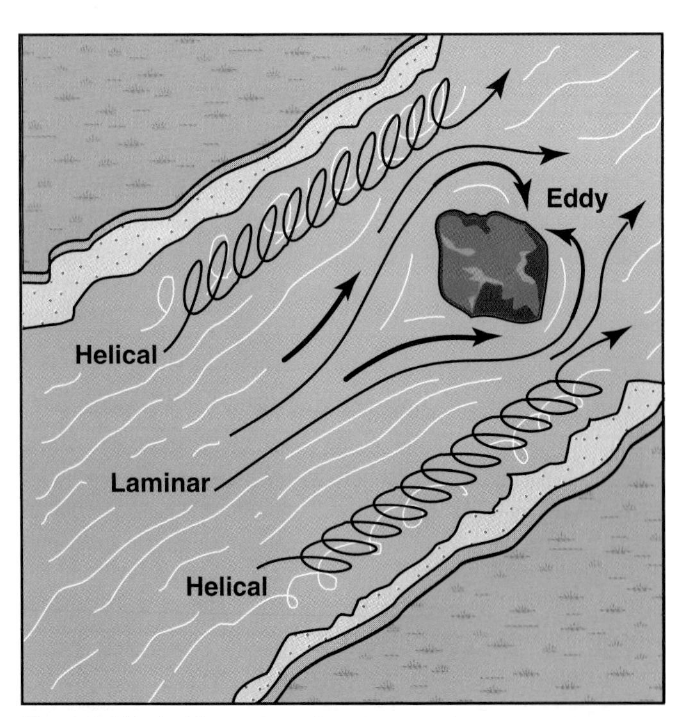

Figure 9.1 Typical flow patterns in a stream or river.

Loads

Water may contain hazards, obstructions, and turbidity that are sometimes called *loads*. *Hazards* are objects of various sizes, such as logs and other debris, that may be floating on the surface, suspended beneath the surface, or moving along the bottom and that may strike victims or rescuers. *Obstructions* are stationary objects in the water, such as rocks or old car bodies, that may or may not extend to the surface and may present an entrapment hazard to victims or rescuers. *Turbidity* may consist of microscopic aquatic organisms, silt, or particulates suspended in the water that obscure vision beneath the surface.

Entrapments

An *eddy* is a segment of water that is moving opposite to the main flow of the current. This movement is usually caused by a rock or other obstruction within the waterway. Its appearance can resemble that of a whirlpool. An eddy can hold a victim or object in a fixed location for extended periods of time because of the opposing forces exerted by the water. This type of flow can also create a hole in the bottom of the waterway in which smaller objects can be deposited.

A *strainer/sweeper* is any type of obstacle that allows water to flow through but traps objects between the water and the obstacle (Figure 9.2). Strainers and sweepers are often formed by fallen trees, construction debris, etc.

Rescuers should approach strainers or sweepers from the side. Either the REACH rescue method or the THROW-to-victim method is recommended; if the ROW method must be used, the rescue vessel should be positioned where it will not be swept away if the hazard breaks free (see Methods of Rescue section later in this chapter).

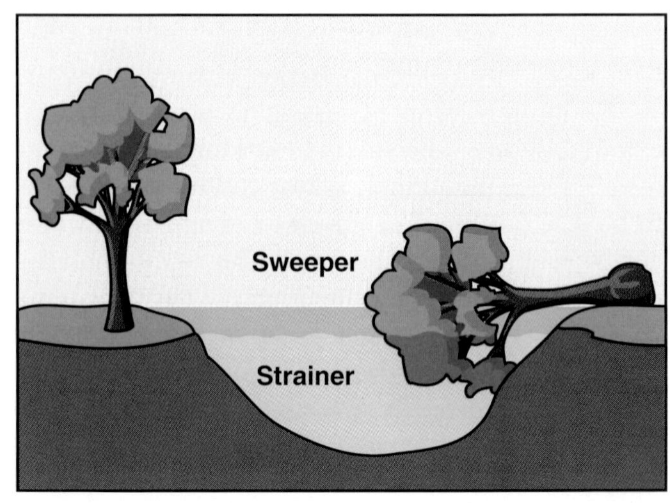

Figure 9.2 Strainers and sweepers are common hazards in flowing water.

Low-Head/Low-Water Dam

The *low-head/low-water dam,* commonly known as the "drowning machine," is one of the more dangerous hazards an aquatic rescuer may encounter (Figure 9.3). The hydraulic action of this dam is virtually impossible to get away from. The rolling action of the water flowing over the dam causes a strong current to return to the point of downwash. An object can be pulled into the downwash, pushed to the bottom, pulled to the surface by the same current, and taken back into the downwash. This cycle can go on indefinitely. Victims caught in the cycle will be continually dunked until they are rescued, manage to break free, or drown.

Rescue attempts should be made from the downstream side only. Rescuers should try to reach out to the victim (REACH method) or throw an aid to the victim (THROW method). The ROW method should be done by specially trained rescue personnel using proper equipment such as a Res-Q-Dek (Figure 9.4). (See Methods of Rescue section later in this chapter.) Also, the use of a helicopter would be one of the safer approaches if the area is accessible.

Figure 9.3 A typical low-head dam. *Courtesy of Calgary (ALTA) Fire Department.*

Figure 9.4 Rescuers operating a Res-Q-Dek. *Courtesy of Calgary (ALTA) Fire Department.*

Polluted Water

The sight of polluted water can deter even the most dedicated rescuer. When dealing with contaminated water, rescuers must protect themselves with proper environmental protection, breathing apparatus, etc.

Cold Water

Several people who have been submerged in water as warm as 70°F (21°C) have been revived after more than 30 minutes because of a phenomenon known as the Mammalian Diving Reflex (MDR). The ability of mammals to be submerged for extended periods without brain or tissue damage is attributed to the body's perception of sudden cold temperature change, usually in the facial region, which concentrates bloodflow to the brain, lungs, and heart.

In drowning accidents, the MDR is much more pronounced in younger people. However, aquatic rescue personnel should treat all victims as recoverable. Because of the possibility of the reflex, rescue personnel should not hesitate to attempt rescue up to at least 60 minutes after submersion. This time is called the *Golden Hour of Aquatic Rescue.*

Cold water immersion gives victims a chance if they drown. However, prior to drowning they are also suffering the effects of exposure. If the water temperature is not equal to or greater than that of the victim, the victim will lose body temperature 25 times faster in water than in air. Rescuers must get victims out of the water as soon as possible, which will also protect the rescuers from the effects of cold-water immersion. Studies have indicated that rapid immersion in cold water can kill a good swimmer in 4 to 5 minutes without proper environmental protection. Table 9.1 shows how long people can be expected to survive in cold water.

TABLE 9.1
Survival Time At Various Water Temperatures

Survival Time	Water Temperature °F	°C
20 to 30 minutes	<40	<5
1½ to 2 hours	40-64	5-20
Indefinitely	>64	>20

Victim Type

Certain types of victims can seriously jeopardize the safety of rescuers, especially victims who are struggling for life and with their fear of the water. Recognizing the types of victims by their behavior will aid rescuers in performing a successful rescue. The following are classifications and characteristics of victim types that rescuers may encounter:

- *Nonswimmer*
  - — vertical in water, not using legs for support
  - — clawing at water with hands
  - — may not respond to instructions, may be unaware of help, may or may not wave for help
  - — panic is evident in their actions

WARNING

There is increased risk to the rescuer because these victims cannot support themselves.

- *Weak or tired swimmer*
  - — arms and legs are used for support
  - — facing direction of help, calling or waving
  - — varying degrees of anxiety are evident
- *Injured/hypothermic swimmer*
  - — same characteristics as weak/tired swimmer
  - — usually holding onto injured area
  - — showing signs of hypothermia
- *Unconscious swimmer*
  - — will usually sink if not wearing personal flotation device (PFD) or life jacket
  - — lying motionless in water
  - — may be either faceup or facedown

EQUIPMENT NEEDED FOR AQUATIC RESCUE

The response district should be surveyed to determine the level of service needed. Areas with an abundance of waterways will need units with more rescue equipment and capability. Every rescue unit should be prepared to at least begin the process of handling a water emergency occurring in its area until additional help arrives.

In areas with significant aquatic rescue potential, all fire apparatus should be equipped with a throw bag and a life jacket or personal flotation device (PFD) for each member assigned to the unit. This equipment will increase the chances of survival for both victims and rescuers. All personnel participating in any water or ice rescue operation should wear a PFD or life jacket (Figure 9.5). Anyone not directly involved in the rescue operation should be kept well back from the shore or the edge of the ice. The following lists equipment needed for various water and ice rescues:

Aquatic Surface Rescues:

- PFDs/life jackets for both rescuer and victim
- Water rescue helmet
- Pike pole
- Ladders
- Throw bag
- Life ring
- Boogie board
- Adequate thermal protection (wet/dry suit with hood, gloves, mask and snorkel, and fins) (Figure 9.6)

Ice Rescues:

- All items listed for surface rescues
- Class III harness
- Lifeline (see Protocol for Ice Rescue section)

Underwater Rescue/Recovery — No Ice:

- Self-contained underwater breathing apparatus (SCUBA) gear
  - — mask and snorkel
  - — fins
  - — diver's knife
  - — regulator
  - — pressure gauge
  - — alternate air source (second regulator)
  - — buoyancy compensator with power inflator

Figure 9.5 A rescuer in a typical PFD. *Courtesy of Calgary (ALTA) Fire Department.*

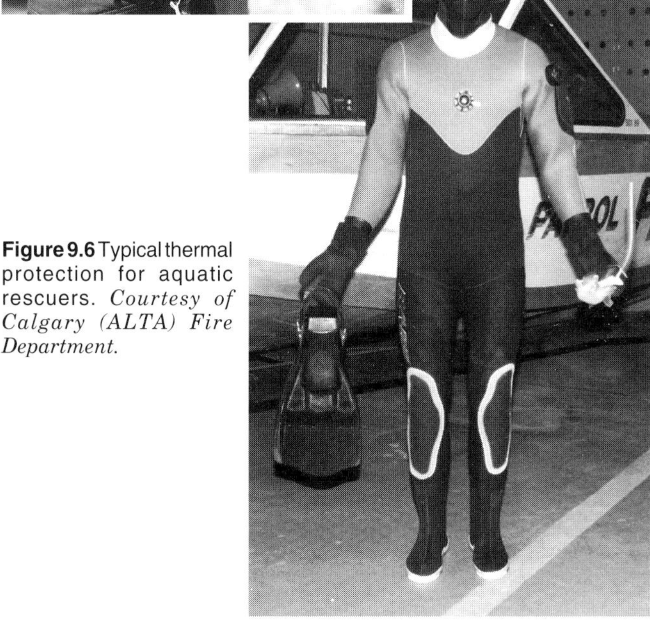

Figure 9.6 Typical thermal protection for aquatic rescuers. *Courtesy of Calgary (ALTA) Fire Department.*

- adequate thermal protection (wet/dry suit)
- depth gauge
- compass
- watch
- tank
- weight belt
- Class III harness

- Spare SCUBA gear
- Diver's warning flag
- Extra dive cylinders
- Dive lights
- Ropes
- Salvage covers
- Backup safety diver

Underwater Rescue/Recovery — Ice:

- All items listed for no-ice rescue/recovery
- Protection from elements (big tent, cabin)
- Toboggan/sled
- Class III harness
- Lifeline
- Ice screws
- Chain saw

Vehicle Rescue:

- All items listed for other aquatic emergencies
- Air lifting bags
- Tow truck

Victim Recovery/Removal From Aquatic Environment:

- PFD/life jacket
- Spinal board
- Basket litter with floats
- Body bag

PREPARATION FOR AQUATIC RESCUE

Preparation for aquatic rescue begins well before the call for help is received. Based on the personal experiences of those involved in the rescue unit, as well as on the experiences of others in the field, general operating guidelines (GOG) must be developed and appropriate equipment acquired. All rescue personnel must be familiarized with the established GOGs as well as with the rescue equipment and its use. They must participate in realistic training exercises of sufficient intensity and frequency to develop and maintain their rescue skills.

When a call for help is received, aquatic rescuers begin to apply the procedures on which they have been trained. Information from the dispatcher will tell rescuers such things as the location of victims (on ice, in swift water), number of victims, type of victims, etc. Processing this information is the beginning of the incident size-up and will help to ensure that the most appropriate resources are dispatched.

DEALING WITH AQUATIC RESCUE INCIDENTS

Once on the scene, rescuers should continue and intensify the size-up process by surveying the scene and questioning any witnesses to the accident. If they are to be successful, rescuers must first develop an operational plan by deciding if they will function in (1) *rescue mode* (some chance of saving human life) or (2) *recovery mode* (no chance of saving human life). They must also consider the *risk / benefit factor* (is the potential benefit worth the risks the rescuers must take) (Figure 9.7). In the recovery mode, there is plenty of time for preparation. In the rescue mode, quick decisions must be made and actions taken because even if a body is submerged, a successful rescue may still be possible because of that "Golden Hour."

The rescue team leader must take control of the incident site, assume the role of incident commander (IC), and establish a highly visible and easily identified command post. All aquatic rescue personnel on the scene are accountable, through the chain of command, to the IC. The IC must gather information, make decisions, oversee operations, and constantly monitor rescuer safety. To maintain a manageable span of control in this multifaceted process, the IC should implement an incident management system to organize and delegate some of these critical functions.

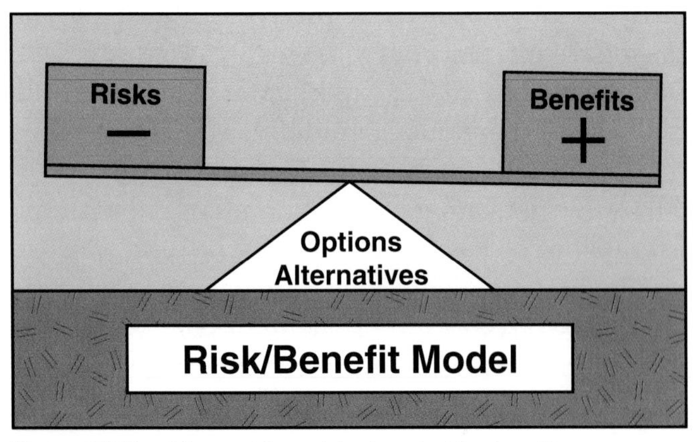

Figure 9.7 The risks must be weighed against the benefits.

SCENE CONTROL AT RESCUE INCIDENTS

Maintaining control of the scene is critical to the successful conclusion of aquatic rescue incidents. To ensure that rescuers can do their jobs without interference, four other groups must be managed: spectators, witnesses, family members, and the media. If not properly handled, any or all of these groups can further complicate an already complex incident.

Even in relatively remote areas, the sound of sirens and the sight of lights flashing on emergency vehicles tend to draw people to the scene. Spectators wandering the scene not only can interfere with the rescue operation, but they may also disturb physical evidence that may be critical to the subsequent accident investigation. Even well-meaning spectators can become additional victims if they are allowed to get too close to the rescue scene. So, while law enforcement personnel are best trained and equipped to handle crowd control responsibilities, someone should be assigned to cordon off the area as soon as possible until law enforcement officers are on the scene.

Those spectators who actually witnessed the accident must also be managed during the incident, and they need to be identified as early in the incident as possible. Even though most witnesses are eager to come forward and tell someone what they saw, it is still worthwhile for someone to ask the assembled crowd if anyone saw what happened. A department representative should be assigned to locate witnesses and to stay with them as long as they are needed for witness interviews.

Some means of dealing with members of the victim's family in these stress-filled, emotional situations should be part of the pre-incident plan. Until another agency, such as the Red Cross, Salvation Army, or a similar organization, can become involved, a department representative should be assigned to stay with any family members on scene. Family members should be isolated from the crowd and kept together in an official vehicle or area with the radio turned off. This will keep them out of the rescuers' way but in a position where they can feel they are a part of the operation. They should be given honest information about what the rescuers are doing and the status of the operation.

Handling the media is sometimes more difficult, but no less important, than handling the family, so they should be dealt with courteously and professionally at all times. News reporters, photographers, and videographers can be kept out of the scene if they are interfering with the operation or if the area has been declared a crime scene by the law enforcement agency having jurisdiction. However, if the media people can get what they need from a public information officer (PIO), they are usually willing to remain close to that individual (Figure 9.8). Setting up a highly visible and easily identified information point early in the incident will often keep media personnel from wandering about the scene.

Figure 9.8 The PIO should keep the media informed about the incident.

PERFORMING THE AQUATIC RESCUE

When confronted with a victim or victims in the water, rescuers should devise and implement a plan of action that will maximize the chances for a successful operation and will minimize the danger to themselves. This planning can be done by determining the type of victim in distress (see Victim Type section given earlier) and by following an appropriate rescue protocol. The protocol is a base upon which to build knowledge, skill, and judgment for all aquatic incidents. As a rescuer moves down the protocol, risk increases to the rescuer.

Protocol For Surface Rescue

METHODS OF RESCUE

Selfrescue. If victims are uninjured when they enter the water, they may be able to rescue themselves if they stay calm. If they fell out of a boat or off of a pier, floating dock, or sea wall, they may be able to hang on to whatever they fell from and make their way to shore or remain in place and call for help. If they were in a vehicle that drove or skidded into the water, they may be able to exit the vehicle through an open window before the vehicle sinks. If the vehicle sinks immediately, the water pressure will keep the doors closed until the interior fills with water. Once this happens, the occupants can usually roll down a window even if the doors will not open.

The methods used in assisted surface rescues are *REACH*, *THROW*, *ROW*, *GO*, and *TOW/CARRY*. However, before attempting to use one or more of these methods, the rescuer must have the knowledge, skill, fitness, and judgment needed to help the person in distress.

Reach. In this method, the rescuer reaches to the victim (making sure to keep center of gravity low), talks to the victim, draws the victim in to shore gradually, and anchors the victim. A pike pole or any other long-handled tool can be used (Figure 9.9). The handle end of the tool should be extended to the victim. Other fire equipment that can be used as aids are ladders, brooms, shovels, fire hose, aerial devices, etc.

Figure 9.9 A rescuer extends a tool handle to a victim. *Courtesy of Calgary (ALTA) Fire Department.*

Throw. In this method the rescuer throws an aid to the victim, ideally with a rope attached (Figure 9.10). However, buoyant aids can be just as effective. Use the same procedure as given in the REACH method. Keep center of gravity low after the throw, talk to the victim, draw the victim in gradually, and anchor the victim.

The throw bag is one of the best aids available to rescuers. Its small, compact size makes it easy to use and store. In use, the bag should be thrown well beyond victims. They should be instructed to grab the rope and roll over onto their back, holding the rope on their chest with their feet up. This helps keep water out of their face and helps keep them from hitting rocks. The rescuer sits down and uses an open belay, with the working side of the rope in the downstream hand so as to not get spun into the water (Figure 9.11).

Figure 9.10 Throwing a lifeline to a victim may be all that is needed. *Courtesy of Calgary (ALTA) Fire Department.*

Figure 9.11 A rescuer in a pendulum belay. *Courtesy of Calgary (ALTA) Fire Department.*

CAUTION: The following rescue techniques should be attempted only by those who have been specifically trained in their application.

Row. When the victim is beyond reaching or throwing aids from solid ground, the ROW method can be used. In this method the rescuer gets a water vessel (rowboat, motorboat, etc.) and goes to the victim. The rescuer should not make physical contact unless absolutely necessary because a panicky victim could pull the rescuer into the water. When close enough to the victim, the rescuer can use a throw aid (Figure 9.12).

Go. In this method the rescuer, wearing the proper environmental protection suit and PFD, enters the water with the victim (Figure 9.13). Before leaving solid ground, the rescuer should make sure that someone is watching from shore to ensure rescuer safety. The rescuer should get an aid (PFD/life jacket, boogie board, towel, etc.) to use in the rescue. The rescuer should also maintain visual contact with the victim from the time the rescuer leaves solid ground until he or she returns to solid ground.

Figure 9.12 Trained rescuers can use small boats in rescue operations.

Figure 9.13 Only those trained for water rescue should attempt rescues in the water. *Courtesy of Calgary (ALTA) Fire Department.*

Upon entering the water, the rescuer should beware of submerged hazards and should *approach* the victim using the head-up front-crawl stroke for speed in reaching the victim. The head-up breaststroke can also be used to conserve the rescuer's energy. From a safe distance, the rescuer should reassess the victim's needs and decide to TOW or CARRY in a defensive position, being ready to fend off the victim if necessary. This defensive position is known as the *reverse and ready* position (Figure 9.14). Returning to shore towing or carrying a victim is hard work, but efforts should be made to keep the victim's face out of the water at all times.

- *TOW*. In this method there is no physical contact between victim and rescuer. The rescuer tows the victim back to solid ground by the victim's clothing or by the aid taken prior to leaving solid ground.

Figure 9.14 The "reverse and ready" position. *Courtesy of Calgary (ALTA) Fire Department.*

- *CARRY*. In this method, physical contact has to be made between victim and rescuer because of the victim being a nonswimmer or a swimmer who is exhausted, injured, or unconscious. A rescuer is at greatest risk whenever there is physical contact between victim and rescuer. Within the CARRY method, there are two classifications:
  — *Assisted*. Implies that the victim can help the rescuer. In the wrist-to-wrist assisted carry, the rescuer stretches the distance between victim and rescuer, which aids in the carry by minimizing water resistance (Figure 9.15).
  — *Controlled*. Implies that the rescuer has complete control of the victim. The rescuer has to work harder in the cross-chest control carry (Figure 9.16). The unconscious control carry is used for long distances in water (Figure 9.17).

Figure 9.15 The victim can sometimes assist the rescuer. *Courtesy of Calgary (ALTA) Fire Department.*

Figure 9.16 The cross-chest control carry. *Courtesy of Calgary (ALTA) Fire Department.*

Figure 9.17 The control carry for unconscious victims. *Courtesy of Calgary (ALTA) Fire Department.*

HELICOPTER RESCUE

As with all other aspects of rescue operations, the key to the successful use of helicopters for water rescue is pre-incident planning that includes realistic exercises to test the plan. If helicopters are available and may be called upon to participate in rescue operations or medical transportation, pre-incident familiarization, planning, and training should done with all personnel who are likely to be involved in helicopter rescue operations. For more information on working with helicopters, refer to DOT publication HS805-703, *Air Ambulance Guidelines*, or the appropriate *Aeronautical Information Publications* of Transport Canada Aviation.

VICTIM REMOVAL

Once rescuers are on solid ground, removing the victim from the water can be difficult because surfaces can be slippery when wet. Good lifting techniques must be used to avoid injury to rescuers or victims. Some examples of shallow water removal techniques are as follows:

- Firefighter's carry (Figure 9.18)
- Underarm, wrist-grasp method; rescuer walks backwards (Figure 9.19)
- Saddleback carry; used if rescuer must walk a long way in water because the water takes the majority of body weight (Figure 9.20)
- Buddy-buddy carry; used only if the victim can help (Figure 9.21)

NOTE: For procedures on lifting injured victims from the water up to a pier or seawall or up a steep incline, see Chapter 4, Rope Rescue.

SPINAL INJURY VICTIM

In open water, a spinal injury victim can make an aquatic rescue much more difficult. The victim may be faceup or facedown, conscious or unconscious, breathing or nonbreathing, and floating or

Figure 9.19 The underarm, wrist-grasp method. *Courtesy of Calgary (ALTA) Fire Department.*

Figure 9.20 The saddleback carry. *Courtesy of Calgary (ALTA) Fire Department.*

Figure 9.18 The firefighter's carry. *Courtesy of Calgary (ALTA) Fire Department.*

Figure 9.21 The buddy-buddy carry. *Courtesy of Calgary (ALTA) Fire Department.*

sinking. The rescuer should take precautions to immobilize the spine if the injury is witnessed or the mechanism of injury indicates possible spinal injury. The following rescue technique is for a witnessed spinal injury victim. To immobilize the victim's head and neck:

Step 1: The rescuer approaches the victim from the side and moves the victim's nearside arm down along the victim's body. The rescuer then places a forearm on the victim's sternum and a hand on the victim's face. The rescuer's little finger rests on the jawline on one side of the face, and the thumb rests on the jawline on the other side. The other fingers grasp the bony structures of the victim's face as firmly as possible. The rescuer's other arm and hand are placed on the victim's spine and the back of the head. The rescuer compresses the victim between both arms (Figure 9.22).

If the victim is facedown, the rescuer then rolls under the victim, turning the victim over slowly (Figure 9.23). Once the victim is turned over, the rescuer monitors the victim's ABCs. If the victim's ears are kept in the water, the water will act as a splint.

Step 2: A second rescuer takes a position on the same side as the first rescuer, just below the victim's waist. The second rescuer grasps both sides of the victim's pelvis and raises the hips to the surface (Figure 9.24).

Step 3: Submerge a spine board under the victim, and place the victim on it.

Figure 9.22 The victim's head and neck are immobilized. *Courtesy of Calgary (ALTA) Fire Department.*

Figure 9.23 A face-down victim must be rolled over. *Courtesy of Calgary (ALTA) Fire Department.*

Figure 9.24 A second rescuer moves in to assist. *Courtesy of Calgary (ALTA) Fire Department.*

VICTIM TURNOVER IN WATER

Turning a victim over will allow a rescuer to monitor the victim's ABCs while still in the water. When there is nothing to indicate a spinal injury, this technique can be used to turn a facedown victim over in shallow water and to turn a victim in deep water into a tow or carry position.

Step 1: The rescuer approaches from above the victim's head. The rescuer reaches out and grasps the victim's right wrist with his or her right hand or grasps the left wrist with the left hand, palm up (Figure 9.25).

Step 2: The rescuer moves backward quickly, turning his or her hand palm down while lifting the victim's wrist and rotating the victim's body in the process. As the victim starts to turn, the rescuer places his or her other hand over the victim's other arm and under the victim's back (Figure 9.26).

Step 3: With the hand that held the victim's wrist, the rescuer opens the victim's airway and assesses the victim's ABCs (Figure 9.27).

Figure 9.25 The rescuer grasps the victim's wrist. *Courtesy of Calgary (ALTA) Fire Department.*

Figure 9.26 The rescuer rotates the victim's body. *Courtesy of Calgary (ALTA) Fire Department.*

Figure 9.27 The rescuer opens the victim's airway. *Courtesy of Calgary (ALTA) Fire Department.*

Protocol For Ice Rescue

The steps in performing an ice rescue are designed to be as simple as possible because the rescuer has other factors to consider. One of those other factors is the unpredictability of the ice. The strength of ice is directly related to how it is formed. Just because ice is thick does not mean that it is strong, and *the victim in the water has demonstrated that the ice is weak*. The presence of chemicals in the water, underwater vegetation, warm-water discharges from industrial facilities, and variable water currents all make it very difficult to estimate ice strength.

CAUTION: Until they have donned a life jacket/PFD or environmental/thermal protection suit (dry suit), rescuers should stay off the ice.

Another factor with which ice rescue personnel must contend is the weather and its effect on those involved in the incident and on the scene. The victim will almost certainly be suffering the effects of hypothermia, so having an advanced life support unit on scene to start immediate patient care is critical. Because of poor footing on the ice, rescuers should wear a helmet, crampons or other traction devices, and proper environmental dress. Using a basket stretcher will make transporting the victim over the ice much easier.

Yet another factor for ice rescuers to consider is that the victim may not be able to be of any help. With frozen hands, the victim may not be able to grasp a rope or other aid; and with heavy, wet clothing, the victim may even have difficulty keeping his or her head above water. With immersion in ice water, the body's temperature can drop dramatically, and the victim's chances of survival may depend on how quickly he or she can get out of the water and into a warmer environment. Once out of the water, the victim's wet clothing must be removed, and treatment for hypothermia must be begun immediately. The ice rescue protocol is as follows:

- Instruct the victim *not* to try to get out of the water until you say to.
- *REACH*. This technique can be implemented only when the victim is close to solid ground and is responsive and able to hold on to an aid.
- *THROW*. This technique allows the rescuer to span more distance while remaining on solid ground. The victim must be responsive and able to hold on to the aid.
- *GO*. This technique is used when the victim is either too far from solid ground to use REACH or THROW or is incapable of grasping an aid.

If the GO technique must be used, some means of spreading the rescuer's weight should be used if available. A fire department ladder, inflatable boat, or other inflatable device can be used for this purpose. Rescuers using inflatable devices on the ice should be tethered to those devices. If none of these items are available, the rescuer will have to use the one-rescuer ice rescue technique. The steps in this rescue technique are as follows:

Step 1: Rescuer dons the proper environmental protection suit (dry suit), hood, gloves, and Class III harness (Figure 9.28).

CAUTION: A lifeline monitored by other rescuers should be attached to the harness and to an anchor point on solid ground.

Step 2: Rescuer crawls out onto the ice, keeping body weight spread as evenly as possible (Figure 9.29).

Step 3: Rescuer slips into ice hole at the victim's side, being careful not to break the ice to which the victim is clinging (Figure 9.30).

Step 4: Rescuer places a lead around victim's torso and attaches it to lifeline with a carabiner (Figure 9.31).

Step 5: Rescuer gives the universal sign for OK (one hand on head) (Figure 9.32).

Step 6: Other rescuers monitoring the lifeline pull victim and rescuer out of the hole and to solid ground (Figure 9.33).

Figure 9.28 A rescuer wearing proper environmental protection for ice rescue. *Courtesy of Calgary (ALTA) Fire Department.*

Figure 9.29 A rescuer crawls out onto the ice. *Courtesy of Calgary (ALTA) Fire Department.*

Figure 9.30 The rescuer enters the water with the victim. *Courtesy of Calgary (ALTA) Fire Department.*

Figure 9.31 The rescuer places a lead around the victim's torso. *Courtesy of Calgary (ALTA) Fire Department.*

Figure 9.32 The rescuer signals "OK." *Courtesy of Calgary (ALTA) Fire Department.*

An alternative to this procedure invloves the use of a rescue platform, if available (Figure 9.34). This device allows the rescuer to approach the victim's location at a full run and to stay out of the water during the entire operation.

Figure 9.33 The rescuer and victim are pulled toward shore. *Courtesy of Calgary (ALTA) Fire Department.*

Figure 9.34 A rescuer moves a rescue platform across the ice. *Courtesy of Angel-Guard Products, Inc.*

Protocol For Vehicle Rescue

Many problems associated with a rescue from a vehicle in the water can be avoided if rescuers first carefully evaluate the situation and develop a definite plan before attempting the rescue. In developing the plan, rescuers should decide how they will function by classifying the operation as described earlier: rescue mode or recovery mode. They should also do a quick risk/benefit analysis. By questioning witnesses and/or by surveying the scene, the following questions should be answered in the process of formulating the operational plan for a vehicle rescue:

- Where and in what condition are the victims?
- Is the vehicle partially submerged?
- Is the vehicle fully submerged? If so, call for a water rescue/dive team.
- What is the likely condition of the vehicle based on the mechanisms of the accident (vehicle hit tree, went through guardrail, length of skid marks, etc.)?
- What are the water conditions (swift, cold, etc.)?

The location and condition of the victims will determine whether the operation should be conducted in the rescue mode or recovery mode. If the vehicle is known to have been submerged for several hours, rescuers should function in the recovery mode, and the risks to themselves should be kept to a minimum. If victims are still inside a recently submerged vehicle, immediate intervention in the rescue mode is required.

CAUTION: Only those who have been trained in aquatic rescue techniques should be allowed to enter the water.

Depending on how far from shore the vehicle is, the GO or the ROW and GO methods may be needed. The same quick action may also be needed if the victim is on the vehicle and it is about to submerge, but the REACH, THROW, ROW and GO methods may be needed depending upon the victim's condition.

The condition of the vehicle can provide information needed to help determine the appropriate mode and method of operation. A badly damaged vehicle that is partially or fully submerged may indicate that the operation should be conducted in the recovery mode, especially if the vehicle has also been submerged for some time. If evidence at the scene indicates that the vehicle rolled over several times before coming to rest in the water, the decisions would probably be different than if the vehicle simply skidded into the water. While there are no specific rules for deciding when an operation should be conducted in a rescue or recovery mode, rescuers should always give potential survivors the benefit of any doubt. While the decision must be based on the information available at the time, it must take into account the "Golden Hour of Aquatic Rescue" and the fact that within fully submerged vehicles enough air can be trapped to allow trapped occupants to breathe for some time. As always, the chances for a successful rescue must be weighed against the risks the rescuers must take in the attempt.

Finally, water conditions will also affect the planning of an operation. Water conditions may have an effect on the mode of operation selected — cold water may suggest the "Golden Hour," and swiftly flowing water will present a set of challenges different than those presented by lakes, ponds, or reservoirs. Vehicle stabilization may be more difficult in swift water but also more important to a successful operation. A strong current may also affect the choice of rescue methods and the water vessel used, and the current may limit the variety of aids that can be used effectively.

In the rescue mode, if there are victims still inside a submerged or partially submerged vehicle, some form of the GO method will be indicated. The following are steps in the GO method:

Step 1: In swift water, approach the vehicle from the side of the stream to avoid being pinned by the current (Figure 9.35).

Step 2: If there is no immediate danger to victims, try to stabilize vehicle on solid ground before removing victims.

Step 3: Remove victims and put them in PFD/life jackets before bringing them to solid ground.

Step 4: Arrange for a tow truck to remove vehicle from water.

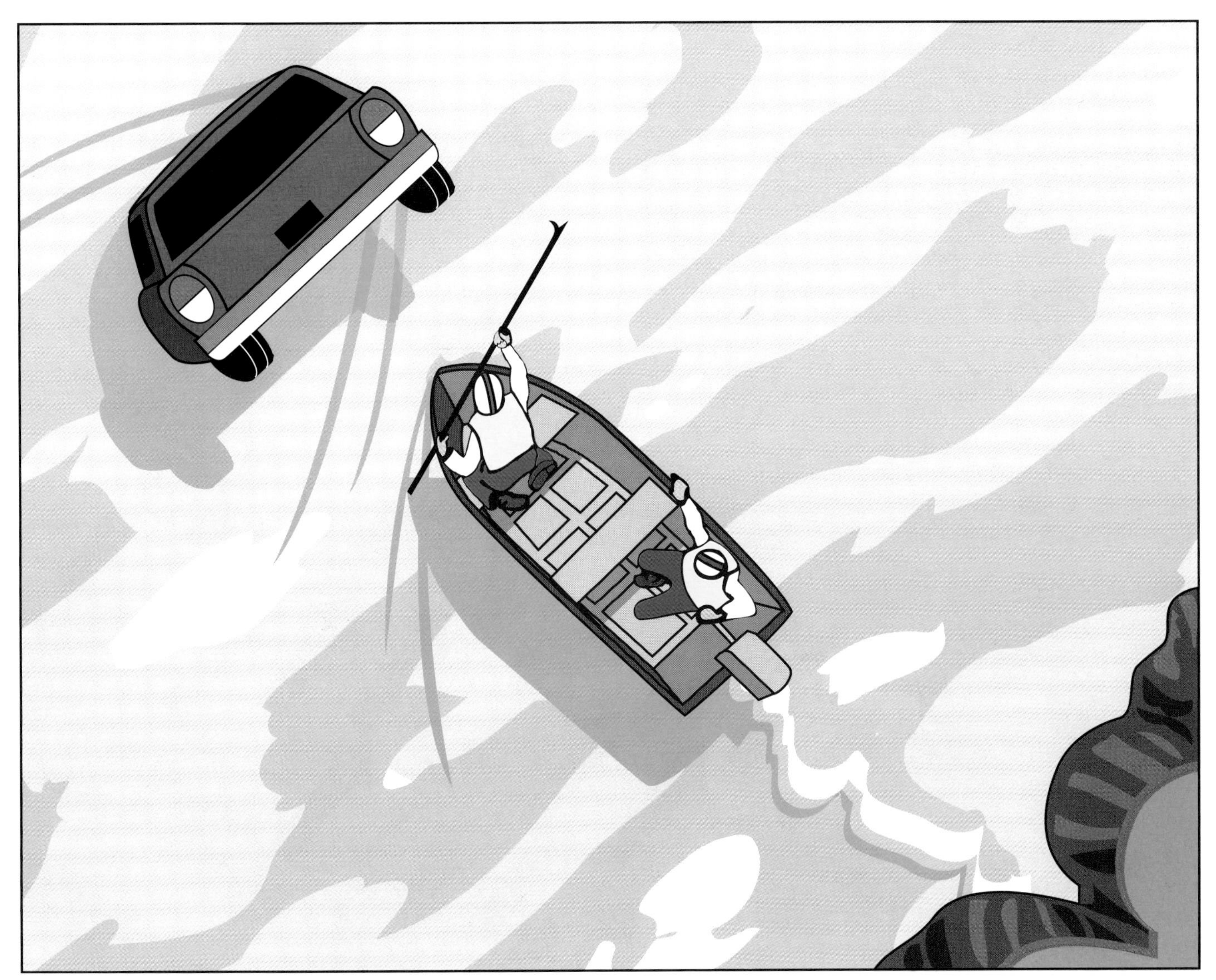

Figure 9.35 Rescuers approach from the side.

Surf Rescue

Even strong swimmers sometimes drown in the surf because they are not familiar with two very dangerous water currents: undertow and riptides. Both of these phenomena result from submerged sandbars that form parallel to the shore line in the surf zone.

Undertow occurs when a massive amount of water that has become trapped on the landward side of a sandbar returns to the sea. The water flows seaward through narrow breaks in the sandbars. The large volume of water flowing through these narrow openings creates very powerful currents that can pull even strong swimmers out to sea. This seaward flow continues even as waves on the surface are moving more water landward; thus, the name *undertow*. In an undertow, people standing in the surf can have their feet swept toward the sea while their upper body is thrust landward by the crashing surf. This knocks them off their feet, and the undertow moves them swiftly out to sea.

Riptides are extremely strong currents that form as the surf recedes and flows laterally toward the openings between sandbars. These currents, too, can sweep people off their feet and quickly out to sea (Figure 9.36).

One of the worst mistakes swimmers make when caught in either of these currents is to try to swim directly back toward shore. This futile attempt to overcome the effects of the current simply tires them out, often resulting in their being drowned. Swimmers should swim with the current,

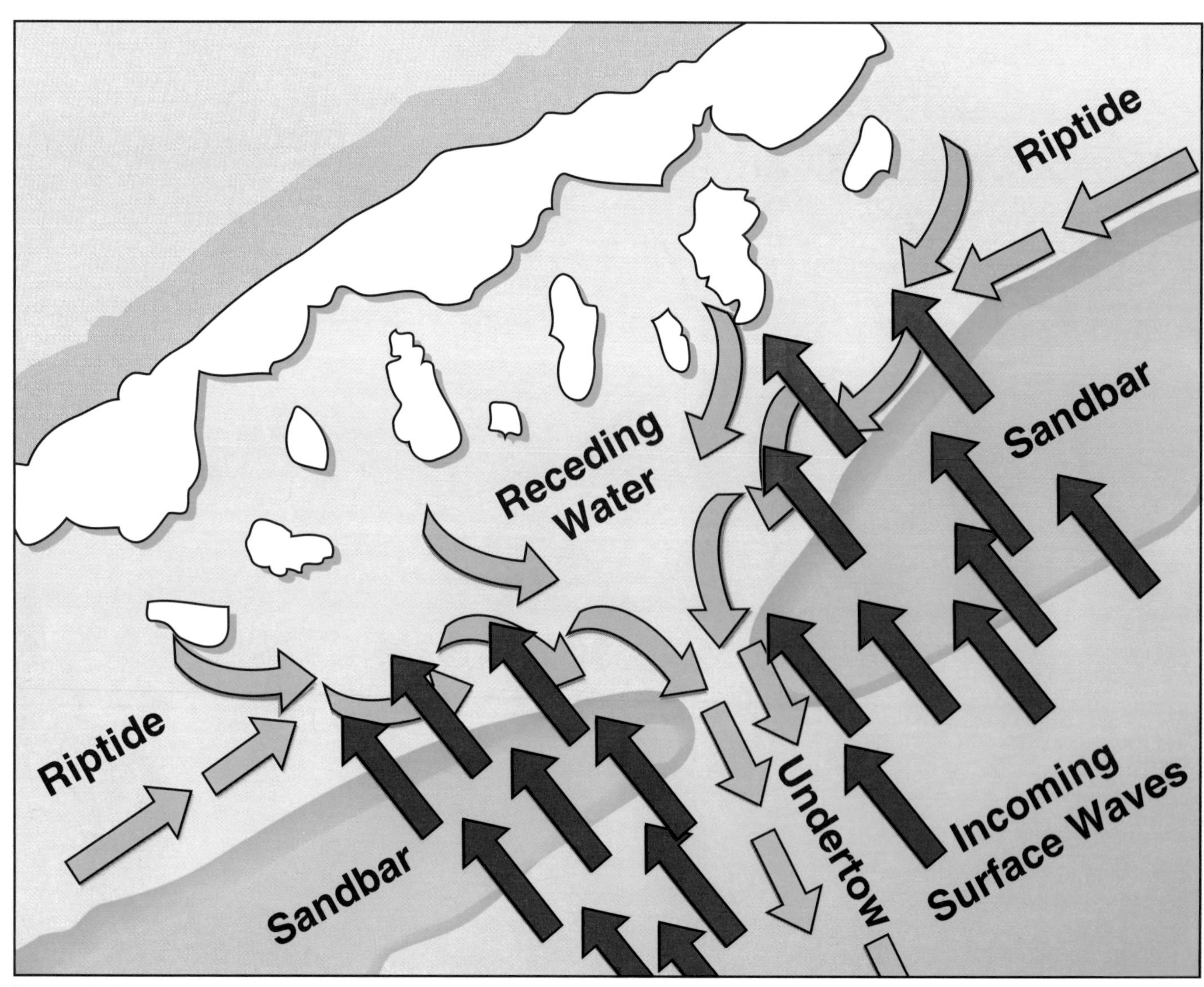

Figure 9.36 Typical undertow and riptide flow patterns.

but angling off to one side. This conserves their energy and allows them to swim back to shore once they are free of the current.

Rescue of swimmers near the shoreline is best done from a small boat beyond the surf zone, although such watercrafts are not always available. Lifeguards on public beaches are often equipped with surfboards or paddleboards for rescue purposes, and some seaside fire departments have acquired them also.

CAUTION: Only strong swimmers who have had aquatic rescue training should attempt to rescue swimmers from the surf.

The rescuer should be equipped with a PFD and should have one for each victim. The rescuer should also be tethered to the boat or to shore, and this may require 500 to 600 feet (154 to 185 m) of polypropylene line.

Recovering drowning victims from the surf is often very difficult, especially in foul weather, as the wave action impairs efforts to do a controlled, systematic search. The increased wave action also stirs up the sand, and the resulting turbidity reduces underwater visibility to near zero. The lack of visibility makes the use of SCUBA gear of little use. If a search is deemed advisable, a human chain of personnel linked elbow to elbow should be formed. The chain begins in knee-deep water and extends out as far as the last person in the chain can stand with his or her head above water (Figure 9.37). They should then walk in unison parallel to the shore for one-quarter of a mile (400 m) in both directions from the last-seen point.

SUBMERGED VICTIM RECOVERY OPERATIONS

Submerged victim recovery can be a very time-consuming operation that involves determining the *last-seen point*, applying *submerged victim recovery factors* in conducting a search, and monitoring *personnel safety*. The main reason for the protracted nature of these operations is that there is no longer any need for haste, and the emphasis shifts to conducting a slow, methodical operation that will minimize the risks to the personnel involved. Everyone involved should be thoroughly familiar with the operational plan. Every step in the operation should be supervised and coordinated by an aquatic team leader in accordance with the plan.

Determining The Last-Seen Point

The last-seen point is the starting point of the search for a submerged victim. Having an accurate last-seen point can greatly facilitate a quick recovery. The last-seen point can be determined in two ways: by deduction — physical evidence on shoreline or in water — and by the statements of witnesses.

DEDUCTION

If a boater, fisherman, or swimmer has failed to return when expected and is considered to be missing after a reasonable but fruitless search by family and others, rescuers should proceed on the assump-

Figure 9.37 A human chain searches the surf zone for a victim.

tion that the missing person is in the water. If the missing person's fishing gear, outer clothing, or other personal effects are located on the shore, it can serve as one form of last-seen point. An ever-expanding series of semicircular search patterns should radiate out from that point (Figure 9.38).

If clothing, an empty boat, or other items belonging to the missing person are found floating in the water, they can also serve as a form of last-seen point. If the evidence is found along the shoreline, the same pattern as described above can be used. If found some distance from shore, a circular pattern should be searched from that point (See Applying Submerged Victim Recovery Factors section).

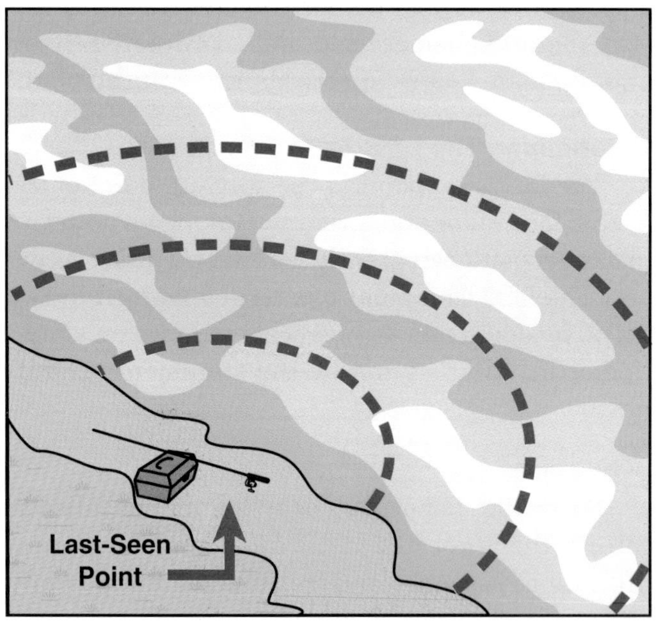

Figure 9.38 Shore-based search patterns.

STATEMENTS OF WITNESSES

A rescuer's uniform can be both a symbol of authority and a source of comfort for people in distress. Whether interviewing a victim's family or other witnesses, rescuers should be sensitive to people's feelings in these stressful situations, and they should always conduct themselves in a calm, professional manner. Rescuers should speak calmly, asking clear, concise questions. When dealing with the family, it is good to refer to the victim by his or her first name, and rescuers should always be honest with the family about what is being done. However, even when asked, rescuers should refrain from speculating about the eventual outcome of the operation.

All witnesses should be asked for identification, including information on how they can be reached in case they need to be called back to the scene. Every witness should be considered credible and their information used. Small children should be interviewed gently and at eye level. Witnesses should be separated from each other during questioning to avoid them being influenced by each other. Each witness should be asked to take rescuers back to the exact point from which they last saw the victim. If there are a number of witnesses, several rescuers can conduct separate interviews simultaneously. By carefully noting what each witness says and by plotting on a sketch of the scene the point at which each witness last saw the victim, a consensus last-seen point may be established.

It may also be necessary to reenact the accident. This can be done by placing a reference object of the same size (a swimmer in appropriate thermal protection or a boat) on the water. The witness then directs the rescuer to the exact point where the victim was last seen. Rescuers can mark this as a credible body location by using a marker buoy or by aligning at least two points on the near shore with landmarks on the opposite shore.

Applying Submerged Victim Recovery Factors

Research into countless submerged victim recoveries over many years has revealed that drowning victims will generally sink to the bottom of any waterway. However, a dry drowning (air in lungs) may allow the body to be moved some distance by the current in flowing water. Air trapped in the victim's clothing can also contribute to movement of the body. In wet drownings (lungs full of water), especially those in still water, the body is more likely to be almost directly below the last-seen point.

In still water, the body should be found on the bottom within a radius equal to the water's depth from the last-seen point. For example, if the water is 20 feet (6.2 m) deep, the body should be within a 40-foot (12.4 m) circle below the last-seen point (Figure 9.39). In flowing water, the search should begin at the last-seen point and progress downstream.

If the body is not located during the search, it is likely to refloat eventually. Refloat is caused by

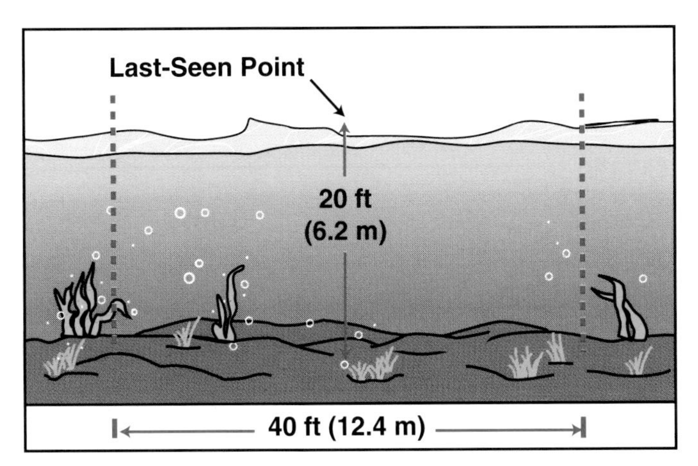

Figure 9.39 Typical search area based on water depth.

gases produced by bacterial action in the victim's intestinal tract making the body buoyant. Water temperature and the contents of the victim's digestive system will affect the timing of refloat from a few hours to a few days. Although unusual, it is also possible that the body may never refloat.

Monitoring Rescuer Safety

As part of the incident management system, the IC should appoint an incident safety officer whose sole responsibility is to monitor the safety of the rescuers involved in the incident. The safety officer must have the authority to stop all or any part of the operation and order the withdrawal of any involved personnel if necessary to protect the rescuers. The safety officer should ensure that all rescuers are properly attired and equipped for the existing and anticipated conditions before they are allowed to leave shore. The safety officer should be equipped with binoculars if necessary and should be in constant communication with the rescue team leader(s), usually by radio.

Conducting The Search

The area to be searched should be marked off and all efforts concentrated within that area. The search must be thorough, covering the entire surface of the bottom within the designated search area. The search may be done with trained divers or with grappling hooks or drags. In shallow water, probing with pike pole handles can also be effective. As with the rest of the operation, the success of the search phase will depend to a large extent on developing and following a plan.

DIVERS

A growing number of fire departments and rescue companies are adding dive teams to their aquatic rescue capabilities. Not having to wait for another agency to respond with their dive team gives these agencies an opportunity to take advantage of the "Golden Hour of Aquatic Rescue" and to effect successful underwater rescues at a higher rate than ever before. However, diving under emergency conditions without proper training can be extremely dangerous. Dive rescue accidents are usually caused by poor judgment, lack of skill, and lack of knowledge.

Qualified rescue divers are trained specialists who must have the skills and equipment necessary to deal with some of the worst underwater conditions possible. These conditions may include zero visibility, entanglement hazards, swift water, and ice. However, it is beyond the scope of this manual to teach rescuers how to use SCUBA equipment. Dive rescue training can be provided by organizations certified to teach these skills.

Chapter 9 Review

Directions

The following activities are designed to help you comprehend and apply the information in Chapter 9 of **Fire Service Rescue**, Sixth Edition. To receive the maximum learning experience from these activities, it is recommended that you use the following procedure:

1. Read the chapter, underlining or highlighting important terms, topics, and subject matter. Study the photographs and illustrations, and read the captions with each.
2. Review the list of vocabulary words to ensure that you know the chapter-related meaning of each. If you are unsure of the meaning of a vocabulary word, look up the word in the glossary or a dictionary, and then study its context in the chapter.
3. On a separate sheet of paper, complete all assigned or selected application and review activities before checking your answers.
4. After you have finished, check your answers against those on the pages referenced in parentheses.
5. Correct any incorrect answers, and review material that was answered incorrectly.

Vocabulary

Be sure that you know the chapter-related meanings of the following words:

- aquatic *(267)*
- current *(268)*
- downwash *(269)*
- drag *(285)*
- estuary *(268)*
- hazard *(268)*
- helical flow *(268)*
- hypothermia *(270)*
- inlet *(267)*
- laminar flow *(268)*
- obstruction *(268)*
- turbidity *(268)*

Application Of Knowledge

1. Choose from the following water and ice rescue procedures those appropriate to your department and equipment, or ask your training officer to choose appropriate procedures. Mentally rehearse the procedures, or practice the chosen procedures under your training officer's supervision.
   - Perform REACH, THROW, ROW, GO, and TOW/CARRY assisted surface rescues. *(268)*
   - Use the firefighter's carry to remove a victim from the water. *(276)*
   - Use the underarm, wrist-grasp method to remove a victim from the water. *(276)*
   - Use the saddleback carry to remove a victim from the water. *(276)*
   - Use the buddy-buddy carry to remove a victim from the water. *(276)*
   - Immobilize a spinal injury victim's head and neck while in the water. *(277)*
   - Turn a spinal injury victim over in the water. *(277)*
   - Perform REACH, THROW, and GO ice rescues. *(278)*
   - Perform a one-person ice rescue. *(279)*
   - Recover victims from a submerged or partially submerged vehicle. *(281)*
   - Perform a surf rescue. *(282)*
   - Search for and recover a drowning victim from the surf. *(282, 283)*
   - Locate and recover a submerged victim. *(283, 284)*
2. Survey your jurisdiction for possible water rescue locations (low-head dams, open water, public swimming areas and pools, bridges, popular ice-fishing areas, etc). Check your department's pre-incident plans to ensure that each of these potential rescue areas has been addressed. *(Local protocol)*

Review Activities

1. Identify the following acronyms and abbreviations:
   - ABCs *(267)*
   - GOG *(271)*
   - IC *(272)*
   - MDR *(269)*
   - PFD *(270)*
   - PIO *(273)*
   - SCUBA *(270)*
2. Identify the following and explain their relevance to water/ice rescue:
   - Air lifting bags *(271)*
   - Basket litter with floats *(271)*
   - Body bag *(271)*
   - Boogie board *(270)*
   - Class III harness *(270)*
   - Crampons *(278)*
   - Eddy *(268)*
   - Grappling hook *(285)*
   - Ice screw *(271)*
   - Last-seen point *(283)*

- Low-head dam/low-water dam *(269)*
- SCUBA outfit *(270, 271)*
- Spinal board *(271)*
- Strainer/sweeper *(268)*
- Throw bag *(270)*
- Water rescue helmet *(270)*

3. List the four basic components of a successful rescue. *(267)*
4. List the four general types of aquatic rescue situations. *(267, 268)*
5. Discuss currents, loads, and entrapments that rescuers may find in aquatic environments. *(268)*
6. Explain the hydraulic actions of low-head dams that earn them the name "drowning machines." *(269)*
7. Outline the guidelines for attempting to rescue a victim trapped in the hydraulics of a low-head dam. *(269)*
8. Explain the mammalian diving reflex phenomenon. *(269)*
9. Explain what is meant by the *Golden Hour of Aquatic Rescue. (269)*
10. List general cold water survival times for victims immersed in water of the following temperatures: *(Table 9.1, 269)*
    - <40°F/<5°C
    - 40°F to 64°F/5°C to -2°C
    - >64°F/>20°C
11. Describe the characteristics of the following types of water rescue victims: *(270)*
    - Nonswimmer
    - Weak or tired swimmer
    - Injured/hypothermic swimmer
    - Unconscious swimmer
12. List the equipment needed for the following water and ice rescues:
    - Aquatic surface rescues *(270)*
    - Ice rescues *(270)*
    - Underwater rescues/recoveries (no ice) *(270, 271)*
    - Underwater rescues/recoveries (ice) *(271)*
    - Partially or completely submerged vehicle rescues *(271)*
    - Victim recovery/removal from aquatic environment *(271)*
13. Explain how rescue teams should prepare for aquatic rescues/recoveries. *(271)*
14. Explain what is meant by *risk/benefit factor. (272)*
15. Describe how to maintain control of the scene, deal with the victim's family, and handle the media in aquatic rescue incidents. *(272)*
16. Briefly describe each of the following aquatic rescue methods:
    - Self-rescue *(273)*
    - Reach *(273)*
    - Throw *(273)*
    - Row *(274)*
    - Go *(274)*
    - Tow *(275)*
    - Assisted carry *(275)*
    - Controlled carry *(275)*
17. Explain factors that make it difficult to estimate ice strength. *(278)*
18. Explain how the weather and the victim's condition can complicate ice rescue. *(278)*
19. List general ice rescue protocol. *(278)*
20. List questions rescuers should ask and factors to consider when formulating an operational plan for rescuing victims from a submerged or partially submerged vehicle. *(280)*
21. Distinguish between an undertow and a riptide. *(282)*
22. Explain actions swimmers should take when caught in an undertow or riptide. *(282, 283)*
23. Discuss the factors that often make it difficult to rescue drowning victims from the surf. *(283)*
24. Describe a technique that may be used to search for a victim submerged in the surf. *(283)*
25. Explain how the rescuer can use deduction and/or statements from witnesses to determine the last-seen point from which to begin aquatic victim recovery operations. *(283, 284)*
26. Describe the search patterns that should be used from the following last-seen points: *(284)*
    - On shore
    - Floating in water along shoreline
    - Floating in water some distance from shore
    - Accident reenactment
27. Distinguish between a dry drowning and a wet drowning. Explain why rescuers make this distinction. *(284)*
28. Explain what causes a drowned body to refloat. *(284)*
29. List the responsibilities of the safety officer at the scene of an aquatic rescue/recovery. *(285)*

Questions And Notes

10

Trench Rescue

Chapter 10
Trench Rescue

INTRODUCTION

Trench construction occurs in virtually every city and town in North America, and in many jurisdictions it occurs almost daily somewhere within their boundaries. With all this excavation going on, cave-ins are bound to happen, and they do. According to National Institute for Occupational Safety and Health (NIOSH) statistics, an average of 60 workers die each year in trench cave-ins. NIOSH figures also show that 77 percent of those fatalities are construction workers; the remaining 23 percent are workers in other fields, including the fire/rescue service. Many of those killed in trench incidents are would-be rescuers who fail to stabilize the trench before they enter it and who then become additional victims when the trench caves in on them. Knowing how to make the trench safe to enter and taking the time to do it give both the victim and the rescuer the best chance for survival.

This chapter discusses the applicable regulations, the composition of soil, and the recognized soil types. It explains the reasons for trench failure, the various types of protective systems in use, and the hazards to be expected in the trench environment. The skills and equipment needed in most trench rescue incidents are described. Also covered are the tactical considerations that are appropriate in trench rescue incidents.

BACKGROUND

Regulations

All work done in excavations in the United States is covered by the OSHA regulations contained in 29 CFR 1926, Subpart P — Excavations, Sections 650-652. Written primarily for the construction industry, these regulations also apply with equal force to emergency workers such as firefighters and rescue squad members.

Definitions

In Section 650, a *trench* is defined as "a narrow excavation (in relation to its length) made below the surface of the ground. In general, the depth is greater than the width, but the width of a trench (measured at the bottom) is not greater than 15 feet (4.6 m)."

According to the standard, a *cave-in* means "the separation of a mass of soil or rock material from the side of an excavation, or the loss of soil from under a trench shield or support system, and its sudden movement into the excavation, either by falling or sliding, in sufficient quantity so that it could entrap, bury, or otherwise injure and immobilize a person."

SOIL

Soil is the natural mineral material of which the earth's crust is composed. It is found in a wide variety of forms and compositions depending on where it is located. Soils vary widely from marine and riparian environments to fertile valleys, deserts, hills, and mountains. To better understand the nature of trenches and other excavations, rescuers need to have some knowledge of the various soil types with which they may have to contend.

Composition Of Soil

There are many different types of soil in nature, but OSHA recognizes five basic types in its system of classification based on stability. In descending order of stability, the types are rock, cemented soils, cohesive soils, granular soils, and loams.

ROCK

The most stable of all soil types is rock, a natural solid mineral material. However, OSHA does not automatically assume rock to be stable and neither should rescuers (Figure 10.1). Because rock is a solid crystalline mass, its hardness may require blasting to allow excavation. Fissures created in the rock by blasting reduce its stability. Also, sedimentary rock is composed of layers (strata) that are natural fissures (Figure 10.2). If the layers are inclined toward the excavation, they can be dangerously unstable, especially if water is seeping between the layers.

Figure 10.1 Even solid rock may not be stable.

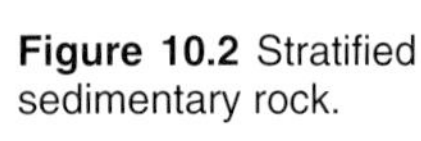

Figure 10.2 Stratified sedimentary rock.

CEMENTED SOILS

Sometimes described as being "hard as a rock," cemented soils are almost as stable. Limestone is perhaps the best known of all cemented soils, and it has the ability to be dissolved and to recement itself. The calcium carbonate of which it is primarily composed dissolves in water, migrates through the soil, then precipitates out, and chemically bonds with the soil in a cemented matrix. However, rescuers must be aware that cemented soils can lose stability through fissuring and when dissolved by water.

COHESIVE SOILS

Cohesiveness in soil is the result of friction between the particles and chemical bonding. While cohesive soils tend to stick together, their natural bonding is not nearly as strong as that of rock or cemented soils. Composed primarily of clay, cohesive soils depend for stability on the type of clay, the moisture content, and the amount of granular soil with which they are mixed.

GRANULAR SOILS

Granular soils are characteristically grainy. Whether the soil is considered to be gravel, sand, or silt depends primarily on the size of the particles. The stability of granular soils is dependent upon the angularity of the grains, the amount of natural cements present, and the moisture content.

LOAMS

Loams are mixtures of cohesive soils and granular soils. A sandy loam would be composed primarily of sand; clay loam would have a preponderance of clay, etc. The stability of loams depends on the amount of each of the constituents and the moisture content.

Weight Of Soil

The weight of soil is usually underestimated by those who have not studied it. Even the loose soil of which many trenchwall failures are composed is incredibly heavy. Depending upon its composition and moisture content, a cubic foot (0.03 m^3) of soil can weigh from 90 to 110 pounds (41 kg to 50 kg). Therefore, a cubic yard (0.76 m^3) of soil can easily weigh well over a ton. Considering that most trenchwall failures involve the movement of several cubic yards of soil, the weight on a fully buried victim may turn a rescue into a body recovery operation. This situation underscores the absolute necessity for rescuers to be adequately protected if they are to avoid becoming additional victims in a subsequent cave-in.

Classification Of Soil

To simplify the field classification of soils, the regulations further classify soils into three general types — A, B, or C — based on their stability according to several criteria. These criteria are:

- Cohesiveness
- Presence of fissures
- Presence and amount of water
- Unconfined compressive strength

- Presence of layering
- Evidence of prior excavation and vibration.

Rescuers should have some knowledge of these soil types in order to fully appreciate the potential for subsequent trenchwall failures in the situations to which they are called.

TYPE A SOILS

Type A soil is considered to be the most stable and the type least likely to fail. Because of its relative stability, excavations in Type A soil are assumed to be the safest in which to work. The regulations define Type A soils as those with an unconfined compressive strength of 1.5 tons per square foot (tsf) (144 kPa) or greater. *Unconfined compressive strength* means the load per unit area at which the soil will fail in compression. Examples of cohesive soils are clay, silty clay, sandy clay, clay loam, and in some cases, silty and sandy clay loam. Cemented soils, such as caliche and hardpan, are also considered Type A.

However, regardless of the mineral content of the soil in a particular location, other factors may affect its stability and thereby put it in a classification of lesser stability. No soil is Type A if it meets *any* of the following criteria:

- It is fissured.
- The soil is subject to vibration from heavy traffic, pile driving, or similar activities.
- The soil has been previously disturbed.
- The soil is part of a sloping layered system where the layers dip into the excavation on a slope of four horizontal to one vertical or greater.
- The material is subject to other factors that would require it to be classified as a less stable material.

These exclusions from Type A reduce the likelihood of any soil being classified as Type A on most construction sites. Therefore, most soils in the situations to which rescuers are called are likely to be either Type B or Type C.

TYPE B SOILS

Although less stable than Type A soils, Type B soils are still relatively stable and are representative of native soils in many parts of North America. To be classified as Type B, soils must meet the following criteria:

- Cohesive soil with an unconfined compressive strength greater than 0.5 tsf (48 kPa) but less than 1.5 tsf (144 kPa)
- Granular cohesionless soils including angular gravel, silt, silty loam, sandy loam, and in some cases, silty and sandy clay loam

  NOTE: Those performing soils classification tests should remain skeptical of soils in this category. While it is true that some granular soils, such as crushed rock that is used to bed railroad tracks, can be called Type B due to the friction generated by the sharp edges of the rock, most granular soils are Type C. One type of silt called a *loess* is characterized by its ability to maintain steep slopes. This ability is due to both its angularity and its high calcium (cement) content. Granular soils that appear to qualify as Type B should be confirmed by a soils engineer or geologist.
- Previously disturbed soils except those that would be classified as Type C soil
- Soil that meets the unconfined compressive strength or cementation requirements of Type A but is fissured or subject to vibration
- Dry rock that is not stable
- Material that is part of a sloped, layered system where the layers dip into the excavation on a slope less steep than four horizontal to one vertical, but only if the material would otherwise be classified as Type B

On a typical construction site, most soils fall into the Type B category unless they are predominantly granular or saturated, in which case they are more likely to be Type C. However, because the trench has already caved in, rescuers should consider the soil to be Type C and proceed accordingly.

TYPE C SOILS

The least stable of the three soil types, Type C soils are nonetheless quite common throughout North America. In addition to the deserts of the

Southwest, most beaches are composed of Type C soils. To be classified as Type C, soils must meet the following criteria:

- Cohesive soil with an unconfined compressive strength of 0.5 tsf (48 kPa) or less
- Granular soils including gravel, sand, and loamy sand
- Submerged soil or soil from which water is freely seeping
- Submerged rock that is not stable
- Material in a sloped, layered system where the layers dip into the excavation on a slope of four horizontal to one vertical or steeper

Figure 10.3 A rescuer obtains a fresh soil sample.

Figure 10.4 The sample should be visually inspected.

Figure 10.5 The trench should be checked for signs of fissuring.

Soils Testing

When contractors and/or rescuers must decide the safest and most efficient way for workers to function in an excavation, OSHA requires them to make that determination based on a system of soils classification. To determine which classification applies to the soil at any particular site, it must be tested by someone competent to make the analysis. The classification should be based on at least one visual and one manual test performed out of the excavation (not in it) on freshly excavated samples (Figure 10.3). Rescuers will quickly perform some of the following tests when developing a plan of action on a trench rescue incident:

VISUAL TESTS

The visual analysis of an excavation takes into account the site in general, the soil adjacent to the trench, the soil forming the trench walls, as well as soil samples taken from within the trench.

Soil that has been removed from the trench should be inspected for cohesiveness by observing the particle sizes (Figure 10.4). Soil composed primarily of fine-grained material is considered to be cohesive; that which is composed of coarse-grained sand or gravel is considered to be too granular to be cohesive. Cohesive soils remain in clumps; granular materials do not, and they break up easily.

The area around the trench, as well as the trench walls should be inspected for tension cracks and evidence of material spalling off the vertical trench walls (Figure 10.5). These cracks could be signs of fissures that would suggest dangerously unstable soil. The area should also be inspected for evidence of underground utility structures that would indicate previously disturbed soil (Figure 10.6).

The trench walls should be inspected for evidence of layering. If found, the layers should be examined to determine whether they slope toward the trench and if so, to estimate the degree of slope.

The area in and around the trench should also be inspected for signs of surface water or water seeping from the trench walls (Figure 10.7). The presence of water seepage adds another factor of instability.

Finally, the general area around the trench should be inspected for any potential sources of vibration that could make the trench walls unstable (Figure 10.8).

Figure 10.6 A typical underground utility enclosure.

Figure 10.7 Water in the trench is a sign of instability.

Figure 10.8 Vehicular traffic nearby can add vibration to the site.

MANUAL TESTS

Manual tests of soil from the trench walls are done to provide additional quantitative and qualitative evidence of the soil's composition and to assist in the proper classification of the soil. First, a moist or wet ball of the soil should be tested for *plasticity*. It is rolled out into a string or thread as little as ½ inch (3.1 mm) in diameter (Figure 10.9). Cohesive materials can be rolled into threads without crumbling. For example, if a ½-inch (3.1 mm) piece of "thread" at least 2 inches (50 mm) in length can be held up by one end without breaking, it is cohesive (Figure 10.10).

In a test of *dry strength*, if a sample is dry and crumbles into individual grains or a fine powder with little or no pressure, it is considered to be granular. If the soil is dry and falls in clumps that break up into smaller clumps but if the smaller clumps can only be broken up with some difficulty, the soil may be clay in some combination with gravel, sand, or silt. If the soil is dry, the clumps do not break up into smaller clumps, the clumps can only be broken up with some difficulty and if there is no visual evidence of cracks or fissures, it may be considered to be unfissured.

Another manual test, *thumb penetration*, can be used to estimate the unconfined compressive strength of cohesive soils. In this test, a freshly excavated clump of soil is pressed with the thumb (Figure 10.11). Type A soils can be indented with thumb pressure but can only be penetrated by the thumb with very great effort. Type C soils can be easily penetrated several inches with the thumb and can be molded with light finger pressure. Type B soils fall somewhere between these extremes.

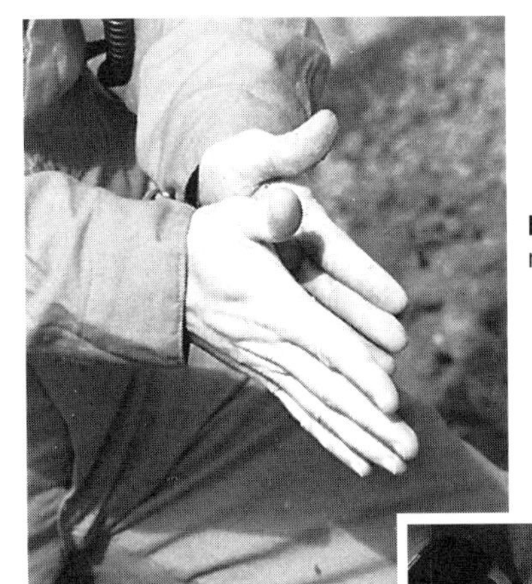

Figure 10.9 The soil sample is rolled into a thread.

Figure 10.10 An example of cohesive soil.

Figure 10.11 A rescuer tests for compressive strength.

OTHER TESTS

Other estimates of the unconfined compressive strength of soils can be obtained by using a pocket penetrometer or a hand-operated shear vane. However, these tests are beyond the scope of this manual and are not considered to be practical tests that rescuers might perform in assessing the stability of a trench.

CAUTION: Because rescuers would not have been called to the scene unless there had already been at least one cave-in, they must assume that the trench walls remain dangerously unstable.

REASONS FOR TRENCH FAILURE

Trench failure can result from a number of different causes or, more likely, from a combination of causes. The most frequent causes of trenchwall cave-ins are the effects of vertical/lateral forces, soft pockets, layered soils, saturated soils, vibration, and surcharge loads.

Vertical/Lateral Forces

The vertical and lateral forces at work in an excavation are the natural force of gravity on a vertical wall that lacks the lateral support that it formerly had. Even relatively stable soils are subject to sudden failure when they have both space in which to move laterally and the motive force of gravity (and the weight of objects on the surface) pushing downward. Sometimes the only trigger needed for cave-in is some amount of vibration added to this.

Another aspect of the conditions that contribute to the failure of trench walls is the depth of the excavation. The general rule is that the deeper an excavation, the more likely it is to fail. Given an average weight of about 2,700 pounds per cubic yard (1 215 kg per m^3), the weight of the unsupported trench walls can generate incredible forces.

Soft Pockets

This condition is most often the result of the soil having been previously disturbed. Where the soil has been previously excavated and backfilled, it is rarely as compact or cohesive as the undisturbed native soil surrounding it.

Layered Soils

As mentioned earlier, when soils are formed in layers, especially layers that incline into the excavation, the chances of trenchwall failure are greatly increased. If there is evidence of water seepage between the layers, the likelihood of a cave-in is even greater.

Saturated Soils

Water has a tremendous effect on soil strength. The rule of thumb is the more water present in the soil, the more the soil will behave like a liquid. When soils become saturated, the cementation and friction bonding of the individual particles begin to disappear, and the particles start to float apart. Thus, the solid mass is transformed into a fluid mass. Moisture also increases the weight of the soil, which may exceed the soil's cohesive strength.

However, ground moisture can also contribute to the failure of trench walls in other ways. If the water in the soil freezes, the soil may *appear* to be more stable because its surface is quite hard. In reality, the soil is even more unstable because the ice crystals separate the soil particles making it softer, and when it thaws, the melted ice lubricates the soil.

If a trench remains open for an extended period, and the length of this period varies greatly depending upon the original moisture content of the soil and atmospheric conditions, the soil can dehydrate. In some granular soil types that are relatively stable when moist, dehydration can reduce cohesiveness and stability to a dangerous point.

Vibration

The effects of vibration are easily underestimated. It is almost invariably present on construction sites and must be taken into consideration when developing a rescue plan. Vehicular traffic (including emergency vehicles), pile driving, and vibratory compaction can seriously reduce trenchwall stability. Both high-frequency and low-frequency vibrations can reduce the stability of an excavation, particularly when moisture is present. Rescuers must assume that every trench wall is suspect and plan accordingly.

Surcharges

Surcharge loads, generated by the weight of anything in close proximity to the excavation, must be seen by rescuers as conditions that increase trenchwall instability and may contribute to subsequent cave-ins. Common sources of surcharge loads are the weight of the spoil pile, the weight of nearby objects, the weight of rescuers looking into the trench, the weight of materials and equipment, vibration, and undermining.

WEIGHT OF SPOIL PILE

Contractors sometimes place the spoil pile too close to the edge of the excavation and/or make the pile too big. OSHA requires that it be placed 2 feet (0.6 m) from the edge and not be over 2 feet (0.6 m) high. The decision of whether to move all or part of a spoil pile must take into account the amount of time involved and the effects of vibration generated by the equipment used to move the spoils.

WEIGHT OF NEARBY OBJECTS

In addition to increasing surcharge loads, structures, concrete slabs, and pavement retain moisture under them, further increasing cave-in probability. Trees and poles add the potential of falling hazards as well as surcharge load problems. Bracing, cribbing, or removing such objects may be considered.

WEIGHT OF MATERIALS AND EQUIPMENT

According to OSHA, the edge of a trench is defined as a horizontal distance from the lip of the trench equal to the vertical depth of the trench. The weight of materials at the edge of the excavation is restricted to the equivalent of a 2-foot (0.6 m) tall spoil pile when timber shoring is used. When either timber or aluminum hydraulic shoring is used, equipment near the trench is restricted to a total of 20,000 pounds (9 t).

UNDERMINING

Undermining of the trench walls can result from overdigging or from areas of the wall sloughing off into the bottom of the trench. This makes the overhanging portions of the trench walls very likely to fail.

TYPES OF TRENCH FAILURE

Depending upon the type of soil involved, the mechanism of collapse, and other variables, trenchwall failures can take different forms. The most common types of trenchwall failure are: lip cave-in, wall slough-ins, and wall shear.

Lip Cave-In

One or both lips of a trench may cave in, often due to excessive surcharge loads too close to the trench lip (Figure 10.12).

Wall Slough-In

One or both walls may slough-in at the bottom of the wall leaving an overhanging portion above (Figure 10.13).

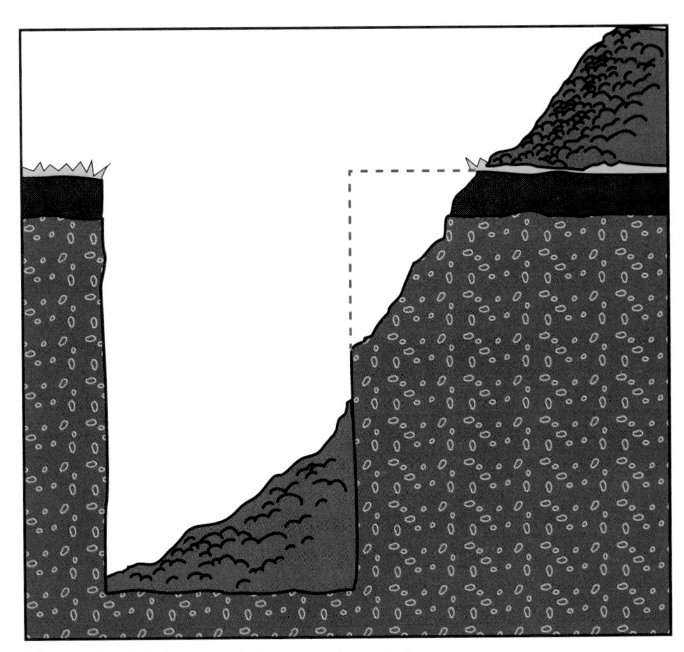

Figure 10.12 The trench lip sometimes fails.

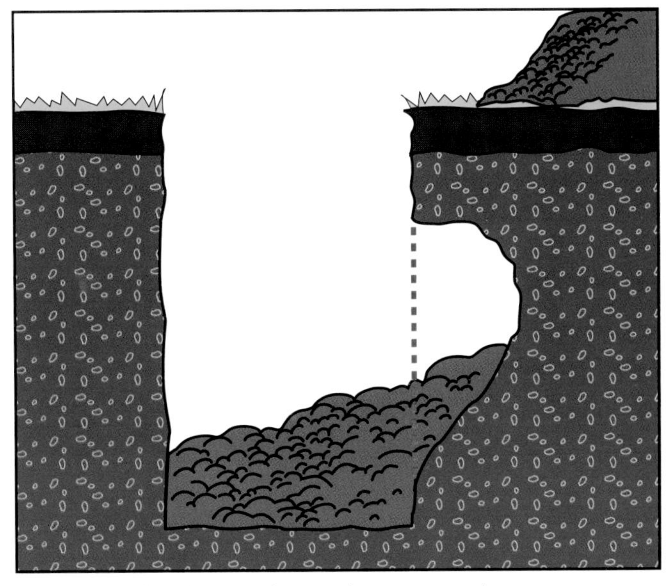

Figure 10.13 Slough-in can leave a dangerous overhang.

Wall Shear

One or both walls can shear away and fall into the bottom of the trench (Figure 10.14).

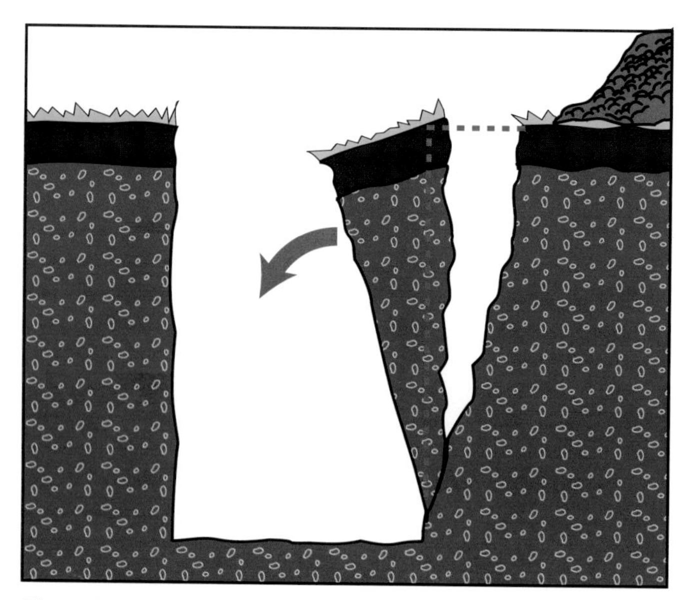

Figure 10.14 Wall shear can happen suddenly and without warning.

PROTECTIVE SYSTEMS

When employees must work in trenches deeper than 4 feet (1.3 m), contractors are required by OSHA to protect the workers from cave-ins as defined at the beginning of this chapter. There are three ways in which the workers can be protected: sloping/benching, shoring, and shielding.

Sloping/Benching

Both sloping and benching (see following sections) are effective methods of preventing cave-ins. In different ways, both of these methods decrease the angle of the trench walls to a point where they are unlikely to collapse. However, because both of these methods involve the excavation and removal of much more soil than is required to just dig the trench, they are rather costly options. Sloping/benching are also too time-consuming to be practical in most rescue situations.

SLOPING

In this method of protecting employees from cave-ins, both sides of the trench are inclined away from the bottom of the trench (Figure 10.15). The angle of incline required to prevent a cave-in varies with different soil types and environmental conditions, but it is usually less than 34 degrees from horizontal.

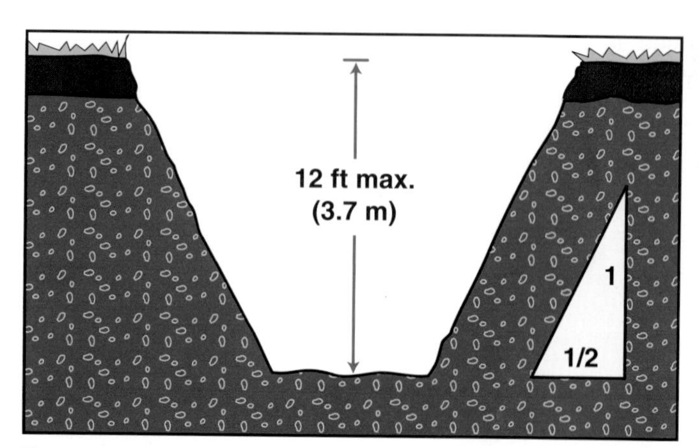

Figure 10.15 Trench walls can be stabilized by being sloped.

BENCHING

In this method, the sides of the trench are excavated to form one or a series of horizontal steps or benches, usually with vertical or near-vertical surfaces between levels (Figure 10.16).

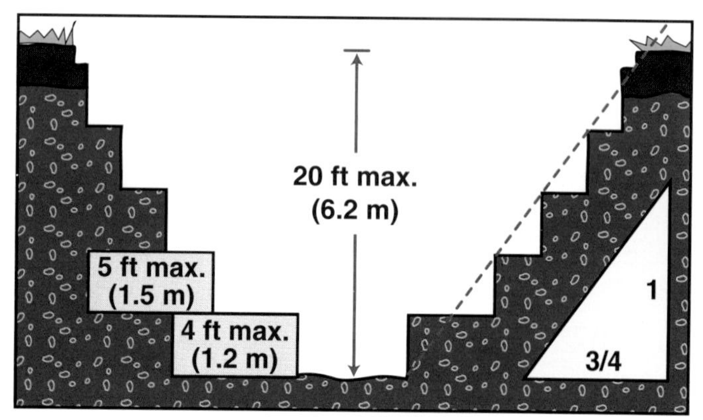

Figure 10.16 Benching is another way to stabilize trench walls.

Shoring

The second means of providing cave-in protection is shoring. In this method, the walls of the excavations are braced with structural components strong enough to prevent a cave-in. Of the numerous ways to shore up excavations, timber shoring is the oldest. However, screw jacks and/or pneumatic or aluminum hydraulic shores may also be used (Figure 10.17). Shoring plans for trenches deeper than 20 feet (6.2 m) must be designed by a registered professional engineer.

Shielding

The third method of protecting workers in excavations is by the use of shielding. Shielding differs from sloping/benching and shoring in that it is not designed to prevent cave-ins. Instead, it is designed to shield the workers from a cave-in should one occur. Most shields consist of two flat, parallel metal walls whose ends are connected by metal cross braces (Figure 10.18). This configuration forms a sort of open box that allows employees to work within it in relative safety. The box is moved along in the trench as work progresses. When conditions allow its use, this method is more cost effective than the other methods and may be the most common method of protecting workers in trenches, especially in pipeline construction.

Figure 10.17 Aluminum hydraulic shores stabilizing a trench.

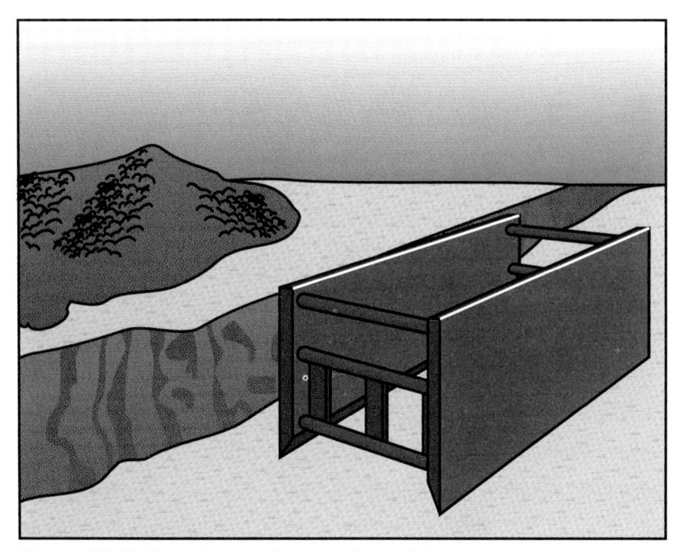

Figure 10.18 One of several styles of trench boxes.

TRENCH RESCUE HAZARDS

Trench rescue operations can be some of the most hazardous operations that rescuers may be called upon to perform. OSHA considers some excavations to be confined spaces, and most if not all of the hazards that are common to other confined spaces may also be found in trenches. Like all other confined spaces, trenches have certain actual or potential hazards within them. These hazards may take any of a number of different forms or a combination of different forms, but most will fall into one of three major categories: physical, atmospheric, or environmental.

Physical Hazards

The physical hazards in trenches are those that are created by the lack of structural integrity of the trench walls and/or hazardous objects within it.

SECONDARY COLLAPSE

The probability of a secondary or subsequent collapse of the trench walls is present as long as the trench remains open. The fact that a wall has already caved in is ample evidence that the walls are unstable. Conditions such as water seeping into the trench can increase the likelihood of a secondary collapse; however, the things that have the most effect on the stability of the trench walls are surcharge loads and vibration.

Surcharge loads. As mentioned earlier, the weight of the spoil pile and other objects near the trench can increase the likelihood of a secondary collapse. The weight of any object within a distance equal to the depth of the trench, including construction and rescue vehicles, contributes to the surcharge load.

Vibration. Vibration from movement of personnel and equipment near the trench, which may have caused the initial collapse, can also contribute to or trigger a secondary trenchwall collapse. If water is present in the soil, vibration can cause or increase the process of liquefaction causing a drastic reduction in cohesion.

Signs of instability. Everyone on scene, but especially the safety officer, should be alert for signs of an impending secondary collapse. Some of the more common signs of trenchwall instability are as follows:

- Bulges in the trench wall, especially near the bottom
- Horizontal cracks or fissures in the trench wall
- Loose chunks falling from the trench wall
- Water seeping into the trench
- Loose debris spontaneously falling from the lip of the trench without apparent cause

UNSTABLE DEBRIS

Loose and unstable debris in the trench or near the lip is prone to sudden and unexpected movement if disturbed, even if just by vibration, and it may fall on rescuers in the trench. Debris, which may have been produced by the excavation or by the initial collapse, can also create a trip hazard for rescuers in and around the trench.

UNSUPPORTED UTILITIES

Utility conduits that cross an open trench can present a number of hazards. Their weight can increase the likelihood of a trenchwall failure. The stresses caused by being unsupported can cause water and/or sewer lines to leak into the trench or the surrounding soil decreasing its cohesiveness. Electrical conduits can add the possibility of a shock hazard for rescuers.

Atmospheric Hazards

Atmospheric hazards in trenches can be created in a number of different ways and may take different forms. However, the potential for the existence of an atmospheric hazard makes it critically important that the atmosphere within the trench be sampled with properly calibrated instruments prior to entry and continuously throughout the operation (Figure 10.19).

Figure 10.19 A rescuer samples the atmosphere within a trench.

OXYGEN DEFICIENCY

According to OSHA, the atmosphere within a confined space, such as a trench, is considered to be oxygen-deficient whenever the percentage of oxygen drops below 19.5 percent. The most common cause of oxygen deficiency in trenches is the oxygen being displaced by an accumulation of heavier-than-air gases in the trench. Except for methane, all petroleum-based vapors and gases are heavier than air and tend to accumulate in low places such as trenches. Other gases, some highly toxic, are also heavier than air.

FLAMMABILITY

OSHA regulations also say that if the atmosphere within a trench contains a flammable gas, vapor, or mist in excess of 10 percent of its lower flammable limit (LFL), it is considered hazardous. This can result from a broken natural gas pipe in the trench, or it can result from a volatile flammable liquid seeping in from a contaminated aquifer.

TOXICITY

Because the atmosphere within a trench tends to be relatively static, toxic gases and vapors that might otherwise be blown away by the wind can remain at harmful concentrations in the trench. Whether the source is biological activity, such as in a sewer trunk, or a liquid residue that is vaporizing, rescuers must be careful to not add themselves to the list of victims.

Environmental Hazards

The environmental hazards in trench rescue situations are those conditions in and around the trench that make the rescue more difficult and/or that place the rescuers or trapped victims in jeopardy. Conditions such as extremes of temperature, high noise levels, water accumulations, and dust can make rescues slower and more difficult and can compromise the safety of rescuers and victims.

TEMPERATURE EXTREMES

Whether high or low, extremes of temperature can threaten both victims and rescuers. Victims trapped in cold, wet spaces for extended periods can suffer from hypothermia. Likewise, extremely hot, dry environments can bring on dehydration and other heat-related illnesses.

NOISE

Because the confines of a trench change the way sound behaves, noises generated within the trench tend to stay in the trench making communication difficult. Sounds produced near but outside the trench may not be heard in the trench. High noise levels created by rescue equipment and personnel working outside the trench can obscure sounds within the trench such as the sounds of creaking shoring members being stressed by soil movement.

WATER ACCUMULATIONS

Moisture is a common phenomenon in trenches. Seepage can accumulate in the bottom of the trench making footing unsure. It can also decrease trenchwall stability by reducing soil cohesion. Moisture can also add unwanted humidity to hot, still air in a trench, or it can hasten the onset of hypothermia in cold situations. Wet, slippery ground can slow rescue efforts due to poor footing and can create a serious fall hazard for rescuers.

CAUTION: Broken water pipes can allow water to enter the trench, and if the water is not stopped, it can threaten rescuers and trapped victims with drowning.

DUST

Airborne dust can remain suspended for some time in the still air in a trench, especially very soon after a trenchwall failure. Also, rescuers working in and around a trench can stir up large amounts of dust, which can obscure vision and make breathing difficult.

EQUIPMENT

As in the other rescue disciplines, the equipment most often needed in trench rescue incidents falls into two broad categories: personal protective equipment (PPE) and other equipment, sometimes referred to as *team gear*. Most of the equipment needed in trench rescue incidents is exactly the same as that needed for other types of rescues.

Personal Protective Equipment

The PPE needed for most trench rescue incidents is no different than that described in Chapter 6, Structural Collapse Rescue. Rescuers still need to be protected from physical, atmospheric, and environmental hazards associated with working in and around open trenches.

Team Gear

The team gear that is peculiar to trench rescue incidents is primarily related to the unique shoring requirements of the trench environment (see Shoring Systems section later in this chapter). Otherwise, the equipment needed for trench rescue is much the same as that needed for structural collapse and confined space rescue incidents (see Chapters 6 and 8).

While they may be used to some extent in structural collapse incidents, trenching tools (short-handled shovels) and canvas or metal buckets are sometimes critical to the success of trench rescues (Figure 10.20). They are used to remove loose soil from the bottom of a trench while searching for or freeing a buried victim.

Figure 10.20 Typical trenching tools.

Ground ladders are also used in trench rescues (Figure 10.21). They are used as a way of getting into and out of a trench.

Heavy equipment, sometimes from outside resources, is also sometimes needed in trench rescue incidents (Figure 10.22).

Figure 10.21 Ground ladders are a critical part of trench safety.

Figure 10.22 Heavy equipment from another agency may be needed. *Courtesy of Joel Woods.*

RESCUE SKILLS

To avoid becoming additional victims themselves, rescuers must know how to make a collapsed trench safe to enter. This *will* involve sampling and monitoring the atmosphere within the trench and *may* involve ventilating it. It *will* involve using one or more of the protective systems, described earlier, in the area where the victim is or is believed to be. Rescue personnel need hands-on training to be able to apply these technical skills correctly.

Atmospheric Monitoring

As described earlier, the atmosphere within a trench can be either oxygen-deficient or contaminated with flammable or toxic fumes or vapors, or both. If there is any reason to suspect that any of these conditions exist in the trench, the atmosphere should be sampled as described in Chapter 8, Confined Space Rescue. If it is found to be either oxygen-deficient or contaminated, it will have to be mechanically ventilated before rescuers are allowed to enter the trench. If the source of the contamination cannot be eliminated in a reasonable amount of time, it may be necessary to continue the ventilation process throughout the rescue operation.

Shoring Systems

Shoring systems designed to stabilize the walls of a trench are different from the shoring systems described in the structural collapse chapter. OSHA regulations provide a series of charts specifying the dimensions and spacing of shoring members to guide contractors and rescuers in shoring their excavations. Rescuers should learn to construct shoring systems that are fully compliant with the regulations for Type C soils — that is the approach taken in this section.

In the absence of a rescue vehicle loaded with timbers, plywood, and other shoring devices and equipment, rescuers may have to construct shoring systems from whatever materials are available on site. Most, but not all, excavation contractors will have mechanical shoring devices on site that can also be used. The types of shoring systems used in trench rescues are those constructed entirely of timber and those that combine the use of timber with screw jacks, pneumatic shores, or aluminum hydraulic shores (see following sections).

Before any type of shoring system is installed in a trench, the area around the trench should quickly be cleared of loose debris, and the area stabilized.

Stabilization is done by laying sheets of plywood on the ground along the lip of the trench (Figure 10.23). On the side of the trench with the spoil pile, there usually is not enough room between the spoil pile and the lip of the trench to allow sheets of plywood to be used, so planks should be placed along that edge (Figure 10.24). If there is room, planks can be placed side by side to provide as much stable walk space as possible.

Figure 10.23 Plywood helps to stabilize the lip of the trench.

Figure 10.24 Planks may be used if space is limited.

TIMBER SHORING

Timber shoring can be very labor intensive and time-consuming to construct. However, well-trained and properly equipped rescue personnel can construct a timber shoring system very quickly. The type of timber shoring system most often used in trench rescue incidents consists of 2- x 12-inch planks nailed to 4- x 8-foot sheets of plywood that are then set against opposite walls of the trench. The upright planks are held firmly apart by horizontal timbers, often called *cross braces* (Figure 10.25). In this system, only the upright planks and the horizontal timbers contribute to the strength of the system; the plywood, called *sheeting*, only prevents rocks, soil, and other loose debris from falling into the trench. However, even for this purpose, 1⅛-inch CDX or ¾-inch arctic white birch (Fin-Form) plywood is required.

In this type of system, the planks should be 10 to 12 feet (3 m to 4 m) long. They should be nailed to the centerline of the plywood, with 1 to 2 feet (0.3 m to 0.6 m) of the plank extending beyond each end of the plywood (Figure 10.26). Using duplex (foundation) nails makes later removal easier. A length of utility rope should be attached to each piece of sheeting to aid in placing them in the trench and in removing them from the trench later (Figure 10.27). Each piece of rope should be at least as long as twice the distance across the trench.

The depth of both sides of the trench should then be measured (Figure 10.28). This will help to determine the vertical spacing of the horizontal cross braces. Measuring both sides of the trench is necessary because the cross braces need to be

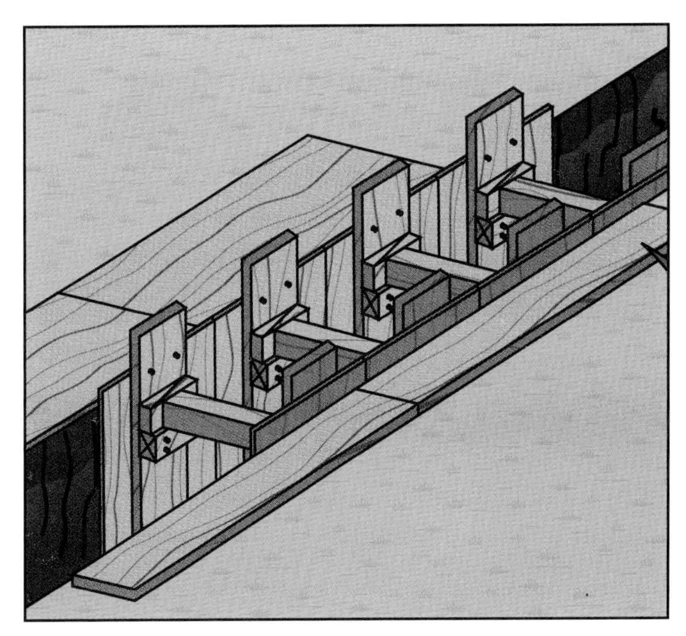

Figure 10.25 Typical timber cross bracing in place.

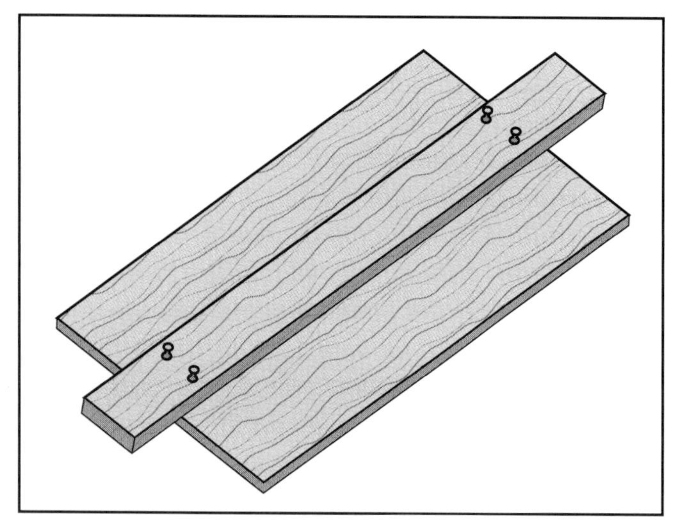
Figure 10.26 The planks extend past the ends of the plywood.

Figure 10.27 Rope handles make sheeting easier to place.

Figure 10.28 Both sides of the trench must be measured.

installed as close to level as possible, and if the trench floor is not level, the cleats or hangers on which the braces will rest will have to be placed at different heights on the opposing uprights. The cross braces must be installed no more than 4 feet (1.2 m) apart, measured center to center, and there must be one within 2 feet (0.6 m) of the trench floor and one no more than 1.5 feet (0.5 m) below the top of the trench (Figure 10.29). Because the depth of the trench may not exactly fit this spacing, the vertical distance between the cross braces may have to be *less* than 4 feet (1.2 m), but the spacing of the top and bottom braces should always be as stated.

Wooden cleats, pieces of 2- x 4-inch lumber about 12 inches (300 mm) long, or metal joist hangers, should be nailed to the uprights at the right levels so that the cross braces will be level when installed (Figure 10.30). If joist hangers are used, they are used on one side of the trench only. The reason the trench-depth measurements are important is so that the cleats can be nailed to the uprights *before* the panels are put in the trench. That way, the shock and vibration of hammering is not transferred to the trench walls as it would be if the nailing were done after the uprights were in the trench. Because each of the cleats or hangers will only have to bear the weight of one end of a cross brace, two nails each should be sufficient.

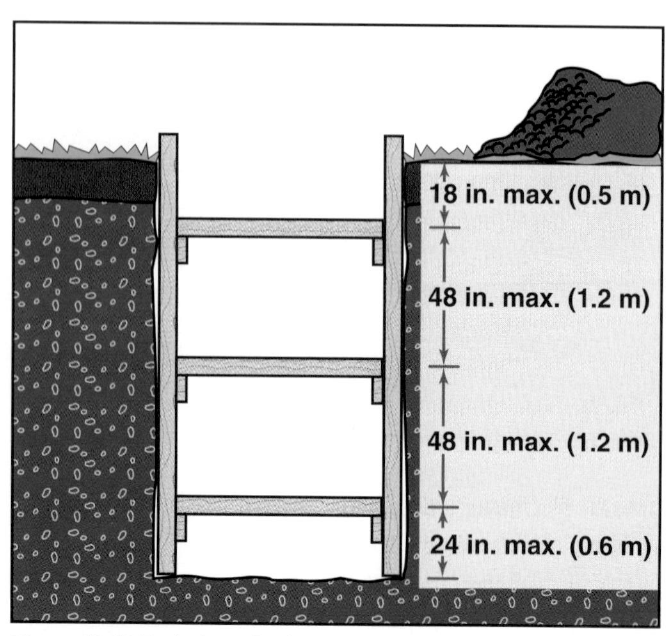

Figure 10.29 Typical spacing between cross braces.

Figure 10.30 A wooden cleat and a metal joist hanger.

Once the cleats or hangers are attached, the panels are ready to be placed in the trench. From the side of the trench that does not have the spoil pile, the first panel is lowered into the trench with the upright facing down. The bottom of the upright comes to rest at the bottom of the trench on the opposite side (Figure 10.31). If ropes are attached to the top of the panel, they are then tossed across the trench to rescuers on the other side; if not, the top of the panel can be pushed with a pike pole. The rescuers on the other side of the trench pull on the ropes until the first panel is upright and snug against the wall of the trench, and they hold it in position (Figure 10.32). With the upright facing up, the second panel is lowered into position opposite the first one and is also held in place (Figure 10.33).

Figure 10.31 The first panel is lowered facedown into the trench.

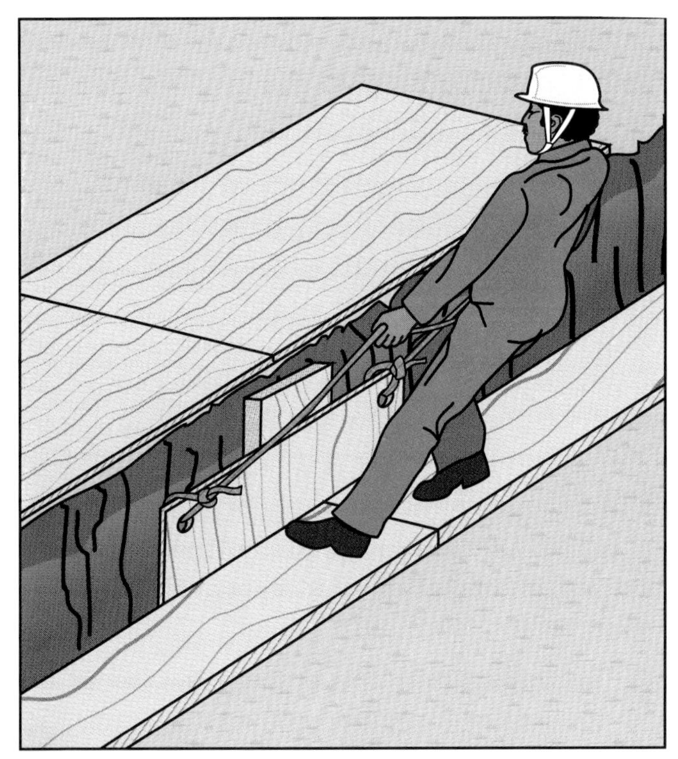

Figure 10.32 The panel is held in position.

Figure 10.33 A second panel is lowered faceup into position.

A ladder is then lowered into the trench at the edge of the panels with at least 3 feet (1.0 m) of the ladder above the lip of the trench (Figure 10.34). It should be positioned on the opposite side from where the second set of panels will be placed. A rescuer can then descend the ladder, no farther than waist deep into the trench, to measure the distance between the uprights at the level of the top brace. A 4- x 4-inch timber is then cut about 1 inch (25 mm) shorter than that distance if cleats and wedges are being used, or about 10 inches (250 mm) shorter if adjustable brackets are being used. With a utility rope tied near each end, the first cross brace is lowered onto the top cleats or the joist hanger (Figure 10.35). If cleats are used, opposing 4- x 4-inch wedges are then driven in at one end to tighten the cross brace between the uprights (Figure 10.36). The other end of the cross brace should be toenailed in place. If an adjustable bracket is used, the foot of the bracket should be screwed out to tighten the brace between the uprights (Figure 10.37). There is no need to toenail the other end of the brace in the joist hanger.

NOTE: Do not overtighten the first cross brace because this may cause one or both uprights to kick out at the bottom.

The rescuer may now descend the ladder far enough to measure for the second cross brace (Figure 10.38). The second brace is then cut and low-

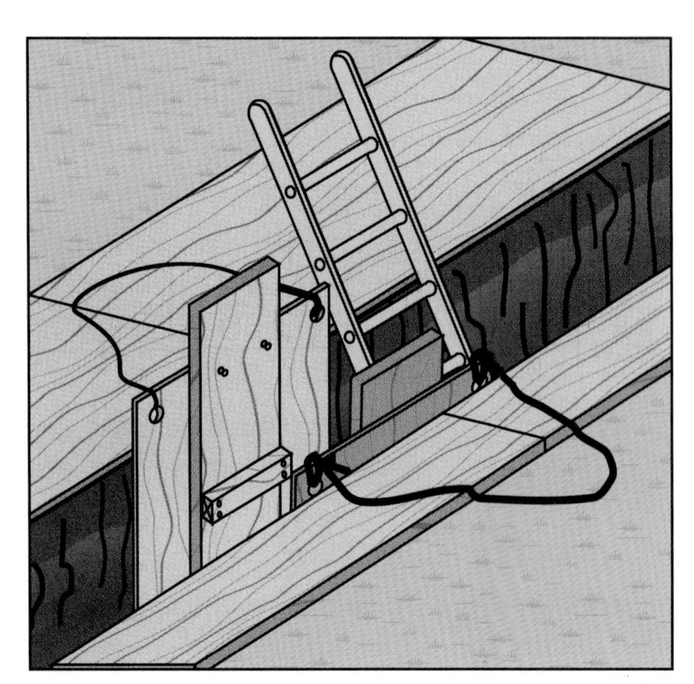

Figure 10.34 A ladder is placed in the trench.

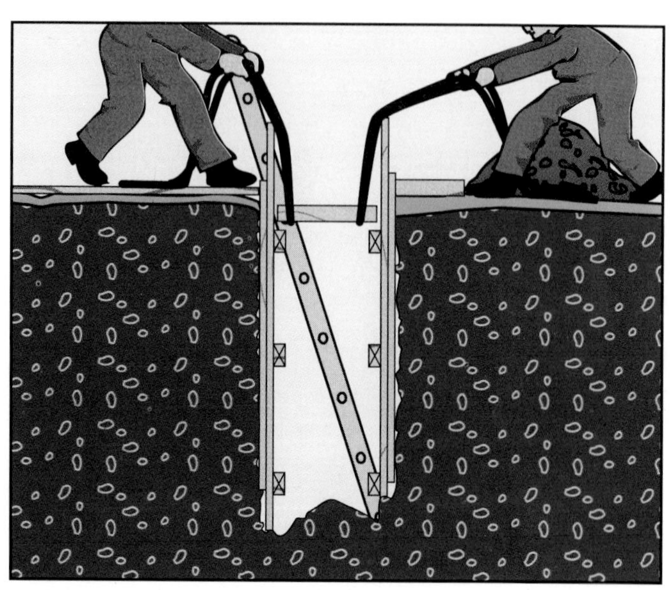

Figure 10.35 A cross brace is lowered into position.

Figure 10.36 Wedges are driven in to tighten the cross brace.

Figure 10.37 A rescuer tightens an adjustable bracket. *Courtesy of Phoenix (AZ) Fire Department.*

Figure 10.38 A measurement for the second brace must be taken. *Courtesy of Phoenix (AZ) Fire Department.*

ered into place just as the first one was. The second brace is then tightened as before. This process is continued until all cross braces are installed between the first two uprights. Then, working from the top down, each cross brace should be tightened further to fully compress the panels against the trench walls. If wedges are used, they should be toenailed in place once they are fully set.

The process can then be repeated for a second pair of panels to create a safe zone within the trench. Once the second pair of panels has been lowered into the trench, a second ladder should be placed alongside these panels (Figure 10.39).

NOTE: If the trench is more than 10 feet (3 m) deep, rescuers may have to call in a technical rescue team, and it will have to install heavier shoring material.

In the absence of plywood, the sheeting can be constructed of tightly butted planks placed vertically along both walls of the trench. The planks are held in place by horizontal timbers called *wales*. The wales are spaced vertically at the same intervals as the cross braces (Figure 10.40). However, constructing this type of shoring system is often too inefficient for contractors to use, and it may also be too time-consuming to be practical as a trench rescue procedure.

Regardless of which system for tightening the cross braces is used, once the last of the braces between two uprights is in position, the tightness of all the braces should be checked and adjusted if needed.

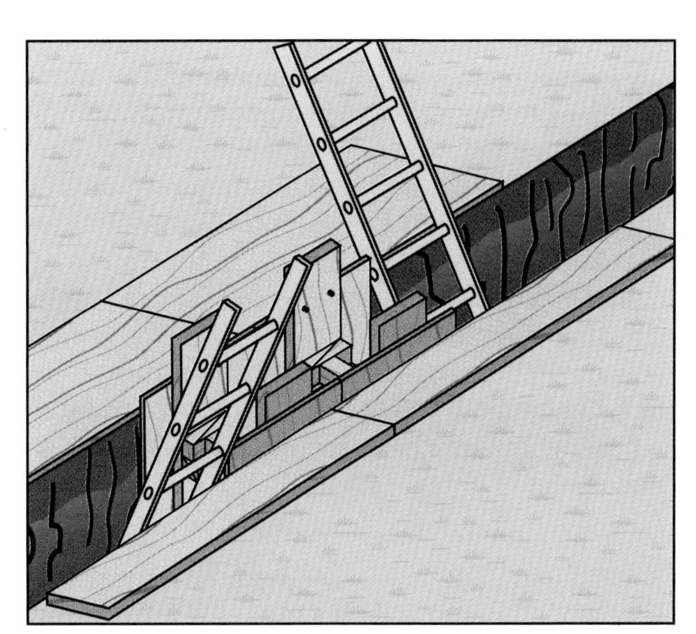

Figure 10.39 A second ladder is placed in the trench.

Figure 10.40 Plank sheeting is held in place with wales.

SCREW JACKS

Because of their ease of application, durability, and relatively low cost, screw jacks have replaced wooden cross braces in many shoring applications. Not to be confused with the type of screw jacks discussed in Chapter 6, this device consists of a swivel footplate with a stem that is inserted into one end of a length of 2-inch (50 mm) steel pipe (not to exceed 6 feet [1.8 m] in length) and a swivel footplate with a threaded stem that is inserted into the other end of the pipe (Figure 10.41). An adjusting nut on the threaded stem is turned to vary the length of the jack and to tighten it between opposing members in a shoring system. In trenches wider than 6 feet (1.8 m), wooden cross braces or large aluminum hydraulic shores have to be used.

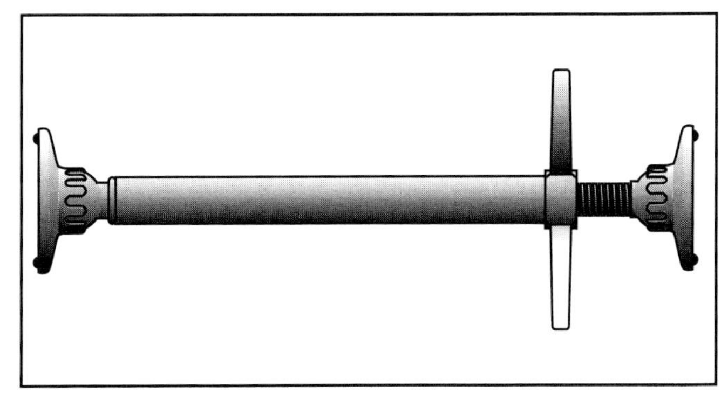

Figure 10.41 A typical trench screw jack.

When used as part of a trench shoring system, properly installed screw jacks simply replace the wooden cross braces; otherwise, the rest of the shoring system is constructed almost exactly as previously described. The only real difference is that when the distance between the uprights has been determined, the pipe should be cut 8 inches (200 mm) shorter than that measurement. Because the jacks are tightened with the adjusting nut, there is no need for the wedges sometimes used with wooden cross braces.

PNEUMATIC SHORING

Usually made of aluminum and sometimes called *air shores* or *rescue struts*, these devices are very similar to screw jacks except in the way in which they are lengthened to apply pressure against two opposing uprights. Pneumatic cylinders consist of a solid metal shaft within a cylinder, with a movable footplate attached (Figure 10.42). The cylinder has a connection through which air is introduced to extend the shaft. Like the screw jacks discussed earlier, air shores serve the same purpose as wooden timber cross braces in trench shoring systems. Air shores also have a positive locking collar used to keep the unit from retracting when the air pressure is released. When in position between two uprights, compressed air or nitrogen is introduced into the middle shore first to prevent the top or bottom of the panel from kicking out. The unit expands to the point of refusal, and the collar automatically (on some models) locks in place when the pressure is released; otherwise, the collar must be locked manually. The unit is then a fixed mechanical cross brace. The top shore is then installed within 1.5 feet (0.5 m) of the top of the trench.

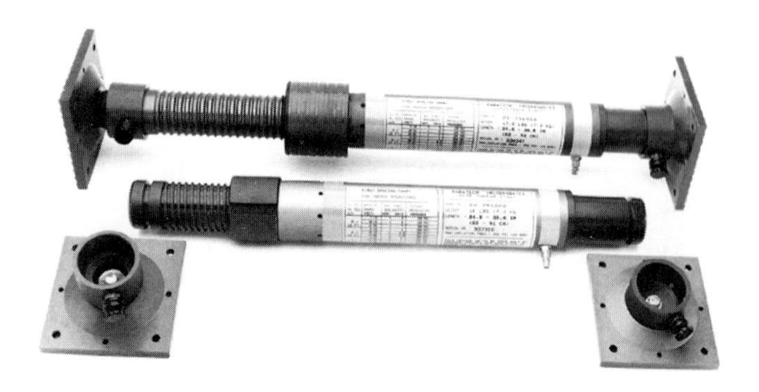

Figure 10.42 Typical pneumatic shores. *Courtesy of Paratech Incorporated.*

Finally, the bottom shore is installed within 2 feet (0.6 m) of the bottom of the trench.

The operating range of most pneumatic shores is from 100 to 350 psi (700 kPa to 2450 kPa). This range can be supplied by properly regulated SCBA cylinders, cascade systems, or air compressors.

ALUMINUM HYDRAULIC SHORES

Very similar in appearance and application to pneumatic shores, these devices are extendable aluminum cylinders operated by hydraulic pressure (Figure 10.43). When applied as a single unit, they serve the same function as pneumatic shores or wooden timber cross braces. However, aluminum hydraulic shores can also be used as part of fully self-contained shoring systems called *speed shores.*

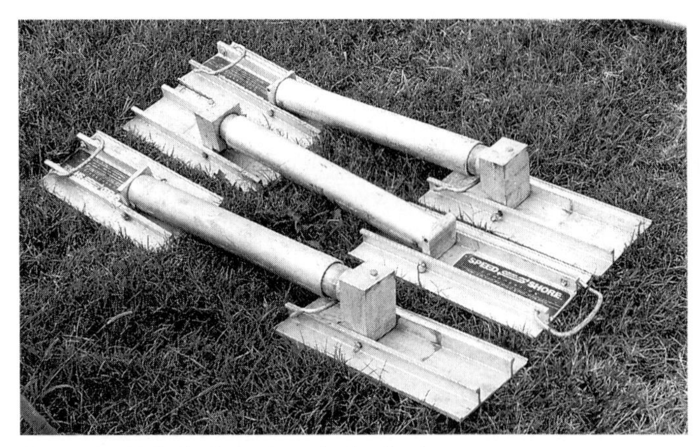

Figure 10.43 Typical aluminum hydraulic shores.

SPEED SHORES

In this configuration, both ends of two aluminum hydraulic shores are connected to aluminum rails up to 7 feet (2.2 m) long (Figure 10.44). Each connection between the shores and the rails is capable of moving 90 degrees so that the system can be collapsed or opened fully (Figure 10.45). The two cylinders are interconnected by a hydraulic line so that they are extended or retracted simultaneously (Figure 10.46). This allows these shoring systems to be installed in the trench without workers or rescuers having to be in the trench before they are installed.

When lowered into a trench, the top of the rail is positioned even with the top of the trench, and that automatically puts the top cylinder within 1.5 feet (0.5 m) of the top of the excavation, as required

Figure 10.44 Typical speed shores.

Figure 10.45 A speed shore is placed in a trench.

Figure 10.46 The interconnecting line is clearly visible.

by the OSHA regulations. As with any other shoring system, the maximum vertical spacing between cylinders is 4 feet (1.2 m), and the bottom cylinder should be no higher than 2 feet (0.6 m) from the bottom of the excavation.

TACTICAL CONSIDERATIONS

The tactical considerations involved in any trench rescue may be broken down into four distinct phases: assessment on arrival, pre-rescue operations, rescue operations, and termination.

Phase I: Assessment On Arrival

The assessment of the situation that must be made by the first-arriving unit can be subdivided into a primary and a secondary assessment.

PRIMARY ASSESSMENT

The primary assessment actually begins with the initial dispatch and continues during response and after arrival. The first-due officer must begin to formulate a mental picture of the situation based on the information provided in the dispatch, the time of day, the weather, and traffic conditions during response to the scene. This mental picture will be further refined and enhanced during the initial size-up.

Information gathering. The process is continued once on the scene by attempting to talk to the reporting party (RP) and/or other witnesses. Of critical importance in the primary assessment is gathering any available information about the number, condition, and location of victims:

- Have all workers been accounted for?
- How many victims are there?
- Is their location known?
- Are they fully or partially buried?
- How much time has elapsed since the cave-in?
- What has been done so far?

Decision making. The answers to these questions help the first-in officer make the first critical decision: Can the units on scene or en route handle the situation, or do additional units need to be called? If more resources are needed, they must be requested immediately to get them on scene as soon as possible (Figure 10.47). If he or she has not done so already, the first-in officer should assume formal command of the incident because the answers to the initial questions also form the basis for the incident action plan that the IC must begin to develop at this point.

Scene control. If the information gathered during the primary assessment confirms that a legitimate rescue emergency exists, the area surrounding the collapsed trench should be cordoned off as described in Chapter 2, Rescue Scene Management (Figure 10.48). Which areas are designated as the hot, warm, and cold zones depend on the size, nature, and complexity of the collapse situation. The boundaries of the zones will be affected by the hazards present.

Figure 10.47 The IC requests additional resources.

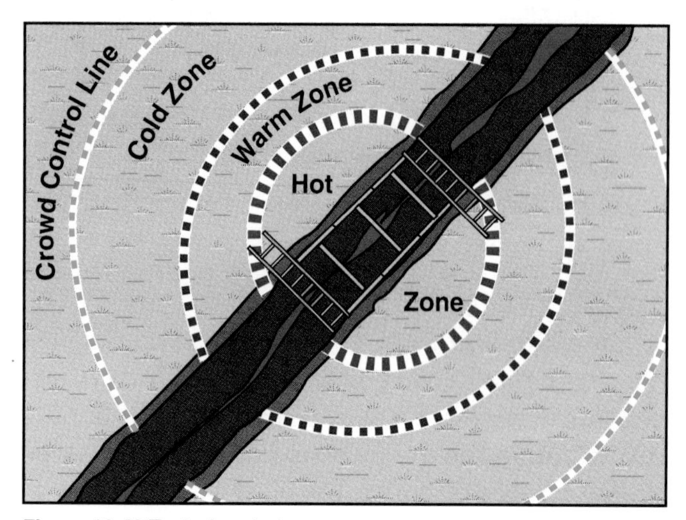

Figure 10.48 Typical control zones.

SECONDARY ASSESSMENT

The secondary assessment involves some reconnaissance of the scene to gather information about the trench, its condition, and the surcharge loads. All information gathered during both the primary and secondary assessments help determine the mode of operation.

Type of soil. Rescuers may not have the time to precisely type the soil; however, observing the spoil pile and the exposed trench wall can give a good visual indication of the type of soil involved. Squeezing a handful of soil from the top of the spoil pile may show how cohesive the soil is. Both of these assessments will suggest the type and amount of shoring that will have to be installed to make the trench safe for rescuers to work in. As mentioned earlier, rescuers should proceed as if they are working with Type C soil.

Condition of trench. The condition of the trench must also be assessed. Some of the questions that must be answered are as follows:

- What type of cave-in occurred?
- Did one or both walls collapse?
- What type of shoring will be needed?
- Will surcharge loads have to be moved?
- Are there hazards in and around the trench?

The answers to these questions provide the IC with more information on which to base decisions about what additional resources to request. Additional shoring equipment and materials may be needed. Portable pumps may be needed to keep the trench from filling with water.

Mode of operation. Finally, all of the information gathered during the primary and secondary assessments confirms the nature and extent of the rescue problem and helps the IC finalize the incident action plan. The information also helps the IC make one of the most important decisions affecting the action plan: whether it is reasonable to think that the victims are alive and that the operation should be conducted as a rescue or that victims could not be expected to have survived, and therefore the operation should be conducted as a body recovery.

Phase II: Prerescue Operations

During this phase, which may last from a few minutes to a few hours, all of the things necessary to make the rescue operation as safe as possible must be done. This process includes finalizing the incident action plan, gathering the necessary resources, monitoring and managing the atmosphere in the trench, and making the trench stable enough to safely enter.

INCIDENT ACTION PLANS

Most trench rescue operations are sufficiently localized that the incident action plan need not be in writing, but there *must be a plan*. Of primary importance is that the plan be finalized and communicated through channels to everyone involved in the operation.

While the original plan should be flexible enough to accommodate a certain amount of the unexpected, *a backup plan should be available* in case something completely unforeseen occurs to invalidate the original plan. If the information gathered during the primary and secondary assessments was somehow inaccurate or misleading or if some subsequent occurrence, such as a secondary collapse, changes the situation significantly, a secondary or backup plan should be ready to be implemented. If it suddenly becomes necessary to rescue the rescuers, that plan of operation should be ready.

GATHERING RESOURCES

Resources consist of personnel and equipment, and both are critically important to the success of the operation. If there are too few rescue personnel or if the personnel are insufficiently trained to perform as needed, the best equipment in the world will not get the job done. Likewise, the most highly trained and motivated rescuers will not be able to do what is necessary if they do not have the tools and equipment they need.

The resources gathered at the scene should reflect the incident action plan. But as stated earlier, if both the personnel and equipment in the initial response are insufficient for the rescue problem at hand, the IC must request additional resources as soon after arrival as possible. The sooner these resources respond, the sooner they will arrive on scene where they are needed. If the IC is initially unsure about the type and/or amount of equipment that will actually be needed, he or she should call for everything that *might* be needed. Resources that prove to be unnecessary can be returned to quarters while still en route or after they arrive on scene.

Personnel. Depending upon the nature and extent of the rescue problem, the number of rescue personnel needed varies. However, even a relatively simple rescue of one victim from a trench cave-in can involve eight to ten rescuers: a two-person rescue team, a two-person backup team, a safety officer, an incident commander, and three or four support personnel to handle shoring materials and operate on-scene equipment. This list does not include EMS personnel that may be needed to treat and transport the victim to a medical facility. Obviously, as the number of victims and the complexity of the rescue problem increase, the number of rescue personnel needed will increase proportionately.

Equipment. The amount and types of equipment needed also varies with the nature and extent of the rescue problem. The relatively simple scenario previously described could require a rescue unit, a shoring unit, a couple of engine and/or truck companies, or an engine and a truck company (depending upon staffing levels), plus a command vehicle and an ambulance. More complex incidents could also require specialized units such as lighting units, air units, communications units, or hazardous materials units. In addition to fire/rescue equipment, there may also be a need for specialized equipment, such as forklifts or cranes, from outside sources.

ATMOSPHERIC MONITORING

The atmosphere in the trench should be sampled before rescuers are allowed to enter. The area should first be checked for oxygen, then for flammables, and finally for toxic gases (Figure 10.49). The information obtained by sampling the atmosphere will help determine the need for mechanical ventilation. The atmosphere within the trench should be repeatedly monitored while victims or rescuers remain there. If the readings change for the worse after rescuers have entered the trench, rescuers should withdraw immediately until a safe atmosphere has been restored.

Figure 10.49 The atmosphere in the trench is checked.

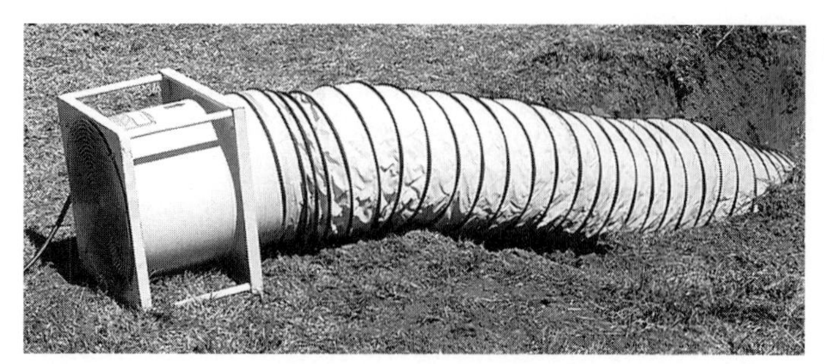
Figure 10.50 Fresh air being blown into a trench.

Figure 10.51 A rescuer consults a reference manual.

Oxygen concentration. The atmosphere within the trench may be normal or oxygen deficient; it is very unlikely to be oxygen enriched. If the atmosphere proves to be oxygen deficient (less than 19.5 percent), the trench will have to be mechanically ventilated before rescuers are allowed to enter (Figure 10.50).

Flammables. If monitoring devices indicate the presence of flammable gases or vapors in the atmosphere within the trench, rescuers must assume the atmosphere to be flammable or explosive. All nearby sources of ignition should be eliminated, and rescuers must not enter the trench until ventilation has reduced the detectable level of flammable gas or vapor to 10 percent of LFL or below. Check the MSDS, NFPA 49, *Hazardous Chemicals Data*; or NFPA 325M, *Fire Hazard Properties of Flammable Liquids, Gases, and Volatile Solids*, to determine these limits (Figure 10.51).

Toxics. If toxic gases or vapors are detected in the trench, rescue operations must be postponed and the trench mechanically ventilated while the source of the contamination is identified and eliminated. The atmosphere within the trench must be tested periodically until it is found to be within safe limits for the particular contaminant involved. Check the MSDS or the National Institute for Occupational Safety and Health (NIOSH) *Pocket Guide to Chemical Hazards* to determine these limits.

The nature and extent of toxic contamination will also affect the IC's decision to approach the problem either as a rescue or as a body-recovery operation. If it is the latter, there is no longer any need for haste, and every effort must be made to reduce the risks to rescue personnel.

WARNING

Because of the toxicity of some materials found in trenches, rescuers must approach these situations with an appropriate level of caution. The procedures described here should be followed without exception.

VENTILATION

Depending on a number of variables, such as the nature and configuration of the cave-in and the speed and direction of any wind, natural (nonmechanical) ventilation may or may not be effective. If natural ventilation is ineffective, some form of mechanical ventilation will have to be used. Mechanical ventilation in trenches takes two forms: positive and negative. The form chosen should be based on the situation and the equipment available on scene. Mechanical ventilation should be used in concert with any prevailing wind, not against it.

Positive ventilation. In the trench environment, positive ventilation involves blowing fresh air into the trench. Positive ventilation may be necessary to clear the trench of airborne contaminants or simply to provide some airflow within the trench in extremely hot conditions. The rate of airflow into the trench must be sufficient to ventilate it without stirring up dust. This may require intermittent use of the blower or fan.

Negative ventilation. In trench rescue, this form of ventilation involves using an explosion-proof smoke ejector or other exhaust fan with a flexible duct attached. The blower is positioned outside the trench, and the end of the duct is lowered into the trench to draw contaminants out and fresh air in (Figure 10.52).

Figure 10.52 Ducting allows this fan to exhaust contaminants.

PREPARING THE SCENE

Even after the necessary resources have been gathered on scene and the atmosphere within the trench has been determined to be safe to enter, other steps are still necessary to prepare the scene. As mentioned earlier, loose debris should be removed from near the lip of the trench, and plywood and/or planks should be laid on the ground to provide a clear, stable walk space. Preparing the scene may also involve identifying and mitigating hazards in and around the trench and providing fire protection. But preparing the scene for a trench rescue will *always* involve installing a protective system in the trench.

Mitigating hazards. Identifying and mitigating hazards in and around the trench can be an extremely important function. Shutting off leaking gas pipes can reduce the flammability hazard in and around the trench (Figure 10.53). Shutting off leaking water pipes can prevent the trench from filling with water and threatening the victim or rescuers with drowning. It can also reduce the chances of electrical shock by keeping the scene dry; however, under some circumstances it may be necessary to wet down the area to reduce the amount of airborne dust. Supporting exposed utility conduits can also reduce the likelihood of their adding hazards to the operation.

Figure 10.53 Gas leaking into the trench must be controlled.

Fire protection. It may be necessary to deploy one or more charged 1½-inch (38 mm) hoselines if there is an actual or a potential flammability hazard in or around the trench. Firefighters with appropriate fire extinguishers or charged hoselines may be needed to stand by if spark-producing power tools need to be used.

Shoring. As described in the rescue skills section, the trench walls in the area where a victim is, or is believed to be, must be shored up to make it safe for rescuers to work in and around the trench (Figure 10.54).

Ladders. To provide access into the trench and a means of egress from it, ground ladders should be placed at both ends of the protective system as described earlier in this chapter. In any case, rescuers should not have to travel more that 25 feet (8 m) to reach a ladder. The tops of the ladders should extend at least 3 feet (1.0 m) above the lip of the trench.

Figure 10.54 Rescuers quickly construct a shoring system.

Phase III: Rescue Operations

Once all of the preparations have been made, the process of actually locating and removing victims from a trench cave-in can begin. This phase of the operation obviously involves entry into the trench, and that necessitates the use of a personnel accounting system. In addition, this phase involves searching for victims, treating them prior to their extrication, and using retrieval systems.

PERSONNEL ACCOUNTABILITY

The purpose of an accountability system is to ensure that only those who are authorized and properly equipped to enter a collapsed trench are allowed to do so and that both their location and their status are known as long as they remain inside the trench. As is true of the incident action plan, the degree of formality of the personnel accounting system should reflect the nature, size, and complexity of the particular rescue problem at hand.

Some trench rescue incidents only require one or two rescuers in the trench at a time, they do not require respiratory protection while working in the trench, and they are close enough to be in constant visual and verbal contact with the attendant. In these cases, the intent of the accountability system can be met with a minimum amount of information being recorded. Merely writing down the team members' names, adding tracer lines to them, and recording their entry and exit times should be sufficient (Figure 10.55).

SEARCH/RESCUE

Once the shoring is in place, rescuers can enter the trench, but they must stay within the safe zone. They can begin searching for any fully buried victims by probing into the loose soil with the blunt end of a pike pole or other tool handle (Figure

Figure 10.55 Names are recorded before rescuers enter the trench.

10.56). If this does not reveal the victim's location, two rescuers should begin to carefully remove loose soil from the trench bottom starting as close as possible to the area where the victim is believed to be and working from each end toward the middle.

Once the victim is found, all efforts should be directed to removing enough soil to expose the victim's head and upper torso. The victim's vital signs should immediately be checked. If there is any chance of the victim surviving, oxygen should be administered. Whether a victim is fully or only partially buried, the rescuers should quickly remove loose soil using only their hands or trench tools to scoop it into buckets (no more than half full) (Figure 10.57). The buckets are then hoisted out of the trench (away from the rescuers) and emptied away from the trench. Local protocols should be followed in treating and transporting the victim to a medical facility. The victim should be packaged as described in Chapter 4, Rope Rescue, for extrication from the trench.

Another means of quickly removing loose soil from the bottom of a trench is through the use of a vacuum truck. Most cities and towns have these units for picking up leaves and debris from street gutters, but they also have been found to be very effective in trench rescue incidents.

Figure 10.56 A tool handle can help locate a buried victim.

Figure 10.57 Loose soil is scooped into a bucket.

Phase IV: Termination

The termination phase of a trench rescue involves such obvious elements as accounting for all rescue personnel and retrieving pieces of equipment used in the operation. But it also involves less obvious elements such as investigating the cause(s) of the incident, releasing the scene back to those responsible for it, and conducting critical incident stress debriefings (CISD) with members of the rescue teams.

EQUIPMENT RETRIEVAL

Depending on the size, complexity, and length of time involved in the operation, the job of retrieving all of the various pieces of equipment used may be very easy or very difficult and time-consuming. Under some circumstances, it can also be quite dangerous.

Identifying/collecting. The process of identifying and collecting pieces of equipment assigned to the various pieces of apparatus on scene is much easier if each piece of equipment is clearly marked (Figure 10.58). However, it may be necessary for the driver/operators of rescue units and other pieces of apparatus on scene to conduct an inventory of their equipment prior to leaving the scene.

Dismantling. Dismantling a shoring system can be a dangerous process if not done correctly, so it is extremely important that those assigned to this task not become complacent once the rescue has been accomplished. If the process is done correctly, there should be no need to abandon any tools or equipment in the trench. The shoring system should be dismantled in the reverse order of its construction. When dismantling a shoring system, rescue personnel must always work within the safe zone, just as they did when constructing the system.

Figure 10.58 A clearly marked piece of equipment.

NOTE: The shoring system may have to be left in place until OSHA representatives complete their investigation of the incident.

INVESTIGATION

All trench rescue incidents should be investigated at some level. A departmental investigation should be conducted for purposes of reviewing and critiquing the operation. If any employees were injured or killed in the incident, it will also be investigated by the Occupational Safety and Health Administration (OSHA) and perhaps by other entities such as the employer's insurance carrier. Because trench cave-ins rarely involve a crime, except perhaps negligence, law enforcement agencies usually will not be involved in investigating these incidents.

RELEASE OF CONTROL

Once rescuers respond to the scene of a trench collapse, they assume control of the excavation and the immediate surrounding area. Within certain limits, they can deny access to anyone, including the owner of the property and the excavation contractor. Legitimate members of the news media have certain constitutionally protected rights of access, but the interpretation of these rights varies from state to state and from country to country. Rescuers should be guided by local protocols when dealing with media representatives.

The process of releasing control of the excavation back to the responsible party is sometimes not as straightforward as it might seem. The responsible party should be escorted on a tour of the excavation, or as close to it as possible consistent with safety, and should be given an explanation of any remaining hazards. If the trench is still too hazardous to leave unattended, the contractor may be required to post a security guard, erect a security fence around the hazard, or both. Before the site is released, the department may require the responsible party to sign a written release that describes the hazards and stipulates the conditions the responsible party must meet.

CRITICAL INCIDENT STRESS DEBRIEFING

The victims in trench rescue incidents sometimes suffer extremely gruesome and horrific injuries as a result of ill-considered rescue attempts by coworkers. Therefore, the members of the rescue teams and any others who had to deal directly with such victims should be *required* to participate in a critical incident stress debriefing (CISD) process. Because individuals react to and deal with extreme stress in different ways — some more successfully than others — and because the effects of unresolved stresses tend to accumulate, participation in this type of process should not be optional.

The process should actually start *before* rescuers enter the scene if it is known that conditions exist there that are likely to produce psychological or emotional stress for the rescuers involved. This is done through a prebriefing process wherein the rescuers who are about to enter the trench are told what to expect so that they can prepare themselves (Figure 10.59).

If rescuers will be required to work more than one shift in these conditions, they should go through a minor debriefing, sometimes called "defusing," at the end of each shift. They should also participate in the full debriefing process within 72 hours of completing their work on the incident.

Figure 10.59 Rescuers are briefed before they enter the trench.

Chapter 10 Review

Directions

The following activities are designed to help you comprehend and apply the information in Chapter 10 of **Fire Service Rescue**, Sixth Edition. To receive the maximum learning experience from these activities, it is recommended that you use the following procedure:

1. Read the chapter, underlining or highlighting important terms, topics, and subject matter. Study the photographs and illustrations, and read the captions with each.
2. Review the list of vocabulary words to ensure that you know the chapter-related meaning of each. If you are unsure of the meaning of a vocabulary word, look up the word in the glossary or a dictionary, and then study its context in the chapter.
3. On a separate sheet of paper, complete all assigned or selected application and review activities before checking your answers.
4. After you have finished, check your answers against those on the pages referenced in parentheses.
5. Correct any incorrect answers, and review material that was answered incorrectly.

Vocabulary

Be sure that you know the chapter-related meanings of the following words:

- aquifer *(300)*
- cohesive *(292)*
- constituent *(292)*
- dehydration *(296)*
- dry strength *(295)*
- fissure *(292)*
- granular *(292)*
- hypothermia *(301)*
- lateral *(296)*
- loam *(292)*
- loess *(293)*
- marine *(291)*
- matrix *(292)*
- plasticity *(295)*
- riparian *(291)*
- saturated *(296)*
- sedimentary rock *(292)*
- sloughing *(297)*
- soil *(291)*
- spalling *(294)*
- strata *(292)*
- surcharge *(297)*
- undermining *(297)*
- vertical *(296)*
- volatile *(300)*

Application Of Knowledge

1. Choose from the following trench rescue procedures those appropriate to your department and equipment, or ask your training officer to choose appropriate procedures. Mentally rehearse the more complex procedures, or practice these procedures under your training officer's supervision.
   - Sample and monitor the atmosphere in a trench. *(302, 303)*
   - Mechanically ventilate a trench. *(313)*
   - Working as a member of a team, construct a timber shoring system that is fully compliant with OSHA regulations for Type C soils. *(302-308)*
   - Working as a member of a team, install a self-contained aluminum hydraulic shoring system that is fully compliant with OSHA regulations for Type C soils. *(302-308)*
   - Brace a trench shoring system with screw jacks. *(307, 308)*
   - Brace a trench shoring system with air shores. *(308)*
   - Brace a trench shoring system with speed shores. *(309, 310)*
   - Terminate a trench rescue operation. *(315)*
2. Obtain a copy of and read OSHA 29 CFR 126, Subpart P — Excavations, Sections 650-652. *(291)*

Review Activities

1. Identify the following abbreviations and acronyms:
   - CDX *(303)*
   - CISD *(316)*
   - IC *(310)*
   - LFL *(300)*
   - MSDS *(312)*
   - NFPA *(312)*
   - NIOSH *(291)*
   - OSHA *(316)*
   - PPE *(301)*
   - RP *(309)*
   - tsf *(293)*
2. Define *trench* per OSHA 29 CFR 126. *(291)*
3. Define *cave-in* per OSHA 29 CFR 126. *(291)*
4. Discuss the following soil types in regard to their composition and stability: *(292)*
   - Rock
   - Cemented
   - Cohesive
   - Granular
   - Loam
5. State the typical weight range of a cubic foot of soil. *(292)*

6. List the criteria upon which soils are classified into A, B, or C types. *(292, 293)*
7. Distinguish among the following soil types:
   - Type A *(293)*
   - Type B *(293)*
   - Type C *(293, 294)*
8. List the criteria that preclude a soil being classified as Type A. *(293)*
9. List criteria that must be met by Type B soils. *(293)*
10. List criteria that must be met by Type C soils. *(294)*
11. Explain what a rescuer would look for in a visual analysis of the soil at a trench rescue excavation. *(294)*
12. Explain how the following manual soil tests are performed and what they indicate: *(295)*
    - Plasticity manual test
    - Dry strength manual test
    - Thumb penetration test
13. Describe the following causes of trench failure:
    - Vertical/lateral forces *(296)*
    - Soft pockets *(296)*
    - Layered soils *(296)*
    - Saturated soils *(296)*
    - Vibration *(296)*
    - Surcharge loads *(297)*
14. State the OSHA requirements for location and size of the spoil pile. *(297)*
15. State the OSHA definition of *trench edge. (297)*
16. State OSHA weight restrictions for timber shoring and for either timber or aluminum hydraulic shoring. *(297)*
17. Describe the following types of trench failure:
    - Lip cave-in *(297)*
    - Wall slough-in *(297)*
    - Wall shear *(298)*
18. State the *minimum* trench depth at which OSHA requires worker protection from cave-ins. *(298)*
19. Identify each of the following methods of protecting workers from trench cave-ins: *(298, 299)*
    - Sloping/benching
    - Shoring
    - Shielding
20. Describe trench rescue hazards in the following categories:
    - Physical *(299, 300)*
    - Atmospheric *(300)*
    - Environmental *(301)*
21. List common signs of trench wall instability. *(299)*
22. State the most common cause of oxygen deficiency in trenches. *(300)*
23. Provide examples of team gear needed for trench rescue. *(301, 302)*
24. Identify the following materials and equipment used in trench rescue:
    - Air shores *(308)*
    - Duplex nails *(303)*
    - Pocket penetrometer *(296)*
    - Screw jack *(307)*
    - Sheeting *(303)*
    - Speed shores *(308)*
25. Identify the following items and actions specific to trench rescue:
    - Cleat *(304)*
    - Cross braces *(303, 304)*
    - Joist *(304)*
    - Shear vane *(296)*
    - Spoil pile *(297)*
    - Stabilization *(303)*
    - Team gear *(301)*
    - Thumb penetration *(295)*
    - Toenail *(306)*
    - Unconfined compressive strength *(293)*
26. Outline the four phases of tactical considerations for trench rescue. *(309-316)*
27. List questions that should be asked during the primary assessment. *(309)*
28. List two quick soil assessments rescuers can perform during the secondary assessment when they do not have time to precisely type the soil. *(310)*
29. List questions that should be asked during the secondary assessment. *(310)*
30. Explain why most trench rescue operations require no written plan. *(311)*
31. Compare equipment resources that may be required for a simple trench rescue with resources that may be required for a more complex trench rescue. *(311)*
32. List two resources the rescuer can use to determine safe levels of flammable gas or vapor allowed in the trench at the time of rescuer entry. *(312)*
33. List the resource that the rescuer can use to determine safe levels of toxic gases or vapors allowed in the trench at the time of rescuer entry. *(312)*
34. Distinguish between positive and negative mechanical ventilation. *(313)*

35. Outline the steps the IC would take to prepare the scene of a trench rescue. *(313, 314)*
36. Describe ways in which rescuers search for and uncover buried victims. *(314, 315)*
37. Describe cautions and guidelines for dismantling a shoring system. *(315, 316)*
38. Describe the variables for releasing control of the scene back to the owner or other responsible party. *(316)*
39. Discuss critical incident stress, prebriefings, and "defusings." Why should participation in a full critical incident stress debriefing within 72 hours of the incident be mandatory for all involved rescuers? *(316)*

Questions And Notes

11

Special Rescues

Chapter 11
Special Rescues

INTRODUCTION

The preceding chapters of this manual covered a wide variety of rescue situations and skills. However, there are still a few specific situations to which rescuers may be called that either are not covered in those chapters or are not covered in sufficient detail because of their unique nature. The rescue situations covered in this chapter are those involving energized electrical lines and equipment, natural gas, caves and mines, and silos and grain vessels.

RESCUES INVOLVING ELECTRICITY

Rescues involving energized electrical lines or equipment are some of the most common situations to which rescuers are called. But the frequency with which these situations occur should not lull rescuers into a false sense of security — these situations can be extremely dangerous. Improper actions by rescue personnel can result in their being injured or killed instantly. Whenever rescuers respond to any situation involving electricity, they should *always* do the following:

- Assume that electrical lines or equipment is energized.
- Call for the power provider to respond.
- Control the scene.

Electrical Service Components

Regardless of whether electrical power is provided by a private or public entity, there are five major components to the electrical service system: generating facilities, transmission lines, substations, distribution lines, and service connections to customers. Rescue incidents can involve any of these components.

GENERATING FACILITIES

Generating facilities are the facilities where electrical power is generated. They may be hydroelectric dams, oil- or coal-fired boiler installations, or nuclear power plants. These facilities are usually well staffed, well run, and well maintained, so incidents requiring a rescue response are very infrequent. Facility employees are usually well trained in handling electrical emergencies, so that portion of any rescue incidents in these facilities should be stabilized by the time rescuers arrive. Rescues in these facilities may involve extreme heights, elevated temperatures, and radioactive materials, but with the exception of radiation incidents, these rescue situations may be no different from those discussed in earlier chapters. However, rescuers should consult facility personnel about machinery and processes within the facility, and must defer to their expertise when dealing with radiation-related rescues.

One difference between other rescue situations and those in steam-producing power-generating plants is the possibility of high-pressure steam leaks. At 400° to 500°F (204° to 260°C), and up to 2,400 psi (16 800 kPa), these invisible jets of steam can cut a person in two.

If personnel must enter such a facility to effect a rescue, they should be accompanied by plant personnel. If plant personnel are not available to accompany them, a rescuer should precede the rest of the crew with a broomstick or a pike pole to which a cloth has been attached (Figure 11.1). If there is a steam leak, it will cause the cloth to move or billow, or it may cut the wooden handle in two. In either case, rescuers should proceed no farther until the steam leak is stopped by plant personnel.

Figure 11.1 A rescuer checks for high-pressure steam leaks.

TRANSMISSION LINES

Also called "high tension lines" or "trunk lines," *transmission lines* carry the electricity produced at generating facilities to various substations where its voltage is stepped down for customer use. The transmission voltage in these lines can be 330,000 volts or higher. The lines are supported by metal or wooden towers (Figure 11.2). These towers are considerably taller than the utility poles that support the power lines to homes and businesses.

The most common rescue situations involving these towers and lines occur when unauthorized people climb the towers and cannot get down without assistance or when lines fall or towers collapse. While these towers are very strong, if they are struck by a fast-moving vehicle or a heavy metal crane, they can collapse. In any case, unless the fallen line starts a vegetation fire, rescuers should follow the three guidelines given earlier: Assume that the line is energized, call the power provider, and control the scene to prevent others from being injured. If a fire was started, it should be allowed to burn away from the downed line a distance equal to one span between towers before it is attacked (Figure 11.3).

SUBSTATIONS

These facilities are located at various points along the electrical system. Substations use transformers to either step the voltage up from the generating station for input into the transmission system or to step the high voltage from the transmission lines down to levels that are appropriate for local distribution lines. Although few remain in

Figure 11.2 Typical transmission towers.

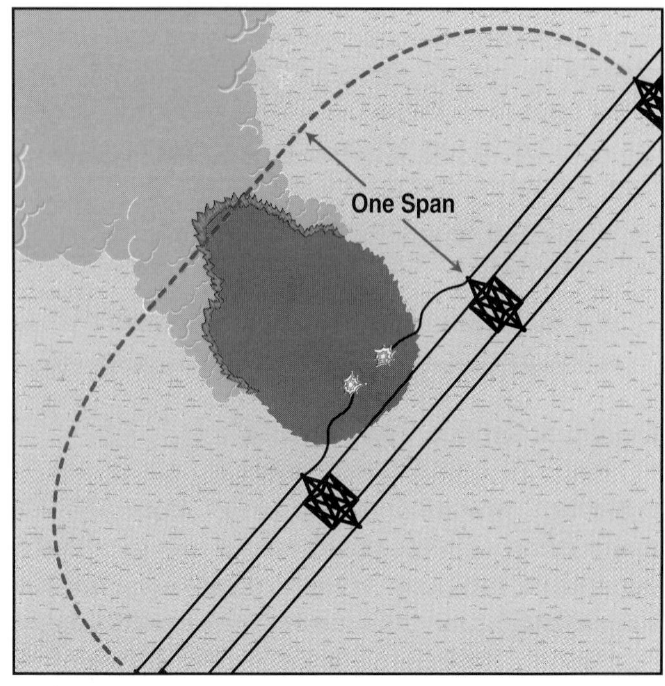

Figure 11.3 Firefighters should stay well clear of energized power lines.

service, transformers filled with oil that contains polychlorinated biphenyls (PCBs), a long-term carcinogen, may still be found in these facilities.

Most substations are open-air facilities with only a fence around them (Figure 11.4). Others are enclosed in buildings, especially in urban areas. The voltages that are common to substation equipment make these facilities much too dangerous for rescuers to enter without the assistance of personnel from the power utility. Rescuers who must enter these facilities should avoid using any metal tools to reduce the possibility of electrocution by contact with energized wiring or equipment.

Figure 11.4 A typical outdoor substation.

Figure 11.5 Typical distribution lines.

Figure 11.6 Pole-mounted transformers.

DISTRIBUTION LINES

Also called "primaries," *distribution lines* carry electricity from substations to the local neighborhoods (Figure 11.5). Voltage in distribution lines (2,400 to 34,000 volts) is generally much lower than in transmission lines. These mid-range voltages are stepped down by pole-mounted transformers to 120 or 440 volts for consumer use (Figure 11.6). Rescue situations involving these lines generally result from electric utility personnel or others coming into contact with these lines aloft or on the ground. There is a remote possibility of these transformers containing PCBs.

In urban areas, distribution lines are commonly placed underground as well as strung between poles. Rescuers should *never* enter an underground utility space without the assistance of utility personnel.

SERVICE CONNECTIONS

From the distribution lines, whether underground or on poles, service connections are made to consumers. Except for accidents that occur in excavations, most rescue incidents involving service connections are the result of someone coming in contact with a drop from a pole to the customer facility (Figure 11.7). A frequent point of contact between people and electricity is at the electric meter (Figure 11.8).

Figure 11.7 A typical customer drop.

Figure 11.8 A typical electric meter.

Handling Electrical Rescues

It is essential for rescuers to remember that they did not cause the accident, that they are not responsible for the victim being in that situation, and that they are not obligated to sacrifice themselves in a heroic attempt to save the victim and especially not in an attempt to recover a body. In fact, it is irresponsible and unprofessional for rescuers to take unnecessary risks that might result in their being incapacitated by an injury and therefore unable to perform the job for which they have been trained. It is not the function of the fire/rescue service to add victims to the situation. Having said this, we can now discuss the safe and proper ways in which rescuers should deal with rescues involving energized electrical wires or equipment.

GENERAL GUIDELINES

Under most conditions, the best course of action is for rescuers to follow the three general guidelines mentioned earlier: (1) assume that wires or equipment are energized, (2) call for the electric utility to respond, and (3) control the scene. Following these guidelines usually stabilizes the situation and prevents others (including rescuers) from becoming additional victims.

SCENE CONTROL

The area surrounding the rescue scene should be cordoned off as described in Chapter 2, Rescue Scene Management (Figure 11.9). Which areas are designated as the hot, warm, and cold zones will depend on the size, nature, and complexity of the rescue situation. The boundaries of the zones may be affected by the presence of other hazards such as spilled or leaking hazardous materials.

Unless absolutely necessary, metal tools should not be allowed into the hot zone, and in no case within 10 feet (3.1 m) of energized wiring or equipment. Rescuers can use AC (alternating current) power locators to determine if power is flowing through a wire, control panel, or other device (Figure 11.10).

USING DIELECTRIC EQUIPMENT

Some fire/rescue units are equipped with dielectric equipment such as "hot sticks," insulated wire cutters, and lineman's rubber gloves (Figure 11.11). However, for this equipment to be safe to use, it must be carefully maintained and stored and requires regular monthly testing. In addition, specific training in the safe handling of this equipment is needed.

CAUTION: Fire/rescue units not meeting these testing and training requirements should not allow their personnel to use this type of equipment.

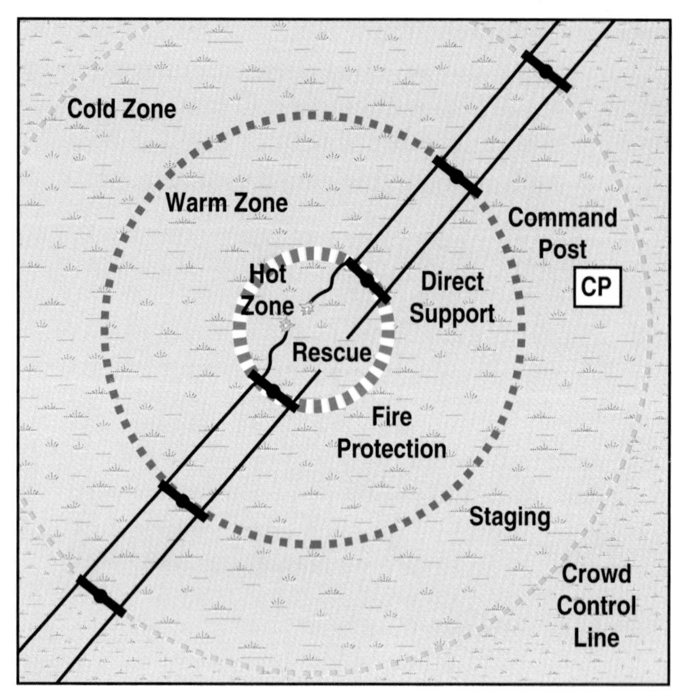

Figure 11.9 Typical control zones.

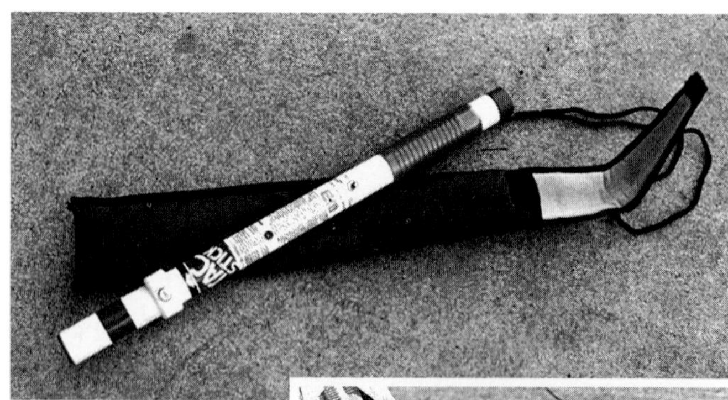

Figure 11.10 One type of AC power locator. *Courtesy of Keith Flood.*

Figure 11.11 A utility worker in lineman's gloves.

However, if all of the maintenance and training requirements are met, this type of equipment can be extremely valuable under the right conditions. The right conditions would be the following: (1) the power has been shut off, (2) the scene is not wet, and (3) someone's life is in jeopardy if the wire is not moved immediately. With a victim in contact with an electrical wire under these conditions, a hot stick can be used to move the wire away from the victim or move the victim away from the wire.

Under the same conditions, insulated wire cutters can be used to cut the wire on both sides of the victim. However, when the wire is cut, it may suddenly recoil in both directions from the cut. Depending upon the gauge of the wire and how violently it recoils, bystanders could be injured if struck by the recoiling wire.

If a victim is in contact with wiring inside the housing of a piece of equipment, the power to the unit can usually be shut off at the circuit breaker panel. If not, a hot stick might be used to pull the victim free, or insulated wire cutters might be used to completely sever the wires providing power to the unit.

WARNING

Standard fire service pike poles with wooden or fiberglass handles are not dielectric equipment and should never be used to move energized electrical wires.

APPROACHING DOWNED WIRES

Electrical wires on the ground can be dangerous without even being touched. Downed electrical lines can energize wire fences or other metal objects with which they come in contact. When an energized electrical wire comes in contact with the ground, current flows outward in all directions from the point of contact. As the current flows away from the point of contact, the voltage drops progressively (Figure 11.12). This is called *ground gradient*. Depending upon the voltage involved and other variables, such as ground moisture, this energized field can extend for several feet from the point of contact. A rescuer walking into this field can be electrocuted because of the differing potentials between the rescuer's feet (Figure 11.13). To avoid this hazard, rescuers should stay away from downed wires a distance equal to one span between poles until they are certain that the power has been shut off.

ENTERING ELECTRIC UTILITY SPACES

When a victim is located inside an underground electric utility vault or manhole, rescuers will have to enter the space either to rescue the victim or to

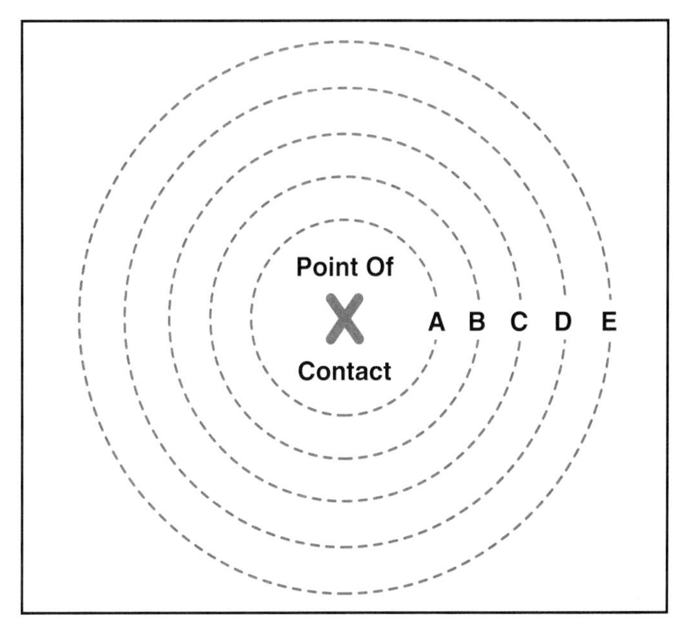

Figure 11.12 Voltage drops as it spreads away from the source.

Figure 11.13 Rescuers must approach downed wires with caution.

recover the body. In addition to following all of the procedures for entering underground confined spaces as described in Chapter 8, Confined Space Rescue, rescuers must also contend with the high-voltage electrical equipment inside. Unless there is some compelling reason to do otherwise, rescuers should apply the general guidelines and wait for utility personnel to arrive.

RESCUES INVOLVING NATURAL/LIQUEFIED PETROLEUM GAS (LPG)

Even though natural gas (methane) and LPG are nontoxic, like carbon dioxide, they can asphyxiate by displacing the oxygen within a confined space. The other significant hazard associated with these gases is their flammability. In their natural state, neither methane nor LPG have an odor. When supplied by a public or private utility, both natural gas and LPG are odorized with ethyl mercaptan to make gas leaks easier to detect.

Natural Gas Service Components

Methane is a natural by-product of petroleum production, and large quantities of the gas are stored in former oil wells that are no longer productive. The gas utility draws gas from this storage as needed and odorizes it before pumping it through large-diameter steel transmission lines to compressor stations. While the industry considers any gas pressure over 125 psi (875 kPa) to be high pressure, the gas being pumped to the compressor stations can be anywhere from 375 to 1000 psi (2 625 kPa to 7 000 kPa).

COMPRESSOR STATIONS

At the compressor stations, the gas pressure is reduced from the high transmission pressure down to a distribution pressure of 30 to 60 psi (210 kPa to 420 kPa) (Figure 11.14). The gas is then transferred to local storage tanks.

Figure 11.14 A gas compressor station. *Courtesy of Keith Flood.*

STORAGE TANKS

Natural gas storage tanks can be any of several different designs and capacities (Figure 11.15). From the storage tanks, gas is distributed throughout the service area, usually through plastic mains, to the individual customers.

Figure 11.15 A floating gas storage tank. *Courtesy of New York (NY) Fire Department.*

SERVICE CONNECTIONS

From the distribution mains, the gas is piped to the individual customer through a small service connection, usually wrapped steel pipe. At the customer's meter, the gas pressure is further reduced to approximately 0.25 psi (2 kPa) for use within the building or facility (Figure 11.16).

Handling Natural Gas Rescues

Rescues involving natural gas are sometimes a result of an accumulation of natural gas exploding within a structure, with or without a subsequent fire. These incidents are often caused by someone attempting to commit suicide by turning off the pilot lights and turning on all of their gas appli-

Figure 11.16 A typical gas meter.

ances. If there is a fire after the explosion, the procedures outlined in Chapter 5, Fireground Search and Rescue, should be followed. The gas-fed fire should not be completely extinguished until the gas is shut off. If there is no fire, the incident should be handled as described in Chapter 6, Structural Collapse Rescue.

Other rescues involving natural gas result from the occupant(s) of a confined space being asphyxiated because natural gas has displaced the oxygen from the space. Because the odorant makes conscious individuals aware of the presence of natural gas, these situations most often develop when people are sleeping. However, in an industrial setting with other strong odors present, victims can be overcome while awake and functioning. Regardless of the cause, this is one situation where rescuers should wear structural turnout gear because of the possibility of the gas being ignited (Figure 11.17). To reduce the likelihood of ignition, rescuers should attempt to shut off the flow of gas and eliminate all possible sources of ignition in the immediate area (Figure 11.18). They should then ventilate the space to make it safe to enter (Figure 11.19).

As in rescues involving other utilities, the gas provider should be called to the scene of rescues involving natural gas. If the release of natural gas was due to a leak in the provider's underground piping, they will have to excavate to clamp the pipe on both sides of the leak because there are few valves in underground gas distribution systems.

Figure 11.17 Rescuers should wear full protection.

Figure 11.18 The flow of gas must be shut off.

Figure 11.19 The space must be ventilated.

LPG Rescues

LPG pipelines can be found anywhere in North America. The handling and distribution of LPG is very similar to that of natural gas, and the gases are very similar except that LPG is heavier than air, so it will tend to collect in depressions and other low areas. The rescue procedures are the same for LPG as for natural gas, and as with any other rescue exposure, the key to successfully handling LPG rescues is pre-incident planning.

CAVE RESCUE

Although firefighters may be called when someone is lost or injured in a cave, they are usually not trained or equipped to perform these rescues. Rescue from caves must be done by those who are familiar with the uniquely hostile environment of a cave and who have the training and equipment needed. Unless they are specially trained to operate in these environments, fire/rescue personnel will usually confine their activities to aboveground support of other cave rescue personnel.

As in most other types of rescue, typical structural turnout gear is usually inappropriate for use in caves and can even be hazardous. Turnouts are often too bulky for the narrow passages, the rubber boots do not provide safe footing, and the helmet is too large and cumbersome.

When called to effect a rescue from a cave, the fire/rescue unit should call their county, state, or provincial Office of Emergency Services to request a cave rescue coordinator. Assistance can also be obtained from the Division of Mine Safety by contacting the nearest District Office for Mine Safety and Health. The district office will assess the need and contact the Chief of Mine Emergency Operations to dispatch skilled cave rescuers.

The following section was prepared for IFSTA by the National Cave Rescue Commission (NCRC). It is included to give rescuers some idea of the techniques and problems involved if called upon to assist in a cave rescue.

Types Of Caves

A rescue company whose district includes caves should have a rescue plan that provides as much information as possible about all the known caves, and that includes standard procedures for calling in trained rescuers. Information concerning local caves is best obtained from the nearest grotto (chapter) of the National Speleological Society. The telephone number of the nearest chapter can be obtained by calling the national headquarters at (205) 852-1300. The members of the local grotto can provide details about local caves that would otherwise take considerable time to research.

SOLUTION CAVES

Solution caves are the most commonly explored and therefore the ones from which there are the most rescues. There are two basic kinds of passages in solution caves: phreatic and vadose. *Phreatic passages* are those formed when the cave was below the water table. Phreatic passages tend to be large, smooth-walled, and mostly level and dry. *Vadose passages* are formed by fast moving streams above the water table. They tend to be very irregular in shape with narrow, fluted passages that often contain active streams. Vadose passages are more difficult to negotiate than phreatic passages.

LAVA TUBE CAVES

Lava tube caves are formed by lava flows on the earth's surface. The lava on the outside of the flow cools and solidifies while the lava in the middle is still molten. The molten lava flows out of the tube leaving a void. These tubes may be stacked atop one another as the result of successive lava flows.

The most significant problem associated with lava tubes is their instability. Almost from the moment the lava core flows from the tube, the tube begins to collapse. This process continues over time until the cave is filled in. Rescuers need brighter lights in lava tubes than in other types of caves because the walls have a rougher, less reflective surface.

TALUS CAVES

Talus caves are formed by fallen rock collected between ridges or at the base of mountains. Although not caves in the strict sense of the word, they have many of the same characteristics of caves, and people can get lost or hurt in them. The passages of a talus cave are the voids between the fallen rocks, and these voids may have streams flowing through them. The irregularity of the passages and the instability of the rocks cause the most problems for cavers and rescuers in talus caves.

Cave Rescue Problem

Most caves have the same basic characteristics that translate into rescue problems: darkness, water, passage irregularities, air movement and temperature, atmosphere, and complexity. The rescuer must be prepared to overcome each of these problems if the victim and the rescuer are to come out of the cave alive.

DARKNESS

The darkness in a cave is absolute, and rescuers must carry lights with them. The rescuer should carry three independent sources of light. The primary light should be a helmet-mounted electric lamp similar to those used in mining (Figure 11.20). A helmet-mounted lamp allows both of the rescuer's hands to be free, which is necessary for negotiating passages, handling the victim, and ascending and descending. The lamp should be able to provide light without needing attention for eight to ten hours. Rescuers should carry spare bulbs and batteries.

The secondary and tertiary light sources can be a good, waterproof flashlight and a candle. There should be spare bulbs and batteries for the flashlight and matches for the candle. The candle can be used while waiting (there may be a lot of waiting during a cave rescue) or while repairing the other light sources. If, for some reason, all light sources are lost or become inoperable, rescuers should stay in one place and not try to move around as this might cause them to fall or to stray even farther from rescue.

Figure 11.20 An intrinsically safe helmet lamp.

Carbide lamps, often used during cave exploration, should not be used during cave rescue. The open flame may burn the victim or the rope, and the carbide must be changed frequently.

WATER

Most caves are natural conduits for water, especially those formed by water. The temperature of the water in caves is usually about 55°F (13°C); immersion can cause the onset of hypothermia within minutes unless the person is protected by a wet suit.

Cavers are sometimes trapped by flooding in a cave. The water can come from a sudden downpour, snowmelt, or other sources, filling the passages and preventing escape. When cut off by rising water, experienced cavers usually seek a high spot downstream from a tight constriction. The constriction limits the flow of water and will often leave an air pocket in which cavers can stay for some time.

In most cases of flooding, rescuers must wait for the water to subside, which can take from a few hours to a few days. Once the ground is saturated, even a slight rain could cause reflooding that could trap the rescuers. Specially trained cave divers will sometimes be needed to rescue cavers trapped by water. The NCRC can get these highly trained scuba divers to the scene quickly.

PASSAGE IRREGULARITIES

Cave passages can be very irregular. They can range from large, smooth-floored "subway tunnel" passages to very low, narrow, sinuous (winding) passages with ankle-deep water on the floor. Some passages are so small that a caver must exhale to squeeze through. People with no history of claustrophobia can suddenly manifest it in these situations.

It is even possible that an injured victim will have to stay underground until the injury heals enough to allow the victim to help while being moved through the narrow, serpentine passages of the cave. If the victim's exact location is known, an alternative is to use the mine rescue drill operated by the Mine Emergency Operations Division of the Mine Safety and Health Administration. The mine rescue drill can cut a 2-foot (0.6 m) diameter shaft through solid limestone at a rate of 50 feet (15 m) per day.

With the exception of lava tube caves, most caves are formed of rock that tends to be stable. Although the floor may be covered with small pieces of rock called *breakdown*, most of these pieces fell hundreds or even thousands of years ago. Because of this stability, shoring and roof jacks are seldom needed in caves.

Vertical drops in caves range from a few feet to more than 500 feet (154 m). Such drops can be negotiated safely only by those skilled in ascending and descending with ropes. Rescuers must also be familiar with cable ladder and belaying (safety rope) techniques and with the use of natural and artificial anchors and their riggings. During a rescue operation is not the time to try to learn these skills.

The irregularities of cave passages can make a rescue very difficult. For example, rescuers might have to lift a stretcher 100 feet (30 m) up a free drop. Then, while hanging on a rope to the side, push the stretcher through a hole 2 feet (0.6 m) wide and 14 inches (400 mm) high. After that, the stretcher might have to pass through several hundred feet of a 10-inch (250 mm) high passage that never exceeds 2 feet (0.6 m) in width.

TEMPERATURE

A cave's temperature is generally the average temperature of the surrounding countryside. In North America, the temperature inside most caves is between 52°F and 60°F (11°C and 15°C). All of the rocks, mud, and water within the cave will eventually assume this temperature, and so will an unprotected victim.

The low temperature inside a cave is the greatest environmental hazard to a lost or injured caver because it can cause hypothermia. A caver loses body heat through conduction into the rocks, mud, and water; through radiation and convection into the surrounding atmosphere; and by respiration. An uninjured caver can keep warm by moving about and by eating carbohydrates, but an injured person is likely to have to lie relatively motionless on the rock, with little means of keeping warm.

Hypothermia caused by conduction, radiation, and convection can be treated by putting the victim into a neoprene exposure bag that covers every part of the body but the face. However, hypothermia aggravated by heat loss through respiration is best treated with a warm-gas inhalator.

ATMOSPHERE

With few exceptions, the air in a cave will be as pure as that of the surrounding area. Unlike mines, caves ventilate themselves very well with changes in barometric pressure. Therefore, rescuers seldom need breathing apparatus, and mechanical ventilation of the cave is usually not necessary. However, some caves may contain carbon dioxide, carbon monoxide, oxides of nitrogen, or other contaminants at dangerous levels.

If carbon dioxide is present, it is usually at an acceptable level. The most common source of carbon dioxide in a cave is decaying vegetation washed in with high water. Another source is the exhalation of many people in a small, cramped space in which there is little or no air movement.

Being heavier than air, carbon dioxide collects in the lowest areas, which can be a problem for a caver trapped there. When a trapped caver is supplied oxygen through a tube, the end of the tube must be below the caver's face; otherwise, the heavier carbon dioxide could suffocate the victim.

Carbon monoxide is not a natural cave contaminant. The most common sources are cavers using small stoves to heat food and water or using gas lanterns for light. Another possible source is the use of gasoline-engine-driven generators and other equipment outside the cave entrance. The exhaust from these engines can be drawn into the cave if they are not positioned downwind from the entrance.

Nitrogen dioxide and other nitrous compounds found in caves are the result of blasting. Blasting is becoming more common as cavers attempt to enlarge passages too small to be crawled through. Also, rescuers must sometimes use explosives to enlarge tight constrictions so that a victim can be brought out. To keep the nitrous oxides as low as possible, only mining-permissible explosives should be used.

If toxic gases are detected in a cave, rescuers must use long-term, closed-circuit SCBA (Figure 11.21). The open-circuit SCBA commonly used by

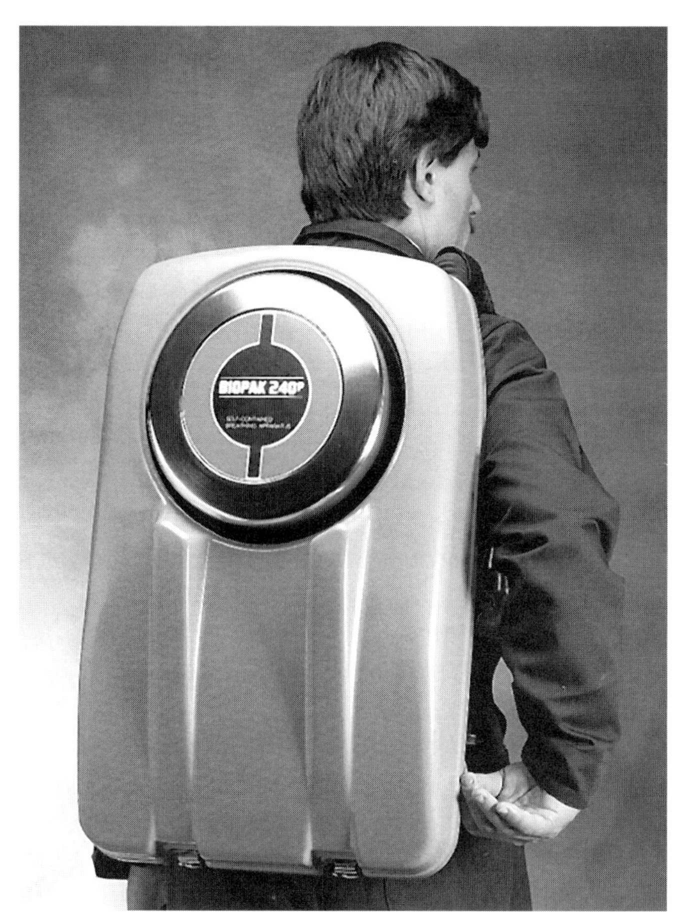

Figure 11.21 A typical rebreather unit. *Courtesy of Biomarine, Inc.*

the fire service does not have sufficient operating capacity for the protracted operations that cave rescues tend to be. An SCBA will also be needed for the victim in these cases.

COMPLEXITY

Many caves are very complex with several levels of maze-like passages and only one or two entrances or exits. Because of this complexity, many cave rescues merely involve finding inexperienced cavers who have become disoriented and lost but are otherwise alright.

Experience has shown that the best way to find lost cavers is to send small teams of rescuers who know the cave system well on a sweep through the most commonly explored passages. If the lost caver is conscious and wants to be found, the first sweep will usually locate them.

If the first sweep is unsuccessful, a more thorough search is made. This time the caver's exploration preferences are taken into account: Does the caver explore low, tight passages? Does he or she prefer climbing? Did the caver wear a wet suit for water passages? Even a very detailed cave map might not show all the passages of the cave system. More adventurous cave explorers might be in passages never before discovered.

Cave Rescue Equipment

Equipment for cave exploration and rescue has evolved over time to deal with the unique cave environment. The basic gear for horizontal travel includes coveralls, boots with lug-tread soles, and a shoulder pack for miscellaneous gear such as extra lights and food. For vertical travel, rescuers need a device such as a rappel rack for descending fixed lines and mechanical ascenders for climbing fixed ropes. Cave rescuers should only use equipment that is their own personal property to be sure of its availability and condition when needed for a rescue.

Rescuers also have to carry team gear into the cave. Team gear includes a stretcher, field telephone with sufficient wire (radios are not reliable underground), and a warm-gas inhalator. If the rescue involves vertical travel, team gear will also include several pulleys and safety cams and several ropes of assorted lengths.

LITTERS

A basket litter may be useful if the passages are large enough, but the passages in many caves are too small for their use. One of the best litters for cave rescue is the flexible plastic litter because of its ruggedness, flexibility, and smooth surface (Figure 11.22). Another form of litter is the drag sheet. Made of conveyor-belt material, the drag sheet is used to drag a victim through low passages.

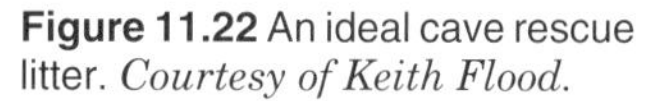

Figure 11.22 An ideal cave rescue litter. *Courtesy of Keith Flood.*

WARM-GAS INHALATORS

Warm-gas inhalators, developed in the early 1970s by Evan Lloyd of Scotland, are the most efficient means of treating hypothermia in the field and should be standard equipment for cave rescue teams. In these units, carbon dioxide is passed over soda lime, generating heat. Compressed breathing air is then passed over the hot soda lime and is warmed to about 115°F (46°C). As the victim inhales, the warmth from the breathing air is absorbed through the lungs into the bloodstream and is distributed throughout the body. This is more effective than warming the victim externally. External warming causes the cold-constricted superficial blood vessels to dilate, which brings more blood to the surface and away from the vital organs. In extreme cases, this situation can even cause death.

CABLE LADDERS

When rescuers must make frequent trips either up or down short drops (50 feet [15 m] or less), cable ladders can be useful. The load limit of a cable ladder is relatively low (600 pounds [272 kg]) and can be exceeded by one rescuer taking a short fall. Because of this, anyone on a cable ladder must use fall protection at all times.

ASCENDING EQUIPMENT

Ascenders used in cave rescue are no different from those described in Chapter 4, Rope Rescue. For more information on Gibbs ascenders, Jumars, and other rope-related equipment, see that chapter.

ROPES

The ropes used for cave rescue are also just as described in Chapter 4. Static kernmantle rope should be used in these and all other life safety applications.

MINE RESCUE

Just as most firefighters are not trained and equipped to carry out search and rescue operations in caves, they are usually not prepared to provide these services in mines. However, just as in cave rescues, fire/rescue personnel can provide aboveground support for the mine rescue teams. The following information is presented to inform fire/rescue personnel about the mine rescue teams required by law and to provide a frame of reference within which mine safety personnel and fire/rescue personnel can develop effective cooperative procedures.

Regulations

The requirements for mine rescue teams are listed in 30 CFR 49.1-10. According to these regulations, all mines must provide a specified minimum rescue capability by training their own teams or by arranging for rescue services from another entity. The basic requirement is that any time miners are working underground, two 5-member rescue teams must be available within 2 hours.

Mine Rescue Teams

These teams, each of which must also have one alternate member, must be "fully qualified, trained, and equipped for providing mine rescue service." To be eligible to serve on a mine rescue team, workers must have been employed in an underground mine for a minimum of one year within the past five years. Members must also be certified by a physician as being "physically fit to perform mine rescue and recovery work for prolonged periods under strenuous conditions."

Rescue Team Training

Prior to serving on a mine rescue team, each member must complete at least 20 hours of initial training on the use, care, and maintenance of the type of breathing apparatus that will be used by the rescue team. In addition, each team member must receive at least 40 hours of refresher training annually, in classes of at least 4 hours per month. Each member must also train for at least 2 hours every 2 months on wearing and using breathing apparatus.

Members must have at least one training session underground every 6 months. Where applicable, they must also train on the use, care, capabilities, and limitations of auxiliary mine rescue equipment. Advanced mine rescue procedures training is also required, including mine ventilation procedures and mine map reading. Any team member who misses more than 8 hours of training in one year is no longer eligible for team membership, unless the missed time is made up. The regulations also set forth the minimum standards for the instructors who deliver this training.

Mine Rescue Equipment

The regulations also require that a specified minimum cache of mine rescue equipment be maintained in one or more designated mine rescue stations. According to the regulations, each mine rescue station must be provided with at least the following equipment:

- Twelve self-contained oxygen breathing apparatus, each with a minimum of 2 hours capacity
- A portable supply of liquid air, liquid oxygen, pressurized oxygen, oxygen-generating or carbon-dioxide-absorbent chemicals in sufficient quantity to sustain each team for 6 hours of operation
- One extra oxygen cylinder (fully charged) for every six self-contained compressed-oxygen breathing apparatus
- Twelve mine-permissible cap lamps and a charging rack
- Two gas detectors appropriate for each type of gas that may be encountered in the mines served
- Two oxygen indicators or two flame safety lamps
- One portable mine rescue communication system or a sound-powered communication system, with at least 1,000 feet (300 m) of communication cable
- Spare parts and tools for repairing breathing apparatus and communications equipment

All of this equipment must be maintained so that it is ready for immediate use. It must be inspected monthly and repaired, if necessary, by someone trained to do so.

Mine Emergency Notification Plan

The regulations also require each underground mine to have a mine rescue notification plan that outlines the procedures for notifying the mine rescue teams when there is an emergency that requires their services. A copy of the plan must be posted at the mine.

SILO/GRAIN VESSEL RESCUES

While the procedures outlined in Chapter 8, Confined Space Rescue, are appropriate for silo and grain vessel rescues, these incidents are different enough from other confined space rescues to warrant their own section. The following information is provided to help fire/rescue personnel be better prepared to deal with these potentially very dangerous incidents.

Silos

Farming, in general, is one of the most hazardous civilian occupations in North America, and accidents involving silos are some of the most lethal. Working in and around silos on a daily basis can cause farm workers to become complacent and careless in these situations, which can be a fatal mistake.

Silo rescues become necessary for any of four common reasons: (1) worker falls into the silo, (2) worker has a heart attack or other episode while in the silo, (3) worker is overcome by toxic gas while in the silo, or (4) worker is caught in the silo unloading mechanism. In any of these situations, the victim may not be able to descend the enclosed chute ladder and will have to be extricated by rescue personnel.

TYPES OF SILOS

The typical farm silo is a large vertical cylinder used to preserve and store livestock feed (Figure 11.23). Silos vary in diameter, height, and type of construction. Most farm silos are either poured concrete, concrete stave, corrugated metal, or wooden stave. Some silos have a roof or cap while others are open at the top.

Another common type of silo is a sealed structure, usually blue in color, with an unloading mechanism built into the bottom and a small vent or access door at the top (Figure 11.24). The vent is designed to be closed to exclude air because these sealed silos are intended to be oxygen-limiting structures.

SILO GAS

Silage is formed by natural chemical fermentation that takes place in the chopped forage shortly after it is placed in the silo. As the silage ferments,

Figure 11.23 A typical farm silo.

Figure 11.24 Access hatch on an oxygen-limiting silo.

a variety of gases, including carbon dioxide, methane, and nitric oxide, are produced. Carbon dioxide can cause suffocation in concentrations of 30 percent or higher, but concentrations this high are unlikely in a silo. Low levels of methane are produced in silos during the first two weeks after silage is stored. However, the nitric oxide that is also formed may combine with oxygen to form nitrogen dioxide, a highly toxic gas known as *silo gas*. If inhaled, nitrogen dioxide — even in very low concentrations — may cause permanent lung damage or death.

Silo gas causes severe irritation of the upper respiratory tract and may cause inflammation of the lungs, with the person feeling little immediate pain or discomfort. The rescuer may inhale silo gas for a short time and notice no immediate ill effects. However, the rescuer may go to bed several hours later and die while sleeping because of fluid collecting in the lungs. If the effects are not fatal, a relapse with symptoms similar to pneumonia can occur one to two weeks after recovery from the initial exposure.

Silo gas is formed at the surface of the silage, but because it is heavier than air, the gas tends to flow down the silo chute into adjoining feed rooms and other low areas at the base of the silo. Silo gas has a strong bleach odor and may appear in the silo as a low-lying amber-, red-, or copper-colored haze.

The greatest danger, especially with corn, occurs one to three days after it is put into the silo, but silo gas can continue to be produced for two to three weeks. Beyond that time it is unlikely that more gas will be produced; however, the gas can remain inside the silo until it is opened for unloading.

WARNING

Silo gas can remain inside unopened silos long after it has stopped being produced. Silos should be well-ventilated before rescuers are allowed to work in or close to them.

RESCUE FROM SILOS

Rescuers must use appropriate caution when rescuing victims in and around silos, especially if the silo was filled during the preceding three weeks. The atmospheric sampling procedures described in Chapter 8, Confined Space Rescue, should be followed without exception. If silo gas is present or suspected, the use of self-contained breathing apparatus is a must. In sealed oxygen-limiting silos, an oxygen deficiency must be assumed, regardless of the time of year or the nature of the rescue.

Rescuers should inspect and evaluate built-in silo ladders before using them because they are often in disrepair and may be unsafe. If ladders are needed, and they probably will be, rescuers should provide their own.

Unlike the crusty surface of spoiled grain (discussed later in this chapter), the surface of the silage is usually firm enough for rescuers to safely walk on. However, as in any other confined space, rescuers entering a silo must wear a retrieval system as described in Chapter 8. There must also be an attendant outside the silo monitoring the rescuers, and a backup team must be immediately available.

Because of difficulties in gaining access, rescuers may have to enter through the top of the silo and be lowered to the victim, using rope systems as described in Chapter 4, Rope Rescue. This will also necessitate the use of a harness or litter for the victim.

A victim found unconscious on the surface of the silage may have lost consciousness as a result of a fall, but it is more likely to be the result of an oxygen deficiency or exposure to a toxic gas, or both. The victim should be lifted from the surface of the silage in order to lower the concentration of toxicants being inhaled, and if allowed under local medical protocols, the victim should be given oxygen. Short of that procedure, the victim should be supplied with pure breathing air.

If any rescuers were exposed to silo gas, they should be treated as additional victims. They should be relieved from duty and immediately transported to the nearest medical facility.

RESCUE FROM SILO UNLOADERS

Some open silos have an electrically operated unloading device that rides on top of the silage. The unloader circles around the silo, removes a thin layer of silage, and blows it down an enclosed chute. This chute often covers a series of doors in the side of the silo.

If the rescue involves a worker trapped in this silage unloading equipment, the presence of silo gas is highly unlikely. There is usually no need for breathing apparatus during the winter months unless the silo is of the oxygen-limiting type.

Residual silage clinging to the silo walls could fall and injure the victim or rescuers. Depending upon the amount of silage involved and its height above the victim, some emergency shoring as described in Chapter 10, Trench Rescue, may have to be done in order to stabilize the situation.

Before any attempt is made to free the victim from the machine, the power to the machine must be shut off and locked out (Figure 11.25). If the power cannot be locked out, a rescuer should be posted at the main switch to prevent anyone from turning the power back on.

A seriously injured victim will probably have to be lifted over the top of the silo in a litter. This will require the use of a rope rescue system as described in Chapter 4, Rope Rescue.

Figure 11.25 Power to the machine must be locked out.

Grain Vessels

Even though there is generally little danger from stored grain, grain bins have been the scene of many deaths and entrapments in flowing grain. Grain transport vessels, commercial elevators, and grain processing facilities have also been involved in numerous flowing grain accidents.

FLOWING GRAIN ENTRAPMENTS

Fully contained masses of grain behave like any other finely divided granular material such as fine gravel or sand. An intentional or accidental loss of containment allows grain to flow like a liquid.

Accidents in flowing grain have three primary sources: flowing grain columns, collapse of horizontal crusts, and collapse of vertical crusts.

Flowing grain columns. The most common grain-related accident is entrapment or suffocation when a worker is drawn into a flowing grain column. When a grain bin is emptied, the grain flows in a funnel-shaped, fluid mass toward the outlet. Grain storage bins on most farms are designed with the outlet located in the center of the bin floor, so the grain flows down the center to an unloading auger.

The flow velocity increases as grain flows from the bin wall at the top of the grain mass into a small vertical column at the center of the bin. This vertical column of grain flows down through the grain mass at nearly the rate of the auger. Almost no grain flows in from the surrounding grain mass.

If a worker falls into the flowing grain, the rate of inflow at the top center of a grain bin is so great that escape is impossible. Once engulfed in the flow of grain, a victim is rapidly drawn to the floor of the bin. Entrapment is similar to being drawn into a whirlpool in water (Figure 11.26). In some cases, the victim is drawn into the auger, which slows or stops the grain flow. Victims are rarely injured by contact with the unloading auger; death comes from suffocation.

CAUTION: Rescuers should never attempt to walk on the surface of contained flax seed as it will not support the weight of a person. It behaves similar to quicksand in that the more the person struggles, the deeper he or she sinks into the mass of seed.

Collapse of horizontal crusts. Entrapments and suffocations can occur if the victim enters a storage bin in which the surface of the grain has become caked due to spoilage. Under these conditions, the surface appears solid. However, it is only a thin layer of crusted grain concealing the void created when grain was removed from below (Figure 11.27). The victim breaks through the crust and is quickly covered by the avalanche of grain collapsing into the cavity.

Collapse of vertical crusts. Victims can be buried beneath the collapse of a wall of free-standing grain. Grain in good condition will assume an angle of 30 degrees to the floor, but spoiled or caked grain can stand almost vertical (Figure 11.28). As grain is removed from the base of a caked mass, the potential for avalanche and engulfment increases. Entrapments of this kind are more common in large grain-handling facili-

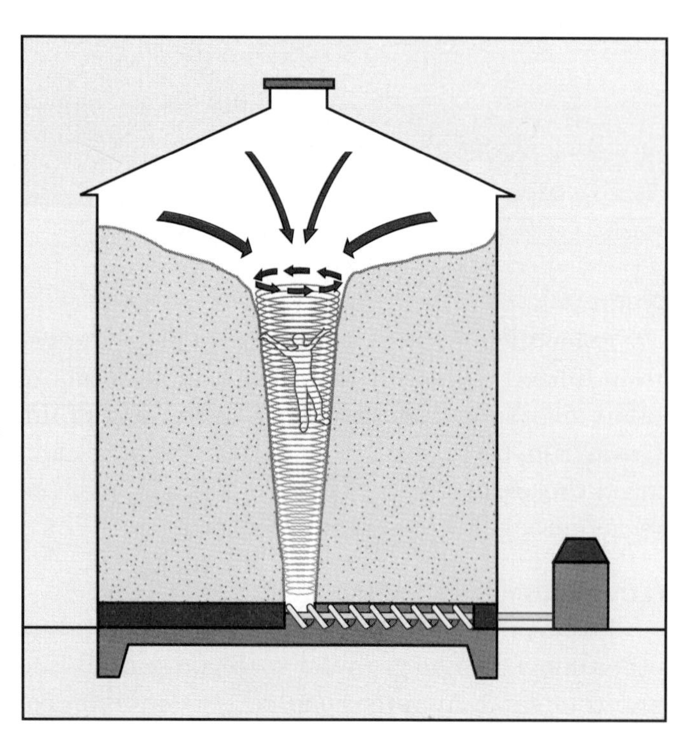

Figure 11.26 Flowing grain can act like a fluid.

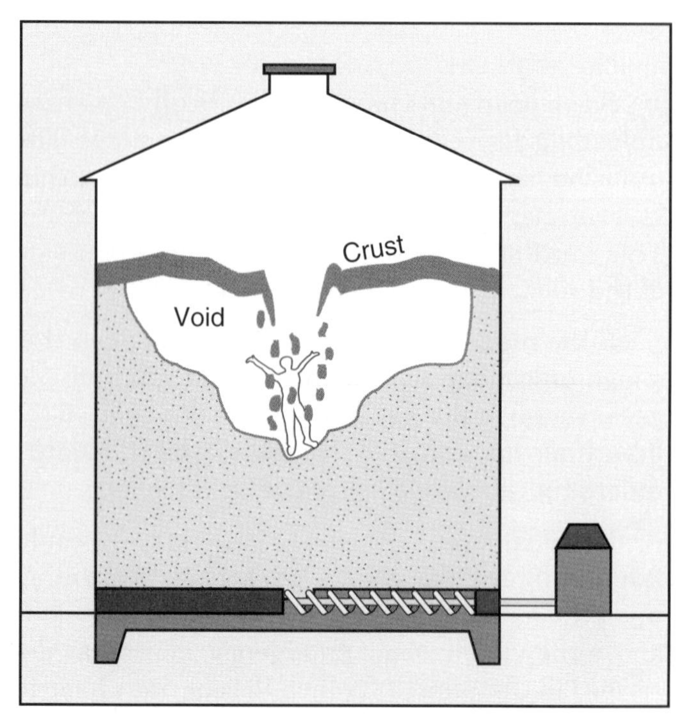

Figure 11.27 A victim in the void below the crust.

Figure 11.28 Vertically crusted grain.

ties than in farm operations; however, they are possible on farms with larger bins.

RESCUE FROM FLOWING GRAIN

Rescuers should always assume that a victim entrapped in grain, even if completely covered, is alive. Successful rescues have taken place in which the victim was completely engulfed in 10 feet (3 m) of grain for more than two hours. However, rescuers must exercise caution in removing the grain in order to avoid injuring the victim further. Whenever entry into the grain bin is necessary, especially if the grain appears to be spoiled or caked, rescuers must follow all of the procedures for confined space entry as described in Chapter 8, Confined Space Rescue.

CAUTION: Dust or mold spores from spoiled grain can cause severe allergic reactions in some people. Respiratory protection should be used if the rescue takes place in a bin of spoiled grain. Paper dust masks are generally sufficient. However, if a rescuer begins to experience shortness of breath, tightness in the chest, or dizziness, he or she should be withdrawn from the grain bin and kept under observation. Reactions to grain dust or mold may be delayed for several hours, and severe reactions often require hospitalization. Rescuers can also react to residues of fumigants used to treat spoiled grain.

Partially covered victim. Two rescuers should be lowered into the bin to reassure the victim and to check his or her ABCs. They should also attempt to attach a harness and lifeline to the victim. The harness and lifeline are only to prevent the victim from sinking further, and no attempt should be made to pull the victim free of the grain. Trying to overcome the tremendous drag created by the grain is likely to cause further injury.

The victim's airway should be cleared of grain. If the victim appears to be having difficulty breathing, oxygen should be administered if allowed under local medical protocols. Sometimes panic and the struggle to work free of the grain, rather than pressure on the chest, cause the victim to have difficulty breathing.

If there is danger of more grain flowing and covering the victim, a shield should be constructed around the victim. A variety of materials have been used successfully as shields, such as a 55-gallon (220 L) drum with both ends removed, sheets of plywood formed into a triangle, and pieces of sheet metal formed into a circle. Once the shield is in place, it should be possible to dig out the victim by scooping the grain from inside the shielded area. Planks or sheets of plywood can be laid on the grain surface outside of the shield to form a stable work platform.

Completely covered victim. If the victim is completely covered, the grain must be removed from the bin in the fastest way consistent with safety. However, rescuers should not start the unloading auger or open the gravity flow gate in an attempt to expedite this process. The victim may be drawn into the auger or become wedged in the gate opening and suffer further injury.

If the bin is equipped with an aeration fan and if someone qualified to operate it is on the scene, the fan should be started as soon as possible. Normally used to circulate air through the grain, the fan can also increase the air available to a completely covered victim under these conditions. Even with the additional air, there is little danger of dust explosions in most farm bins. However, if a grain dryer is an integral part of the aeration fan and if the fan cannot be operated without the dryer, the fan should not be used.

Because of the volume of grain involved and the tendency of the grain to backflow, attempting to dig out a completely covered victim is not likely to be successful. Experience has shown that the grain must be removed from the bin as quickly as possible. The most successful emergency procedure for rapid grain removal is to cut large openings spaced uniformly around the base of the bin. The openings can greatly reduce the time needed to find the victim, and they can be cut in the metal skin of the bin with a rescue saw, air chisel, or cutting torch (Figure 11.29).

CAUTION: Because of the danger of igniting the grain, a cutting torch should be used only when no other cutting tool is available. The lower flammability limit (LFL) of airborne combustible dust can be approximated as a condition in which the dust obscures vision at a distance of 5 feet (1.5 m) or less, so a charged hoseline or a water-type fire extinguisher should be at hand when a cutting torch is used.

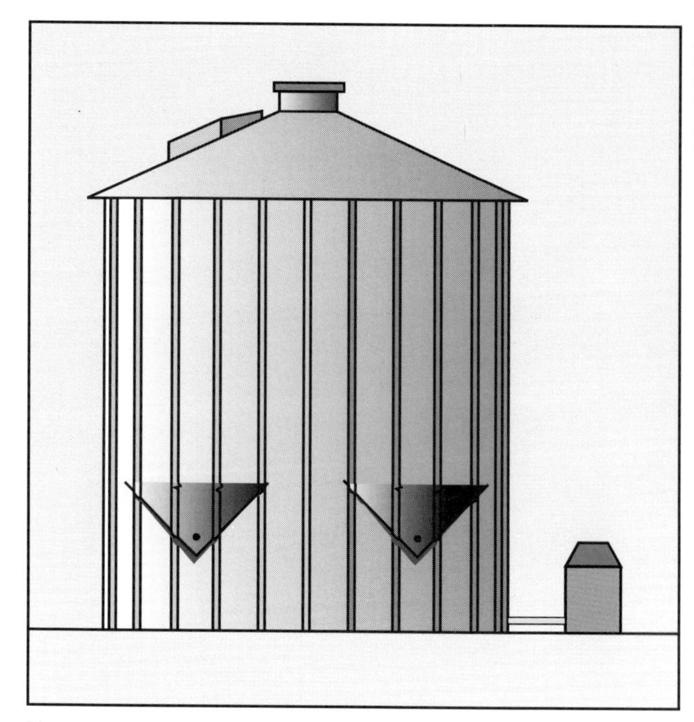

Figure 11.29 Typical vents cut in a grain bin.

The openings should be cut 4 to 6 feet (1.2 m to 1.8 m) above the ground to keep the pile of grain from building up and blocking the flow. When semicircular or V-shaped cuts, 30 to 40 inches (700 mm to 1 000 mm) across, are bent up, they form crude valves with which the flow of grain can be controlled (Figure 11.30). The size of these openings allows a large volume of grain to be removed quickly, while reducing the possibility of the victim becoming wedged in one of them and being injured further.

Figure 11.30 Controlling the outflow of grain.

CAUTION: Rescuers must not stand on top of the grain during this rapid removal process.

Rescuers can observe the top of the grain from the open roof hatch. Once the victim has been uncovered, the openings cut in the bin can be closed to stop the flow of grain and stabilize it. This will allow rescuers to safely enter the bin and effect the rescue or recovery.

GRAIN TRANSPORT ENTRAPMENTS

Many workers, most under 16 years old, have suffocated when they became entrapped in grain inside high-capacity transport vehicles. Trucks equipped with grain beds or gravity dump trailers are most often involved in flowing grain accidents (Figure 11.31). The victims are either buried while loading from a combine or storage bin or are drawn into the flow of grain from the vehicle as it is being unloaded. However, because these vessels are relatively small, entrapments in grain transport vehicles usually present fewer problems for rescuers than those in large, fixed storage facilities.

The same rescue techniques used in fixed grain storage bins are used with transport vessels. After the safety of rescuers, rapid removal of grain from around the victim is the first priority. Cutting open the sides of the vehicle in the same manner as described earlier will rapidly remove most of the grain. As with fixed-storage facilities, the gravity flow openings on bottom-dump vehicles should not be used. Victims can be drawn deeper into the grain or become wedged in the opening.

Figure 11.31 A typical bottom-dump grain trailer.

Chapter 11 Review

Directions

The following activities are designed to help you comprehend and apply the information in Chapter 11 of **Fire Service Rescue**, Sixth Edition. To receive the maximum learning experience from these activities, it is recommended that you use the following procedure:

1. Read the chapter, underlining or highlighting important terms, topics, and subject matter. Study the photographs and illustrations, and read the captions with each.
2. Review the list of vocabulary words to ensure that you know the chapter-related meaning of each. If you are unsure of the meaning of a vocabulary word, look up the word in the glossary or a dictionary, and then study its context in the chapter.
3. On a separate sheet of paper, complete all assigned or selected application and review activities before checking your answers.
4. After you have finished, check your answers against those on the pages referenced in parentheses.
5. Correct any incorrect answers, and review material that was answered incorrectly.

Vocabulary

Be sure that you know the chapter-related meanings of the following words:

- carcinogen *(325)*
- hydroelectric *(323)*
- cordon *(326)*
- odorant *(329)*
- grotto *(330)*
- tertiary *(331)*
- conduit *(331)*
- hypothermia *(331)*
- sinuous *(331)*
- serpentine *(331)*
- claustrophobia *(331)*
- cache *(335)*
- silage *(335)*
- fermentation *(335)*
- forage *(335)*

Application Of Knowledge

1. Choose from the following special rescue procedures those appropriate to your department and equipment, or ask your training officer to choose appropriate procedures. Mentally rehearse the procedures, or practice the chosen procedures as set up and supervised by your training officer.
   - Rescue an unconscious victim from a silage-filled silo. *(336, 337)*
   - Rescue a victim partially covered by flowing grain. *(339)*
   - Rescue a victim completely covered by flowing grain. *(339)*
   - Rescue a victim trapped in a silo unloader. *(337)*
   - Rescue a victim trapped in a grain transport vessel. *(340)*
   - Perform a simulated electricity-related rescue. *(326-328)*
2. Find and record the telephone numbers and contact names for the following cave rescue resources: *(Local protocol)*
   - Nearest cave rescue coordinator
   - District Office for Mine Safety and Health
   - Nearest grotto of the National Speleological Society
3. Survey your community for locations of possible electrical, cave, mine, and silo/grain vessel rescues. Review your department's pre-incident plans in these rescue areas. *(Local protocol)*

Review Activities

1. Identify the following abbreviations and acronyms:
   - NCRC *(330)*
   - LPG *(328)*
   - 30 CFR 49.1-10 *(334)*
   - PCB *(325)*
2. Identify the following electrical service components and terms:
   - AC power locator *(326, Figure 11.10)*
   - Dielectric equipment *(326)*
   - Distribution lines *(325)*
   - Electric meter *(325)*
   - Generating facility *(323)*
   - High tension lines *(325)*
   - Hot sticks *(326)*
   - Polychlorinated biphenyl *(325)*
   - Primaries *(325)*
   - Service connections *(325)*
   - Substation *(324)*

- Transformer *(324)*
- Transmission lines *(324)*
- Trunk lines *(324)*

3. List the three procedures rescuers should always follow when responding to any situation involving electricity. *(323)*
4. List hazards that may be encountered at electrical generating facilities. *(323)*
5. Outline the safe procedure for entering a steam-producing power-generating plant to effect a rescue. *(323)*
6. Determine whether the following statements about electricity-related rescue are true or false:
   - Generating-facility steam can be pressurized up to 2,400 psi (16 800 kPa). *(323)*
   - High-pressure steam leaks are visible as dense, gray-white vapors. *(323)*
   - High-pressure steam leaks can cut a person in two. *(323)*
   - If an energized line starts a fire near a transmission line, it should be allowed to burn away from the downed line a distance equal to one-half span between towers before it is attacked. *(324)*
   - The voltage in transmission lines can be 330,000 volts or higher. *(324)*
   - Transformers filled with oil that contains PCBs have been phased out and can no longer be found at electrical substations. *(325)*
   - Electrical distribution lines carry 2,400 to 34,000 volts of electricity. *(325)*
   - Rescuers should never enter an underground electrical utility space without the assistance of utility personnel. *(325)*
   - The electric meter is a frequent point of contact between people and electricity. *(325)*
   - Fire/rescue units not meeting testing and training procedures for dielectric equipment should not allow their personnel to use this type of equipment. *(326)*
   - Pike poles with wooden or fiberglass handles may be used as dielectric equipment to move electrical wires. *(327)*
7. Explain why rescue personnel should not take unnecessary risks to save victims of electrical and other accidents. *(326)*
8. State the safety guideline regarding metal tools at electrical accidents. *(326)*
9. Describe instances in which dielectric equipment may be used. *(326, 327)*
10. Explain what ground gradient is and how it can be hazardous to rescuers. *(327)*
11. Identify the following gas service components and terms:
    - Compressor station *(328)*
    - Methane *(328)*
    - Liquefied petroleum gas *(328)*
    - Ethyl mercaptan *(328)*
    - Gas storage tank *(328)*
    - Service connection *(328)*
    - Gas meter *(329)*
12. Explain the two primary hazards associated with natural and liquid petroleum gases. *(328)*
13. State what the gas service uses to odorize natural and liquid petroleum gases, and explain why it odorizes these naturally odor-free gases. *(328)*
14. List the gas pressure ranges for the following: *(328)*
    - High transmission pressure
    - Distribution pressure
    - Pressure at customer's gas meter
15. Explain why a gas-fed fire should not be extinguished until the gas is shut off. *(329)*
16. Explain what rescuers should do to reduce the likelihood of ignition in a gas-related rescue. *(329)*
17. Describe how the gas utility must handle a gas leak in underground piping. *(329)*
18. State the primary difference between LPG and natural gas in regard to rescue operations. *(329, 330)*
19. Identify the following cave rescue items and terms:
    - Belaying technique *(332)*
    - Breakdown *(332)*
    - Cable ladder *(334)*
    - Carbide lamp *(331)*
    - Mechanical ascenders *(333)*
    - Mining-permissible explosives *(332)*
    - Nitrous oxides *(332)*
    - Rappel rack *(333)*
20. Explain why fire/rescue personnel usually confine their activities to aboveground support of cave and mine rescue personnel. *(330)*
21. List resources fire/rescue units should call in their county, state, or province when called to effect a cave rescue. *(330)*
22. Distinguish between phreatic cave passages and vadose cave passages. *(330)*

23. Distinguish among the characteristics and hazards associated with solution caves, lava tube caves, and talus caves. *(330)*
24. Discuss ways in which rescuers safely handle the following cave rescue problems:
    - Darkness *(331)*
    - Water *(331)*
    - Passage irregularities *(331, 332)*
    - Temperature *(332)*
    - Atmosphere *(332, 333)*
    - Complexity *(333)*
25. Explain why shoring and roof jacks are seldom needed in cave rescue. *(331)*
26. List basic gear for the following cave situations: *(333)*
    - Horizontal travel
    - Vertical travel
    - Team entry
27. Explain when rescuers should use a neoprene rescue bag and when they should use a warm-gas inhaler to treat cave rescue hypothermia victims. *(334)*
28. State the basic requirement for availability of mine rescue teams. *(334)*
29. Describe training and eligibility requirements for members of mine rescue teams. *(334, 335)*
30. List the minimum equipment that must be provided at each mine rescue station. *(335)*
31. Explain what a mine rescue notification plan is and the regulations regarding it. *(335)*
32. Identify the following terms associated with silo/grain vessel rescues:
    - Nitrogen dioxide *(336)*
    - Forage *(335)*
    - Silage *(335)*
    - Silo unloader *(337)*
    - Oxygen deficiency *(337)*
    - Unloading auger *(337)*
    - Flowing grain column *(338)*
33. Distinguish between open silos and sealed silos. *(335)*
34. Describe how silo gas is formed. *(335, 336)*
35. Describe the physical effects of inhaling silo gas. *(336)*
36. Describe the two ways in which silo gas can be identified. *(336)*
37. List specific hazards associated with rescue from silos. *(336, 337)*
38. List and describe the three primary sources of accidents from flowing grain. *(338)*
39. Explain why rescuers should never attempt to walk on the surface of contained flax seed. *(338)*
40. Distinguish among the hazards associated with accidents caused by flowing grain columns, horizontal grain crust collapse, and vertical grain crust collapse. *(339, 340)*

Questions And Notes

Glossary/Index

Glossary

A

Angle Of Repose
Greatest angle above the horizontal plane at which loose material (such as soil) will lie without sliding.

Atmosphere
Area within a confined space where dust, vapors, mists, or other hazardous materials may exist.

B

Bearing Wall
Wall that supports itself and the weight of the roof and/or other structural components above it.

C

Cave-In
Collapse of unsupported trench walls.

Ceiling Concentration
Maximum allowable concentration of dust, vapors, mists, or other hazardous materials that may exist in a confined space.

Curtain Wall
Nonbearing wall attached to the outside of a building with a rigid steel frame.

D

Dead Load
Weight of the structure, structural members, building components, and any other feature that is constant and immobile.

Dielectric
Nonconductor of dielectric current. Term usually applied to tools that are used to handle energized electrical wires or equipment.

E

Excavation
Opening in the ground that results from digging, and is wider than it is deep.

F

Face
See Wall.

H

Hauling System
Mechanical advantage system that is constructed of rope and appropriate hardware and is designed for lifting a load.

Hot Work
Any operation that requires the use of tools or machines that may produce a source of ignition.

Hydraulic Shoring
Shores or jacks with movable parts that are operated by the action of hydraulic fluid.

I

Inerting
Introducing a nonflammable gas (i.e., nitrogen or carbon dioxide) to a flammable atmosphere in order to remove the oxygen and prevent an explosion.

Intrinsically Safe Equipment
Equipment designed and approved for use in flammable atmospheres. Formerly called "explosion proof."

L

Live Load
Loads within a building that are movable. Merchandise, stock, furnishings, vehicles, occupants, and rescuers are examples of live loads.

Lower Flammable Limit
Lowest concentration of a flammable vapor in air that will support continuous combustion.

O

OSHA
Acronym for Occupational Safety and Health Administration, a division of the U.S. Department of Labor that develops and enforces standards and regulations for occupational safety in the workplace.

Oxygen Deficiency
Insufficient oxygen to support life or flame; 16 percent oxygen is needed for flame production and human life.

Oxygen-Enriched Atmosphere
Area in which the concentration of oxygen is in excess of 25 percent.

P

Permissible Exposure Limit (PEL)
Maximum time-weighted concentration at which 95 percent of exposed, healthy adults suffer no adverse effects over a 40-hour workweek; an 8-hour time-weighted average unless otherwise noted. PELs are expressed in either ppm or mg/m^3. They are commonly used by OSHA and are found in the NIOSH pocket guide to chemical hazards.

Pneumatic Shoring
Shores or jacks with movable parts that are operated by the action of a compressed gas.

Purging
Freeing from impurities, such as ventilating a contaminated space, by introducing fresh air.

R

Respirator
Filter breathing device, *not* an SCBA.

S

Screw Jack
Nonhydraulic jacks that can be extended or retracted by turning a collar on a threaded shaft.

Shear Strength
Ability of a building component or assembly to resist lateral or shear forces.

Sheeting
Generally speaking, wood planks and wood panels that support trench walls when held in place with shoring.

Shoring
General term used for lengths of timber, screw jacks, hydraulic and pneumatic jacks, and other devices that can be used to hold sheeting against trench walls. Individual supports are called *shores, cross braces*, and *struts*.

Skip Shoring
Procedure for supporting trench walls with uprights and shores at spaced intervals.

T

Tilt-Up Construction
Type of construction in which concrete wall sections (slabs) are cast on the concrete floor of the building and are then tilted up into the vertical position. Also known as Tilt-Slab Construction.

Trench
Temporary excavation in which the length of the bottom exceeds the width of the bottom. The term *trench* is generally limited to excavations that are less than 15 feet wide at the bottom and less than 20 feet deep.

Trench Lip
Edge of a trench.

Uprights
Generally speaking, planks that are held in place against sections of sheeting with shores. Uprights add strength to the shoring system. They distribute forces exerted by trench walls and counterforces exerted by shores over wider areas of the sheeting.

W

Wall
Side of a trench from the lip to the floor. Also called the Face.

Index

IFSTA MANUALS AND FPP PRODUCTS

For a current catalog describing these and other products, call or write your local IFSTA distributor or Fire Protection Publications, IFSTA Headquarters, Oklahoma State University, Stillwater, OK 74078-0118. Phone: 1-800-654-4055

AWARENESS LEVEL TRAINING FOR HAZARDOUS MATERIALS
prepares fire, police, EMS, and public utilities to recognize and identify the presence of hazardous materials at an emergency scene. Addresses the requirements in NFPA 472, Chapter 2: Competencies for First Responders at the Awareness Level. includes responsibilities of the first responder, identification systems, types of containers, and personal protective equipment. 1st Edition (1995), 152 pages.

STUDY GUIDE AWARENESS LEVEL TRAINING FOR HAZARDOUS MATERIALS
The companion study guide in question and answer format. (1995), 184 pages.

FIRE DEPARTMENT AERIAL APPARATUS
includes information on the driver/operator's qualifications; vehicle operation; types of aerial apparatus; positioning, stabilizing, and operating aerial devices; tactics for aerial devices; and maintaining, testing, and purchasing aerial apparatus. Detailed appendices describe specific manufacturers' aerial devices. 1st Edition (1991), 386 pages, addresses NFPA 1002.

STUDY GUIDE FOR AERIAL APPARATUS
The companion study guide in question and answer format. 1991, 140 pages.

AIRCRAFT RESCUE AND FIRE FIGHTING
comprehensively covers commercial, military, and general aviation. Subjects covered include personal protective equipment, apparatus and equipment, extinguishing agents, engines and systems, fire fighting procedures, hazardous materials, and fire prevention. It also contains a glossary and review questions with answers. 3rd Edition (1992), 247 pages, addresses NFPA 1003.

BUILDING CONSTRUCTION RELATED TO THE FIRE SERVICE
helps firefighters become aware of the many construction designs and features of buildings found in a typical first alarm district and how these designs serve or hinder the suppression effort. Subjects include construction principles, assemblies and their resistance to fire, building services, door and window assemblies, and special types of structures. 1st Edition (1986), 166 pages, addresses NFPA 1001 and NFPA 1031, levels I & II.

CHIEF OFFICER
explains the skills necessary to plan and maintain an efficient and cost-effective fire department. The combination of an ever-increasing fire problem, spiraling personnel and equipment costs, and the development of new technologies for decision making require far more than expertise in fire suppression. Today's chief officer must possess the ability to plan and administrate as well as have political expertise. 1st Edition (1985), 211 pages, addresses NFPA 1021, level VI.

SELF-INSTRUCTION FOR CHIEF OFFICER
The companion study guide in question and answer format. 1986, 142 pages.

FIRE DEPARTMENT COMPANY OFFICER
focuses on the basic principles of fire department organization, working relationships, and personnel management. For the firefighter aspiring to become a company officer, or a company officer wishing to improve management skills. 2nd Edition (1990), 278 pages, addresses NFPA 1021, levels I, II, & III.

COMPANY OFFICER STUDY GUIDE
The companion study guide in question and answer format. Includes problem applications and case studies. 1991, 243 pages.

ESSENTIALS OF FIRE FIGHTING
is the "bible" on basic firefighter skills and is used throughout the world. The easy-to-read format is enhanced by 1,600 photographs and illustrations. Topics covered include: personal protective equipment, building construction, firefighter safety, fire behavior, portable extinguishers, SCBA, ropes and knots, rescue, forcible entry, ventilation, communications, water supplies, fire streams, hose, fire cause determination, public fire education and prevention, fire suppression techniques, ladders, salvage and overhaul, and automatic sprinkler systems. 3rd Edition (1992), 590 pages, addresses NFPA 1001.

STUDY GUIDE FOR 3rd EDITION OF ESSENTIALS OF FIRE FIGHTING
The companion learning tool for the new 3rd edition of the manual. It contains questions and answers to help you learn the important information in the book. 1992, 322 pages.

PRINCIPLES OF FOAM FIRE FIGHTING
Covers both Class A and Class B fires and foams. Includes information on portable foam extinguishers, foam concentrates, portable foam proportioning equipment, foam and ARFF apparatus, and fixed foam extinguishing systems. Also discusses the latest compressed air foam systems. Provides detailed information on using foam to control structural, wildland, industrial, petrochemical facility, and aircraft fires. 1st Edition (1996).

PRINCIPLES OF EXTRICATION
leads you step-by-step through the procedures for disentangling victims from cars, buses, trains, farm equipment, and industrial situations. Fully illustrated with color diagrams and more than 500 photographs. It includes rescue company organization, protective clothing, and evaluating resources. Review questions with answers at the end of each chapter. 1st Edition (1990), 365 pages.

FIRE CAUSE DETERMINATION
gives you the information necessary to make on-scene fire cause determinations. You will know when to call for a trained investigator, and you will be able to help the investigator. It includes a profile of firesetters, finding origin and cause, documenting evidence,

interviewing witnesses, and courtroom demeanor. 1st Edition (1982), 159 pages, addresses NFPA 1021, Fire Officer I, and NFPA 1031, levels I & II.

FORCIBLE ENTRY
reflects the growing concern for the reduction of property damage as well as firefighter safety. Contains technical information about forcible entry tactics, tools, and methods, as well as door, window, and wall construction. Tactics discuss the degree of danger to the structure and leaving the building secure after entry. Includes a section on locks and through-the-lock entry. 7th Edition (1987), 270 pages, helpful for NFPA 1001.

GROUND COVER FIRE FIGHTING PRACTICES
explains the dramatic difference between structural fire fighting and wildland fire fighting. It discusses the apparatus, equipment, and extinguishing agents used to combat wildland fires. Outdoor fire behavior and how fuels, weather, and topography affect fire spread are explained. Also covers personnel safety, management, and suppression methods. It contains a glossary, sample fire operation plan, fire control organization system, fire origin and cause determination, and water expansion pump systems. 2nd Edition (1982), 152 pages.

FIRE SERVICE GROUND LADDER PRACTICES
addresses NFPA 1001, *Standard for Fire Fighter Professional Qualifications*, (ladders). Ladders provides a broad scope of understanding of ground ladders and their use. Ladders familiarizes you with design construction, maintenance, and service testing. It shows procedures for handling, arising, and climbing ladders and includes methods for special uses. Ladders contains a glossary, review questions, and a sample testing repair form. 9th Edition (1996), 203 pages.

HAZARDOUS MATERIALS FOR FIRST RESPONDERS
covers the objectives for First Responder at the Awareness and Operational levels contained in NFPA 472. Includes information on properties of hazardous materials, recognizing and identifying hazardous materials, personal protective equipment, emergency scene command and control, incident control tactics and strategies, and decontamination. 2nd Edition (1994), 241 pages.

STUDY GUIDE FOR IFSTA HAZARDOUS MATERIALS FOR FIRST RESPONDERS
The companion study guide in question and answer format. 2nd Edition (1994), 253 pages.

HAZARDOUS MATERIALS: MANAGING THE INCIDENT
addresses OSHA 1910.120 and NFPA 472, *Standard for Professional Competence of Responders to Hazardous Materials Incidents.* Provides the reader with a logical, systematic process for responding to and managing hazardous materials emergencies. It is directed toward the haz mat technician, incident commander, the off-site specialty employee, and haz mat response team members. Includes numerous charts, diagrams, scan sheets, checklists, and reference information. Topics include haz mat management system, health and safety, ICS, politics of haz mat incident management, hazard and risk evaluation, decontamination, and more! 2nd Edition (1994)

STUDENT WORKBOOK FOR HAZARDOUS MATERIALS: MANAGING THE INCIDENT
The companion study guide in question and answer format. 2nd Edition (1994)

INSTRUCTOR'S GUIDE FOR HAZARDOUS MATERIALS: MANAGING THE INCIDENT
Provides lessons based on each chapter. 2nd Edition (1994)

HAZ MAT RESPONSE TEAM LEAK AND SPILL GUIDE
contains articles by Michael Hildebrand reprinted from *Speaking of Fire's* popular Hazardous Materials Nuts and Bolts series. Two additional articles from *Speaking of Fire* and the hazardous material incident SOP from the Chicago Fire Department are also included. 1st Edition (1984), 57 pages.

EMERGENCY OPERATIONS IN HIGH-RACK STORAGE
is a concise summary of emergency operations in the high-rack storage area of a warehouse. It explains how to develop a pre-emergency plan, the equipment needed to implement the plan, type and amount of training personnel need to handle an emergency, and interfacing with various agencies. Includes consideration questions, and trial scenarios. 1st Edition (1981), 97 pages.

HOSE PRACTICES
is the most comprehensive single source about hose and its use. The manual details basic methods of handling hose, including large diameter hose. It is fully illustrated with photographs showing loads, evolutions, and techniques. This complete and practical book explains the national standards for hose and couplings. 7th Edition (1988), 245 pages, addresses NFPA 1001.

FIRE PROTECTION HYDRAULICS AND WATER SUPPLY ANALYSIS
covers the quantity and pressure of water needed to provide adequate fire protection, the ability of existing water supply systems to provide fire protection, the adequacy of a water supply for a sprinkler system, and alternatives for deficient water supply systems. 1st Edition (1990), 340 pages.

INCIDENT COMMAND SYSTEM (ICS)
was developed by a multiagency task force. Using this system, fire, police, and other government groups can operate together effectively under a single command. The system is modular and can be used to meet the requirements of both day-to-day and large-incident operations. It is the approved basic command system taught at the National Fire Academy. 1st Edition (1983), 220 pages, helpful for NFPA 1021.

INDUSTRIAL FIRE BRIGADE TRAINING: INCIPIENT LEVEL
assists management in complying with applicable laws and regulations, primarily NFPA 600 and 29 CFR 1910, and to assist them in training those who provide incipient level fire protection for industrial occupancies. It is also intended to serve as a reference and training resource for individual emergency responders.
1st Edition (1995), 184 pages.

FIRE INSPECTION AND CODE ENFORCEMENT
is a comprehensive guide to the principles and techniques of inspection for both uniformed and civilian inspectors.Text includes information on how fire travels, electrical hazards, and fire resistance requirements. It covers storage, handling, and use of hazardous materials; fire protection systems; and building construction for fire and life safety. 5th Edition (1987), 316 pages, addresses NFPA 1001 and NFPA 1031, levels I & II.

STUDY GUIDE FOR FIRE INSPECTION AND CODE ENFORCEMENT
The companion study guide in question and answer format with case studies. 1989, 272 pages.

FIRE SERVICE INSTRUCTOR
explains the characteristics of a good instructor, shows how to determine training requirements, and teach to the level of your class. It discusses principles and procedures of teaching and learning, and covers the use of training aids and devices. Also covers the principles of testing as well as test construction. Included are chapters on safety, legal considerations, and computers. 5th Edition (1990), 326 pages, addresses NFPA 1041, levels I & II.

LEADERSHIP IN THE FIRE SERVICE
was created from the series of lectures given by Robert F. Hamm to assist in leadership development. It provides the foundation for getting along with others, explains how to gain the confidence of your personnel, and covers what is expected of an officer. Includes information on supervision, evaluations, delegating, and teaching. Topics include: the successful leader today, a look into the past may reveal the future, and self-analysis for officers. 1st Edition (1967), 132 pages.

FIRE SERVICE ORIENTATION AND TERMINOLOGY
Fire Service Orientation and Indoctrination has been revised. It has a new name and a new look. Keeping the best of the old — traditions, history, and organization — this new manual provides a complete dictionary of fire service terms. To be used in conjunction with **Essentials of Fire Fighting** and the other IFSTA manuals. 3rd Edition (1993), addresses NFPA 1001.

PRIVATE FIRE PROTECTION AND DETECTION
Discusses ways in which fires may be prevented or attacked in their incipient phase until the fire brigade or public fire protection arrives. Covers information on automatic sprinkler systems, hose standpipe systems, fixed fire pump installations, portable fire extinguishers, fixed special agent extinguishing systems, and fire alarm and detection systems. Information on the design, operation, maintenance, and inspection of these systems and equipment is provided. 2nd Edition (1994).

PUBLIC FIRE EDUCATION
provides valuable information for ending public apathy and ignorance about fire. This manual gives you the knowledge to plan and implement fire prevention campaigns. It shows you how to tailor the individual programs to your audience as well as the time of year or specific problems. It includes working with the media, resource exchange, and smoke detectors. 1st Edition (1979), 169 pages, helpful for NFPA 1021 and 1031.

FIRE DEPARTMENT PUMPING APPARATUS
is the Driver/Operator's encyclopedia on operating fire pumps and pumping apparatus. It covers pumpers, tankers (tenders), brush apparatus, and aerials with pumps. Explains safe driving techniques, getting maximum efficiency from the pump, and basic water supply. It includes specification writing, apparatus testing, and extensive appendices of pump manufacturers. 7th Edition (1989), 374 pages, addresses NFPA 1002.

STUDY GUIDE FOR PUMPING APPARATUS
The companion study guide in question and answer format. 1990, 100 pages.

FIRE SERVICE RESCUE
provides fire and rescue personnel with information and procedures related to victim removal from hazardous or life-threatening situations. Covers management and incident command, rope rescue, fireground search and rescue, rescue situations involving structural collapse, elevators, confined space, water/ice, trench, caves and mines, and grain storage vessels. 6th Edition (1996).

RESIDENTIAL SPRINKLERS A PRIMER
outlines U.S. residential fire experience, system components, engineering requirements, and issues concerning automatic and fixed residential sprinkler systems. Written by Gary Courtney and Scott Kerwood, reprinted from *Speaking of Fire.* Supplements **Private Fire Protection.** 1st Edition (1986), 16 pages.

FIRE DEPARTMENT OCCUPATIONAL SAFETY
addresses the basic responsibilities for a safety officer and the requirements and procedures for a safety and health program. Includes an overview of establishing and implementing a safety program, physical fitness and health considerations, safety in training, fire station safety, tool and equipment safety and maintenance, personal protective equipment, en- route hazards and response, emergency scene safety, and special hazards. 2nd Edition (1991), 366 pages, addresses NFPA 1500, 1501.

SALVAGE AND OVERHAUL
covers salvage operations, equipment selection and care, as well as techniques for using salvage equipment to minimize fire damage caused by water, smoke, heat, and debris. The overhaul section includes methods for finding hidden fire, protection of fire cause evidence, safety during overhaul operations, and restoration of property and fire protection systems after a fire. 7th Edition (1985), 225 pages, addresses NFPA 1001.

SELF-CONTAINED BREATHING APPARATUS
contains all the basics of SCBA use, care, testing, and operation. Special attention is given to safety and training. The chapter on Emergency Conditions Breathing has been completely revised to incorporate safer emergency methods that can be used with newer models of SCBA. Also included are appendices describing regulatory agencies and donning and doffing procedures for nine types of SCBA. Covers NFPA, OSHA, ANSI, and NIOSH regulations and standards as they pertain to SCBA. 2nd Edition (1991), 360 pages, addresses NFPA 1001.

THE SOURCEBOOK FOR FIRE COMPANY TRAINING EVOLUTIONS
provides volunteer and career training officers and company officers with ideas for presenting more than 50 weekly or monthly training sessions. Each session contains information on the standards covered, equipment needed, outlines for the presentations and practical exercises, and a listing of pertinent resources and training materials. The sessions cover basic fire fighting, apparatus operation, company evolutions, indoor sessions for rainy days, and competitive exercises with a practical training value. 1st Edition (1994) 238 pages.

STUDY GUIDE FOR SELF-CONTAINED BREATHING APPARATUS
The companion study guide in question and answer format. 1991, 131 pages.

FIRE STREAM PRACTICES
This carefully written text covers the physics of fire and water; the characteristics, requirements, and principles of good streams; and fire fighting foams. **Streams** includes formulas for the application of fire fighting hydraulics, as well as actions and reactions created by applying streams under a variety of circumstances. The friction loss equations and answers are included, and review questions are located at the end of each chapter. 7th Edition (1989), 464 pages, addresses NFPA 1001 and NFPA 1002.

GASOLINE TANK TRUCK EMERGENCIES
provides emergency response personnel with background information, general procedures, and guidelines to be followed when responding to and operating at incidents involving MC-306/DOT 406 cargo tank trucks. Specific topics include: incident management procedures, site safety considerations, methods of product transfer, and vehicle uprighting considerations. 1st Edition (1992), 51 pages, addresses NFPA 472.

FIRE SERVICE VENTILATION
describes and illustrates the safe operations related to ventilation, products of combustion, elements and situations that influence the ventilation process, ventilation methods and procedures, and tools and mechanized equipment used in ventilation. The manual includes chapter reviews, a glossary, and applicable safety considerations. 7th Edition (1994), addresses NFPA 1001, 311 pages.

WATER SUPPLIES FOR FIRE PROTECTION
acquaints you with the principles, requirements, and standards used to provide water for fire fighting. Rural water supplies as well as fixed systems are discussed. It includes requirements for size and carrying capacity of mains, hydrant specifications, maintenance procedures conducted by the fire department, and relevant maps and record- keeping procedures. Review questions at the end of each chapter. 4th Edition (1988), 268 pages, addresses NFPA 1001, NFPA 1002, and NFPA 1031, levels I & II.

CURRICULUM PACKAGES

COMPANY OFFICER
A competency-based teaching package with 17 lessons as well as classroom and practical activities to teach the student the information and skills needed to qualify for the position of Company Officer. Corresponds to **Fire Department Company Officer**, 2nd Edition.

The Package includes the Company Officer Instructor's Guide (the how, what, and when to teach); the Student Guide (a workbook for group instruction); and 143 full-color overhead transparencies.

ESSENTIALS CURRICULUM PACKAGE
A competency-based teaching package with 19 chapters and 22 lessons as well as classroom and practical activities to teach the student the information and skills needed to qualify for the position of Fire Fighter I or II. Corresponds to **Essentials of Fire Fighting**, 3rd Edition.

The Package includes the Essentials Instructor's Guide (the how, what, and when to teach); the Student Guide (a workbook for group instruction); and 445 full-color overhead transparencies.

LEADERSHIP
A complete teaching package that assists the instructor in teaching leadership and motivational skills at the Company Officer level. Each lesson gives an outline of the subject matter to be covered, approximate time required to teach the material, specific learning objectives, and references for the instructor's preparation. Sources for suggested films and videotapes are included.

INCIPIENT INDUSTRIAL CURRICULUM FOR INDUSTRIAL FIRE BRIGADE TRAINING
designed to aid those in industry and those who teach industrial fire brigade training, this comprehensive curriculum can be used for instructor-led group instruction or for instructor-monitored self-instruction. Includes Instructor's Guide, 106 full-color transparencies, and Student Applications Workbook.

TRANSLATIONS

LO ESENCIAL EN EL COMBATE DE INCENDIOS
is a direct translation of **Essentials of Fire Fighting**, 2nd edition. Please contact your distributor or FPP for shipping charges to addresses outside U.S. and Canada. 444 pages.

PRACTICAS Y TEORIA PARA BOMBEROS
is a direct translation of **Fire Service Practices for Volunteer and Small Community Fire Departments**, 6th edition. Please contact your distributor or FPP for shipping charges to addresses outside U.S. and Canada. 347 pages.

OTHER ITEMS

TRAINING AIDS
Fire Protection Publications carries a complete line of videos, overhead transparencies, and slides. Call for a current catalog.

NEWSLETTER
The nationally acclaimed and award-winning newsletter, *Speaking of Fire*, is published quarterly and available to you free. Call today for your free subscription.

ESSENTIALS
3RD EDITION
5TH PRINTING, 5/96

COMMENT SHEET

DATE ______________________ NAME ______________________________________

ADDRESS ____

ORGANIZATION REPRESENTED ____

CHAPTER TITLE ____________________________________ NUMBER __________

SECTION/PARAGRAPH/FIGURE _________________________ PAGE ____________

1. Proposal (include proposed wording or identification of wording to be deleted), OR PROPOSED FIGURE:

2. Statement of Problem and Substantiation for Proposal:

RETURN TO: IFSTA Editor
Fire Protection Publications
Oklahoma State University
930 N. Willis
Stillwater, OK 74078-8045

SIGNATURE ______________________________

Use this sheet to make any suggestions, recommendations, or comments. We need your input to make the manuals as up to date as possible. Your help is appreciated. Use additional pages if necessary.